T0139980

Advances in Intelligent Systems and Computing

Volume 705

The series "Advances in Intelligent Systems and Computing" contains publications on theory, applications, and design methods of Intelligent Systems and Intelligent Computing. Virtually all disciplines such as engineering, natural sciences, computer and information science, ICT, economics, business, e-commerce, environment, healthcare, life science are covered. The list of topics spans all the areas of modern intelligent systems and computing such as: computational intelligence, soft computing including neural networks, fuzzy systems, evolutionary computing and the fusion of these paradigms, social intelligence, ambient intelligence, computational neuroscience, artificial life, virtual worlds and society, cognitive science and systems, Perception and Vision, DNA and immune based systems, self-organizing and adaptive systems, e-Learning and teaching, human-centered and human-centric computing, recommender systems, intelligent control, robotics and mechatronics including human-machine teaming, knowledge-based paradigms, learning paradigms, machine ethics, intelligent data analysis, knowledge management, intelligent agents, intelligent decision making and support, intelligent network security, trust management, interactive entertainment, Web intelligence and multimedia.

The publications within "Advances in Intelligent Systems and Computing" are primarily proceedings of important conferences, symposia and congresses. They cover significant recent developments in the field, both of a foundational and applicable character. An important characteristic feature of the series is the short publication time and world-wide distribution. This permits a rapid and broad dissemination of research results.

More information about this series at http://www.springer.com/series/11156

Damodar Reddy Edla · Pawan Lingras
Venkatanareshbabu K.
Editors

Advances in Machine Learning and Data Science

Recent Achievements and Research Directives

Editors
Damodar Reddy Edla
Department of Computer Science
and Engineering
National Institute
of Technology Goa
Goa
India

Venkatanareshbabu K.
Department of Computer Science
and Engineering
National Institute
of Technology Goa
Goa
India

Pawan Lingras
Department of Mathematics
and Computing Science
Saint Mary's University
Halifax, NS
Canada

ISSN 2194-5357 ISSN 2194-5365 (electronic)
Advances in Intelligent Systems and Computing
ISBN 978-981-10-8568-0 ISBN 978-981-10-8569-7 (eBook)
https://doi.org/10.1007/978-981-10-8569-7

Library of Congress Control Number: 2018934903

This Springer imprint is published by the registered company Springer Nature Singapore Pte Ltd.
The registered company address is: 152 Beach Road, #21-01/04 Gateway East, Singapore 189721, Singapore

Preface

It is our pleasure to present the reader the proceedings of the first international conference on *Latest Advances in Machine learning and Data Science* (LAMDA 2017) organized during October 25–27, 2017, in the Department of Computer Science and Engineering, National Institute of Technology Goa, India. LAMDA 2017 is the first international conference, which aims at providing an opportunity for scientists, engineers, educators, and students to gather so as to present the state-of-the-art advancements in the fertile domains of machine learning and data science.

Machine learning is a field of computer science that provides computers the ability to learn without being explicitly programmed. Machine learning focuses on constructing computer programs that improves with experience. Machine learning is a fast-growing field, which is widely used in the Web browsing, information filtering on the Web, credit card fraud detection, and many more applications. Data science is an interdisciplinary field about scientific methods, processes, and systems to extract knowledge or insights from data in various forms, either structured or unstructured. It includes data analysis fields such as data mining, machine learning, statistics, and knowledge discovery of data. Data science can be used in domains like pattern and speech recognition, robotics, bioinformatics.

We received 123 papers from different regions, and based on the reviews, technical program chairs selected 48 quality papers for the proceedings, out of which 38 papers have been published in this volume. These papers comprise topics related to machine learning and data science including classification of images, data, videos, clustering techniques, support vector machines, statistical learning, association rules, mining of images, graph mining, Web mining.

At LAMDA 2017, technical interactions within the research community were in the form of keynote speeches, panel discussions, and oral presentations. It encourages open debate and disagreement as this is the right way to push the field forward in the new directions. It provides an environment for the participants to present new results, to discuss state of the art, and to interchange information on emerging areas and future trends of machine learning and data science. It also creates an opportunity for the participants to meet colleagues and make friends who share similar research interests.

The organizers and volunteers in LAMDA 2017 made great efforts to ensure the success of this event. The success of the conference was also due to the precious collaboration of the co-chairs and the members of the local organizing committee. We would also like to thank the members of the program committee and the referees for reviewing the papers and the publication committee for checking and compiling the papers. We would also like to thank the publisher, Springer-Verlag, for their agreement and cooperation to publish the proceedings as a volume of "Advances in Machine Learning and Data Science—Recent Achievements and Research Directives." We wish to extend our gratitude to all the keynote speakers and participants who enabled the success of this year's edition of LAMDA.

Goa, India
Halifax, Canada
Goa, India
October 2017

Damodar Reddy Edla
Pawan Lingras
Venkatanareshbabu K.

Contents

About the Editors

Dr. Damodar Reddy Edla is Assistant Professor and Head in the Department of Computer Science and Engineering, National Institute of Technology Goa, India. His research interests include data mining and wireless sensor networks. He has completed his doctorate from Indian School of Mines, Dhanbad, India, in 2013. He has published more than 35 research papers in national and international journals and conference proceedings. He is editorial board member of five international journals.

Dr. Pawan Lingras is a graduate of IIT Bombay with graduate studies from University of Regina. He is currently a Professor and Director of Computing and Data Analytics at Saint Mary's University, Halifax. He is also internationally active having served as a Visiting Professor at Munich University of Applied Sciences, IIT Gandhinagar; as a Research Supervisor at Institut Superieur de Gestion de Tunis; as a Scholar-in-Residence; and as a Shastri Indo-Canadian scholar. He has delivered more than 35 invited talks at various institutions around the world. He has authored more than 200 research papers in various international journals and conferences. He has also co-authored three textbooks and co-edited two books and eight volumes of research papers. His academic collaborations/co-authors include academics from Canada, Chile, China, Germany, India, Poland, Tunisia, UK, and USA. His areas of interests include artificial intelligence, information retrieval, data mining, Web intelligence, and intelligent transportation systems. He has served as the general co-chair, program co-chair, review committee chair, program committee member, and reviewer for various international conferences on artificial intelligence and data mining. He is also on editorial boards of a number of international journals. His research has been supported by Natural Science and Engineering Research Council (NSERC) of Canada for 25 years, as well as other funding agencies including NRC-IRAP and MITACS. He also served on the NSERC's Computer Science peer-review committee. He has been awarded an alumni association excellence in teaching award, student union's faculty of science teaching award, and president's award for excellence in research at Saint Mary's University.

Dr. Venkatanareshbabu K. is with the Machine Learning Group, Department of Computer Science and Engineering, National Institute of Technology Goa, India, where he is currently an Assistant Professor. He obtained doctoral degree from Indian Institute of Technology Delhi in 2014. He was with Evalueserve Pvt Ltd, as a Senior Research Associate in 2009. He has been a Visiting Scientist with the Global Medical Technologies, California, in 2017. He has been the Principal Investigator of many sponsored research and consultancy research projects in the fields of neural networks and machine learning. He is also actively involved in teaching and project coordination for the Graduate and Post Graduate Program in Computer Science and Engineering Department at the National Institute of Technology Goa. He has authored a number of research papers published in reputed international journals in the areas of neural networks, classification, and clustering.

Optimization of Adaptive Resonance Theory Neural Network Using Particle Swarm Optimization Technique

Khushboo Satpute and Rishik Kumar

Abstract With the advancement of computers and its computational enhancement over several decades of use, but with the growth in the dependencies and use of these systems, more and more concerns over the risk and security issues in networks have raised. In this paper, we have proposed approach using particle swarm optimization to optimize ART. Adaptive resonance theory is one of the most well-known machine-learning-based unsupervised neural networks, which can efficiently handle high-dimensional dataset. PSO on the other hand is a swarm intelligence-based algorithm, efficient in nonlinear optimization problem and easy to implement. The method is based on anomaly detection as it can also detect unknown attack types. PSO is used to optimize vigilance parameter of ART-1 and to classify network data into attack or normal. KDD '99 (knowledge discovery and data mining) dataset has been used for this purpose.

Keywords Intrusion detection system · Adaptive resonance theory 1
Particle swarm optimization

1 Introduction

In recent years, intrusion detection systems have gained attention of the researchers to complement the traditional security systems and enhance the network security aspects of cyber world. Failure to misuse detection is that they detect by comparing with already available attack signatures, i.e. requires patterns corresponding to known attacks. The supervised learning techniques suffer from the downside of requirement for labelled training data to develop a normal behaviour profile.

K. Satpute (✉)
MCOE, Pune, India
e-mail: Khushboosatpute88@gmail.com

R. Kumar
NITK, Surathkal, India
e-mail: kumar.rishik@gmail.com

D. Reddy Edla et al. (eds.), *Advances in Machine Learning and Data Science*,
Advances in Intelligent Systems and Computing 705,
https://doi.org/10.1007/978-981-10-8569-7_1

The previous work proposed by [1] Eldos and Almazyad et al. (2010) provided a good set of training parameters practically for the ART1 obtaining good balance between parameters quality which results in better performance and less time in training In [2] Prothives and Srinoy et al. (2009), integration of adaptive resonance theory and rough set is applied for recognizing intrusion in network. Supervised learning for the fuzzy ARTMAP different synthetic data is learned, and hyper-parameter values and training set sizes are studied in [3] Granger et al. (2007). [4] Al-Natsheh and Eldos et al. (2007) proposed a hybrid clustering system as genetically engineered parameters ART1 or ARTgep based on the adaptive resonance theory 1 (ART1) with a genetic algorithm (GA) optimizer. In [3] Granger et al. (2006), particle swarm optimization eliminates degradation of error. Results obtained with the particle swarm optimization reveal the importance of optimizing parameters. In, [5, 6], Amini et al. applied the UNNID system with intrusion. Results showed ART-1 in 93.5% and ART-2 in 90.7% recognize intrusion attack from normal.

2 Theoretical Background

2.1 *Particle Swarm Optimization*

Swarm intelligence is a population-based stochastic optimization technique where PSO is one of the efficiently developed techniques proposed by Eberhart [7] which is based on bird flocking. The velocity and position of particles are updated using the following equation.

$$v_i^{q+1} = v_i^q + c_1 r_1 (p_i^q - x_i^q) + c_2 r_2 (p_g^q - x_i^q) \tag{1a}$$

$$x_i^{q+1} = x_i^q + v_i^{q+1} \tag{1b}$$

where i = 1, 2, …, n, size of the swarm is given by n, positive constant called acceleration constants are denoted as C1 and C2, and also random numbers are used denoted as r1 and r2.

2.2 *Adaptive Resonance Theory*

Adaptive resonance is neural network that accepts binary inputs introduced by Carpenter and Grossberg [9]. ART neural networks are special because of their special ability to cluster the data dynamically. ART is an unsupervised net and detects similarity between the data. ART is its ability to discover data through vigilance parameter ρ ranging between [0, 1].

3 Proposed Method

The proposed PO-ART1 intrusion detection system has four main phases shown in Fig. 1. Selection of this parameter is important for the performance enhancement of the ART1 so as to efficiently discover data based on similarity measures. Once the optimum vigilance is obtained, the data is classified using ART1 and testing is performed.

Figure 2 shows the randomly generated population is fed into particle swarm optimization algorithm where position of every particle represents the value of vigilance parameter, and using PSO and updating Eq. (1a) and (1b), optimum vigilance parameter is optimized. The ART1 is trained using the internal training strategy, and vigilance for each iteration is applied to evaluate the minimum classification error.

If the vigilance parameter is obtained, then this optimized vigilance parameter is used in training ART1 which categorizes the given data into anomalous data and normal data. Figure 3 shows the architecture of a novel methodology PO-ART1 for network IDS. Stepwise architecture is shown where initially features as input patterns are provided to ART1, and simultaneously randomly generated particles are fed as input to PSO algorithm. Next classification error is evaluated, and ART1 is optimized using PSO. In order to find clusters, following classification error E is used as $E = \frac{Ic}{Tc}$, where Ic is correctly classified and Tc is the total number of patterns. Finally, output is obtained using optimized ART1.

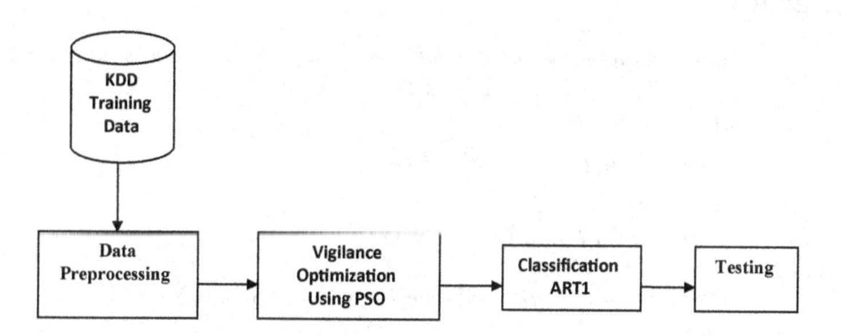

Fig. 1 Block diagram for the intrusion detection system

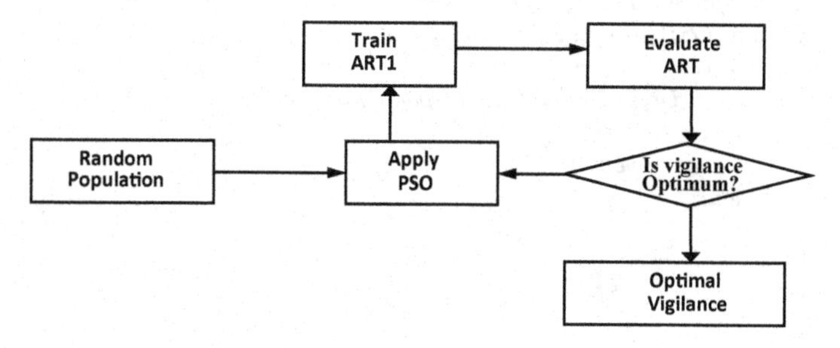

Fig. 2 Selection of vigilance parameter

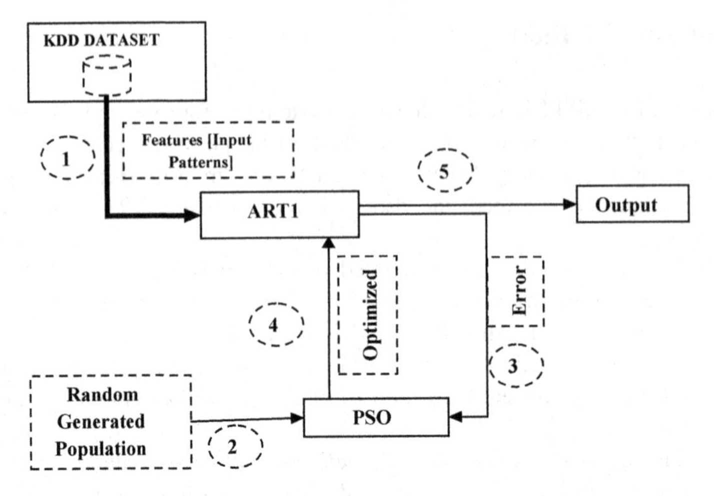

Fig. 3 Architecture diagram for proposed method—PO-ART1

Algorithm 1 shows the pseudo-code of the PSO training strategy applied to optimize ART1 neural networks. It seeks to minimize ART1 classification error $E(x_i^q)$ in the solution space of vigilance parameter values. We define a group of particles.

Algorithm 1

1. *Initialization-maximum number of iteration q and fitness function as E.* *P ,V,* c1 , c2 , *r*1 *and r*2 are set
2. *Start the Iteration with q*=0

while (q $\leq$ *qmax) or E(*P_g^q*)*$\geq$*Edo*

for every particle P do

Train ART1 and Update weights

Compute fitness value E

4 *If E(Cuurent position)<E* (Personal Best Position)

then

particle's personal best position is updated

end

best global fitness is selected from particles

g=Argmin [*E(xi) for every P particle)*

do

Update velocity and Update position

end

increment q

end

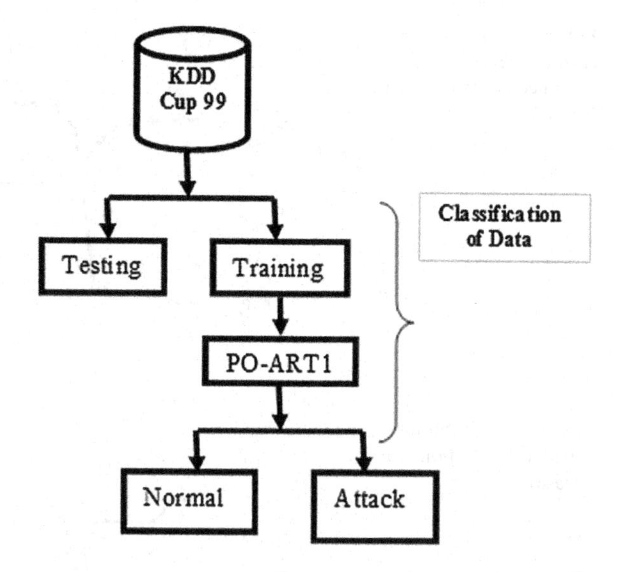

Fig. 4 Classification of Attacks and Normal using PO-ART1

The position of particle i is represented as Xi, and the position of each particle is equal to the value of vigilance parameter ρ, where particle positions are evaluated according to the fitness function being optimized. At first, we set the positions of all particles randomly. Particle is having its previous best position as P I and group global position as Pg. Particle also possesses velocity Vi. Velocity and position of particle i are updated according to PSO Eq. (1a) and (1b). Measurement of any fitness value E $\left(x_i^q\right)$ involves computing classification error for the ART1 network and learn the parameter values at particle position. ART1 has a learning rate $l = 2$, and weight is updated according to the internal training strategy of basic ART1. After selection of optimum vigilance parameter and ART1 performance is enhanced using PO-ART1, the pattern gets classified into two categories as anomaly and normal. Figure 4 shows the classification of data into normal and attack types using PO-ART1.

Table 1 Performance result of PO-ART1-based NIDS and comparison with UNNID ART1

Performance result of PO-ART1 based NIDS			
Proposed method	Detection rate	Accuracy	False alarm rate
PO-ART	98.1	98.8	0.2
Comparison for performance of PO-ART1 and UNNID ART1 [5]			
	True classification Rate (TR)	False Positive Rate a(FPR)	False Negative Rate as (FNR)
PO-ART1	98.8	0.2	0.021
UNNID ART1 [5]	97.42	1.99	0.59

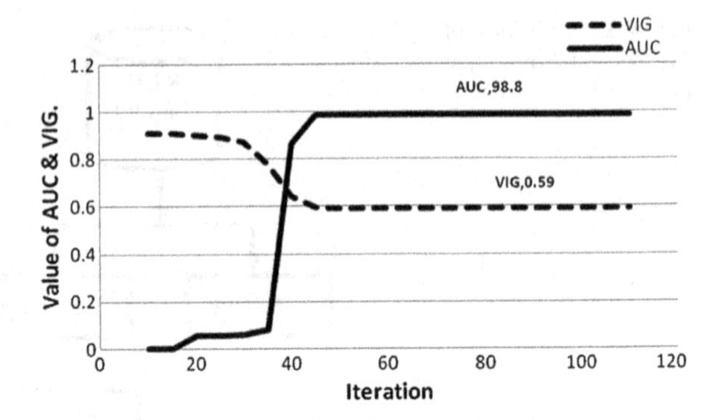

Fig. 5 Comparison of accuracy as (AUC) and vigilance as (VIG) versus iteration

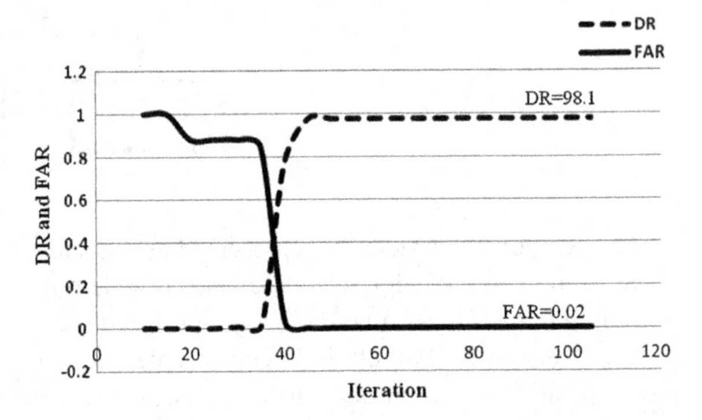

Fig. 6 No. of training cycle versus the detection rate of intrusion

4 Evaluation Criteria

Performance evaluation is based on detection of normal and attack in two classes

- True positive rate (TPR) or detection rate (DR): $\frac{TP}{TP + FN}$
- False positive rate (FPR) or false alarm rate (FAR): $\frac{FP}{TN + FP}$
- Accuracy: $\frac{TN + TP}{TN + TP + FN + TP}$.

As shown in result, performance of optimized adaptive resonance is higher. In Fig. 5, the comparison of the accuracy represented as (AUC) and vigilance represented as (VIG) w.r.t the number of training cycle is shown. Initially with higher vigilance value, very low accuracy is achieved. As shown in graph rate the vigilance parameter gets optimized to 0.59 and then becomes constant, for which higher accuracy of 98.8% is obtained. The performance result of IDS are shown in Fig. 6 where detection rate (DR) and false alarm rate (FAR) with respect to the number of iterations are plotted. After 40 iteration cycles, rate decreases with increasing detection rate is observed which becomes constant after 50 iterations with accuracy as 98.1% with low alarm rate (false) 0.2%.

5 Conclusion

Intrusion detection using an unsupervised neural network ART1 is proposed which is optimized using particle swarm optimization. Compared with the results of conventional ART1 in IDS, experimental results show that accuracy of 98.8% with alarm rate (false) 0.2%. Also the algorithm has very low time and space complexity, is simple and has no complex and difficult computation.

References

1. Eldos, T.M., Almazyad, A.S.: Adaptive resonance theory training parameters: pretty good sets. J. Comput. Sci. **6**, 1443–1449 (2010). https://doi.org/10.3844/jcssp.2010.1438.1444
2. Prothives, K., Srinoy, S.: Member, IAENG. "Integrating ART and rough set approach for computer security" In: Proceedings of the International Multi Conference of Engineers and Computer Scientists, IMECS 2009, vol. I. Hong Kong, March 18–20 (2009)
3. Granger, E., Henniges, P., Sabourin, R., Oliveira, L.S.: Supervised learning of fuzzy artmap neural networks through particle swarm optimization. J. Pattern Recogn. Res. **2**(1), 27–60 (2007)
4. Al-Natsheh, H.T., Eldos, T.M.: Performance optimization of adaptive resonance neural networks using genetic algorithms. In: Foundations of computational intelligence, 2007. IEEE Symposium, 143–148 (2007)
5. Amini, M., Jalili, R., Shahriari, H.R.: RT-UNNID: a practical solution to real-time network-based intrusion detection using unsupervised neural networks. Elsevier Comput. Secur. J. **25**(6) (2006)
6. Amini, M., Jalili, R.: Network-based intrusion detection using unsupervised adaptive resonance theory (ART). In: Proceedings of the Fourth Conference on Engineering of Intelligent Systems (EIS 2004). Madeira, Portugal (2004)
7. Kennedy, J., Eberhart, R.C.: Particle swarm optimization. In: Proceedings of the IEEE International Joint Conference on Neural Networks, p. 1942e1948 (1995)
8. Carpenter, G.A., Grossberg, S.: Adaptive resonance theory. In: Encyclopedia of Machine Learning, Springer US, 22–35 (2010)
9. KDD'99 dataset, University of California, Irvine (1999). http://kdd.ics.uci.edu/databases/kddcup99/kddcup99.html
10. Tavallaee, M., Bagheri, E., Lu, W., Ghorbani, A.A.: A detailed analysis of the kddcup 99 data set (2009) http://nsl.cs.unb.ca/NSL-KDD/

Accelerating Airline Delay Prediction-Based *P-CUDA* Computing Environment

Dharavath Ramesh, Neeraj Patidar, Teja Vunnam and Gaurav Kumar

Abstract Machine learning techniques have enabled machines to achieve human-like thinking and learning abilities. The sudden surge in the rate of data production has enabled enormous research opportunities in the field of machine learning to introduce new and improved techniques that deal with the challenging tasks of higher level. However, this rise in size of data quality has introduced a new challenge in this field, regarding the processing of such huge chunks of the dataset in limited available time. To deal such problems, in this paper, we present a parallel method of solving and interpreting the ML problems to achieve the required efficiency in the available time period. To solve this problem, we use CUDA, a GPU-based approach, to modify and accelerate the training and testing phases of machine learning problems. We also emphasize to demonstrate the efficiency achieved via predicting airline delay through both the sequential as well as CUDA-based parallel approach. Experimental results show that the proposed parallel CUDA approach outperforms in terms of its execution time.

Keywords Machine learning (ML) · Naïve Bayes · GPU · CUDA
Tree reduction

D. Ramesh (✉) · N. Patidar · T. Vunnam · G. Kumar
Department of Computer Science and Engineering, Indian Institute of Technology (ISM), Dhanbad 826004, Jharkhand, India
e-mail: ramesh.d.in@ieee.org

N. Patidar
e-mail: neerajism@cse.ism.ac.in

T. Vunnam
e-mail: vunnamteja@gmail.com

G. Kumar
e-mail: gaurav315kumar@gmail.com

D. Reddy Edla et al. (eds.), *Advances in Machine Learning and Data Science*, Advances in Intelligent Systems and Computing 705,
https://doi.org/10.1007/978-981-10-8569-7_2

1 Introduction

The introduction of new technologies has made the computer science at its best as an emerging field. With the availability of new software paradigms, dependency on computers and machines to perform extreme and computationally exhaustive tasks is growing day by day. For example, the traditional way of file storage on local file systems is replaced by distributing and even more secure cloud storage systems. Among all, the Internet has turned out to be the most resourceful technology ever created for instant sharing of knowledge and resources with the rest of the world with ease. Its popularity around the world can be assessed by the fact that the total number of Internet users has grown by around 82% or almost 1.7 billion in the last 5 years and this number is forecasted to increase up to 4 billion around the year 2020. The main reasons behind this increasing popularity of the Internet are its speed, economic nature and ease of accessibility to the users. The aid of Internet has made the activities easier to a tremendous extent by reducing the communication delay between users in the different parts of the world.

The dependency and conscience to improve our methods by means of research and analysis have led to an immense hike in the rate of data production. As a result of its increased popularity, around 90% of the world's data has come into existence in the past 2 years and Internet has turned out to be the largest contributor among its sources. With the availability of sufficient data for research and analytics, machine learning and related techniques have picked up popularity [1] to achieve some computationally impossible solutions by using well-defined mathematical models. Machine Learning [2] is a term used to define a technique to find a solution or more precisely improve the existing solution gradually by following the process of learning through previous observation or experience without any human intervention. Machine learning is a popular community to solve problems which require human-like instinct and decision-making. With the ease and availability of Big Data, machine learning is used to solve some complex and interesting problems [1] which help in achieving those tasks that normally require some special human assistance, intelligence or decision-making skills to execute successfully. For example, making classification, predictions in advanced robotics, and driver-less cars include such tasks.

1.1 Limitations of Classical ML Methodology

The basic approach to solve the machine learning problem involves two subtasks: (i) training and (ii) testing. Apart from the techniques [1] used, both these subtasks are common in the process of solving any machine learning task. Both these tasks are extremely crucial in solving a machine learning problem. The quality of the training and testing methods eventually decides the overall quality of the model or solution. Initially, training is performed to train or develop a mathematical model/classifier/machine with the help of existing chunks of data in a huge quantity. The

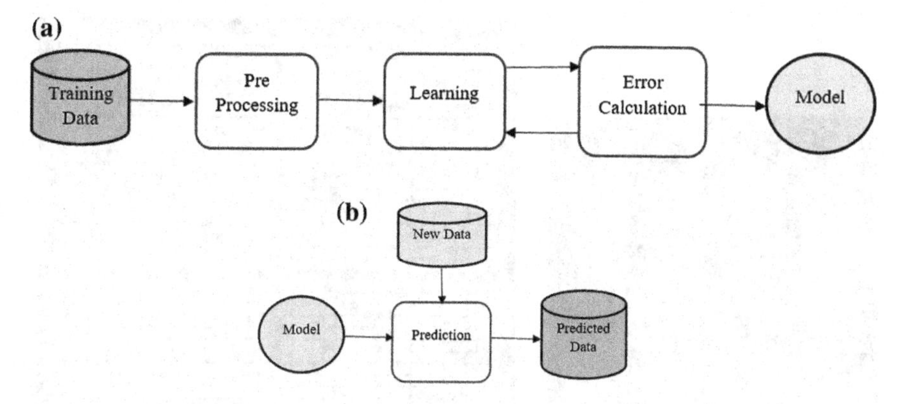

Fig. 1 Machine learning as a process, **a** learning phase, **b** testing phase

vast amounts of data help in improving the accuracy and performance of the model, but processing these chunks of data is another crucial problem. Hence, the overall performance of the ML algorithm is limited by the processing methodologies used to process the data for training the model. The same can be improved to process those huge datasets, so that appreciable results can be achieved within the conserved time limits. A learning phase and training phase of ML process are depicted in Fig. 1a, b.

The second task is testing and analysing the accuracy of the model by using the related test data. This step is crucial as it measures the accuracy and the acceptability of the model. But, the solution must work for regular as well as new problems [2], i.e. to make the solution robust against different possible situations. A suitable ML model can be used to test the model against a large dataset which may or may not belong to dataset. Again, the vastness of the dataset will extend the evaluation period of our model which is completely undesirable. To deal with these kind of problems, several efforts have been made in this field to improve its performance. To over-relate and achieve this problematic instance, in this paper, we discuss a strategy using the GPU-based parallel processing platform named as CUDA, to perform the ML tasks in allowable time limits [3, 4]. This approach works in an efficient way and can make the training and testing strategies workable within the time period scan.

2 Resolving with CUDA Platform

2.1 CUDA

In 2007, NVIDIA, a GPU-designing firm, released the initial version of CUDA—a parallel computing platform which uses the graphical processing units (GPUs) for massive parallelization of computation tasks [4]. It follows general purpose

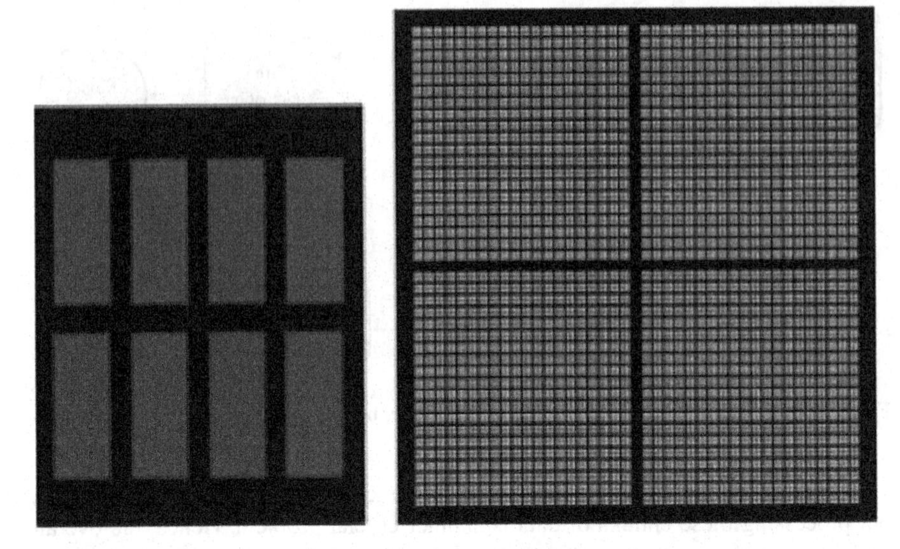

Fig. 2 CPU versus GPU

computing, on graphics processing unit (GPGPU) approaches to perform the general tasks which are computationally intensive and require a lot of time to be executed if performed on the CPU [5]. It allows users to write massively parallel programs using languages like C, C++ and Fortran [6]. The CPUs are optimized for sequential processing and contain only limited numbers of powerful cores, whereas on the other hand, GPUs are optimized for performing parallel tasks and contain thousands of smaller and less powerful cores [7]. The graphical representation of CPU and GPU is depicted in Fig. 2.

2.2 Impact on Solving ML Problems

Solution to the ML problems using traditional systems may turn out to be an expensive technique with the increase in the size of the dataset. Especially, the time required to process such a large set of data increases drastically. Using parallel GPU, computation can turn out to be a perfect solution for solving such large-scale machine learning problems without any extra increase in the machine cost by harnessing the graphical computation power of the system [3, 8].

3 Problem Formulation: Predicting Flight Delay Occurrences

Every year around one-fifth of the airline flights suffer a cancellation or delay which results in critical loss of time and resources to both airlines and passengers. Hence, creating a model which can predict whether a given flight with certain parameters like source, destination, duration will suffer a delay or not can offer huge help to the passengers and airline managers in choosing and managing their flights. Here, we use a dataset from American Statistical Association (ASA) 2009 data expo [9], which includes dataset for all commercial flights within USA from year 1987 to 2008. This dataset includes nearly 120 million flight records. To predict the future occurrences, we consider Naïve Bayes Classifier (NBC). The sequential standard of NBC is described in Sect. 3.1.

3.1 Solution: Sequential Naïve Bayes Classifier

3.1.1 Naïve Bayes Classifier

Naïve Bayes Classifier (NBC) belongs to a class of conditional probabilistic classifiers which assume a major assumption of independence between all features of its dataset to calculate most probable outcome depending on the input given by the user [10]. It uses Bayesian interpretation to measure a degree of belief to identify the class which includes the input features [11]. Given an input in the form of a vector;

$$X = (x_1, x_2, x_3, \ldots, x_n) \tag{1}$$

The NBC will provide the probability P_k for each class, whether the input vector X belongs to the given class C_k, i.e.

$$P(C_k \mid (x_1, x_2, x_3, \ldots, x_n)) \tag{2}$$

Using Bayes theorem, the given probability can be decomposed as;

$$P(B \mid A) = \frac{P(B)P(A \mid B)}{P(A)} \tag{3}$$

The above formulae can be extended to calculate the joint probability considering several parameters for each sample;

$$P(B_i \mid A) = \frac{P(B_i)P(A \mid B_i)}{\sum_{i=1}^{n} P(B_i)P(A \mid B_i)} \quad (4)$$

However, the denominator can be ignored since it will remain same for all the classes [12]. Hence, the classification problem reduces to finding numerator for a given input vector *X*. Using the assumption of independence between the features of the vector *X*, i.e. the probability of occurrence of one feature, is independent of the probability of occurrence of any other feature or collection of features. At the same time, we can simplify the calculation of the numerator probability same as a joint probability.

3.1.2 Normal Distribution

For calculating the probability of occurrence of features which are continuing in nature rather the discrete, we use the normal (also known as Gaussian) distribution function [10] which can be given as follows.

$$P(x_k \mid C_i) = g(x_k, \mu_{Ci}, \sigma_{Ci}) = \frac{1}{\sqrt{2\pi}\sigma_{Ci}} e^{\frac{(x_k - \mu_{Ci})^2}{2\sigma_{Ci}^2}} \quad (5)$$

where $g(x_k, \mu_{Ci}, \sigma_{Ci})$ is the Gaussian (normal) density function for attribute A, while μ_{Ci} and σ_{Ci} are the mean and standard deviation, respectively, given the values for attribute A for C_i. Advantages of using NBC over other classifiers are: (i) simple and easy to test new data, (ii) minimum error rate as compared to other complex algorithms, (iii) high degree of accuracy with lower computational complexity and (iv) compete with more advanced algorithms like SVM, etc.

3.2 *Methodology: Predicting Airline Delays*

3.2.1 Training the Model

To train the model, we start by calculating mean and variance for each relevant parameter in the flight description which belong to either class $C_{Delayed}$ or $C_{Nodelay}$. The calculated values of mean and variance are shown in Table 1.

Using the table similar to above, we calculate the probabilities to find out the numerator for given values of features of flight whose class has to be identified as either delayed or on-time.

Table 1 Sample airline delay database

Sample	Dep. time	CRS dep. time	Arr. time	CRS arr. time	Act elapsed time	CRS elapsed time	Arr. delay	Dep. delay
1.	741	730	912	849	91	79	23	11
2.	729	730	903	849	94	79	14	−1
3.	741	730	918	849	97	79	29	11
4.	729	730	847	849	78	79	−2	−1
5.	749	730	922	849	93	79	33	19
6.	728	730	848	849	80	79	−1	−2
7.	728	730	852	849	84	79	3	−2
8.	731	730	902	849	91	79	13	1
9.	744	730	908	849	84	79	19	14
10.	729	730	851	849	82	79	2	−1

3.2.2 Testing

After performing preprocessing, we use mean and variance calculated in the previous step to find the posterior probabilities for the given test samples. After combining the posterior probabilities of all the features, we get the combined probabilities of that test sample belongs to the same class. For each sample, the class with the highest numerator value can be considered as the most optimum prediction for that sample. After this, we apply the strategy to find the probability for all the flight variables in the test dataset one by one, sequentially. The corresponding algorithm which includes these two steps is shown in Algorithm 1.

Algorithm 1: *NBC Classification algorithm*

Input: *Airline Dataset*
Output: *Set of probabilities for each class*

Step-1. Select the relevant features for each flight to be considered in the flight delay evaluation.
Step-2. Find **Mean** () and **Variance** () for the selected features over each class in the complete dataset.
Step-3. For each input vector, **X** calculate the individual probability of occurrence for each different class.
Step-4. Combine all the probabilities, to calculate the joint probability in the numerator.
Step-5. Select and assign each input vector class which has the highest numerator value.

---

4 Parallelizing the Approach

In this section, we test and improve the performance of the above algorithm using the parallel CUDA computing model [13, 14]. To accelerate the process to find the mean, we use a methodology known as *Tree-Based Reduction Sum*. This technique utilizes the parallel architecture of CUDA [6] to efficiently calculate the mean and variance for a given feature belonging to a particular class. The instance of *Tree-Based Reduction Sum* is depicted in Fig. 3.

In reduction sum, we particularly focus on the tree reduction technique to evaluate the sum of the given large set of values which can produce a speedup nearly equal to *log(n)*. As CUDA has no global synchronization facility to easily communicate with all the threads [13], this limitation can be avoided by using a recursive kernel invocation while adding a small hardware/software overhead in the individual kernel launch [15]. The state of recursive kernel invocation is shown in Fig. 4.

4.1 Parallelizing Testing Phase

To classify the test dataset, we select an optimum block size (i.e. the total number of threads per block) and total number of blocks each containing the threads equal to the block size [16]. After passing the set of mean and variance related to each

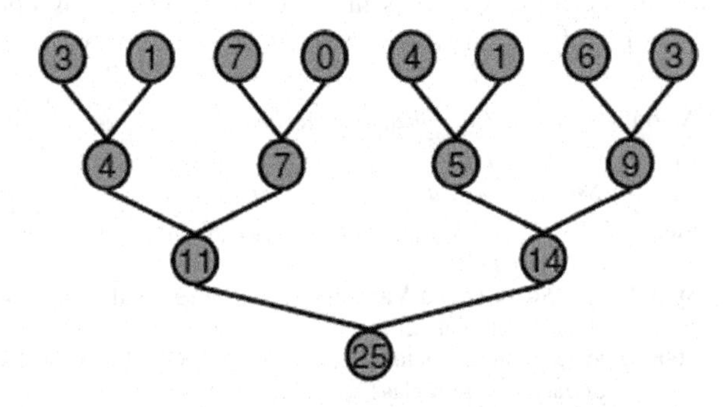

Fig. 3 Tree reduction sum approach

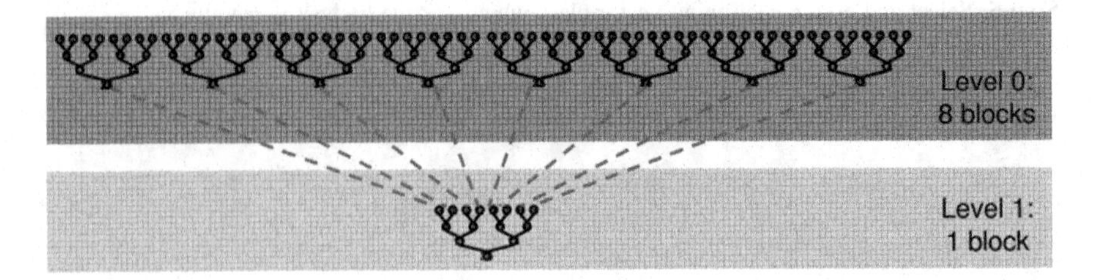

Fig. 4 Recursive kernel invocation

feature in the flight description, we calculate the individual probabilities of the test flight belonging to a particular class one by one for all possible classes [9]. Figure 5 shows the calculation through the NB model. Hence, each block processes the number of flights equal to the number of threads, i.e. one thread per test flight that evaluate the total results. As compared to the sequential processing of each test flight, the parallel CUDA algorithm only takes time equivalent to the number of threads per block and some communication time between the CPU and GPU to distribute [17] the process and recombine the solution. The modified parallel classification algorithm is shown in Algorithm 2.

Algorithm 2: *Parallel NBC Classification algorithm*

Input: *Airline Dataset*
Output: *Set of probabilities for each class*

Step-1. Select the relevant features for each flight to be considered in the flight delay evaluation.
Step-2. Find **Mean** () and **Variance** () using the reduction sum technique for the selected features over each class in the complete dataset.
Step-3. Divide the dataset into set of blocks which includes flight vectors equal to number of threads. For each input vector, **X** calculate the individual prob ability of occurrence for each different class.
Step-4. Combine all the probabilities, to calculate the joint probability in the numerator.
Step-5. Select and assign each input vector class which has the highest numerator value.

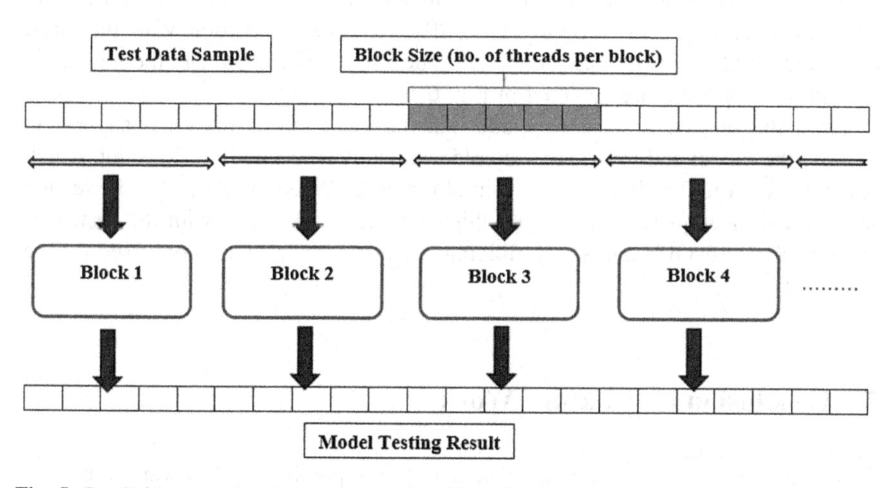

Fig. 5 Parallel test results calculation through NB model

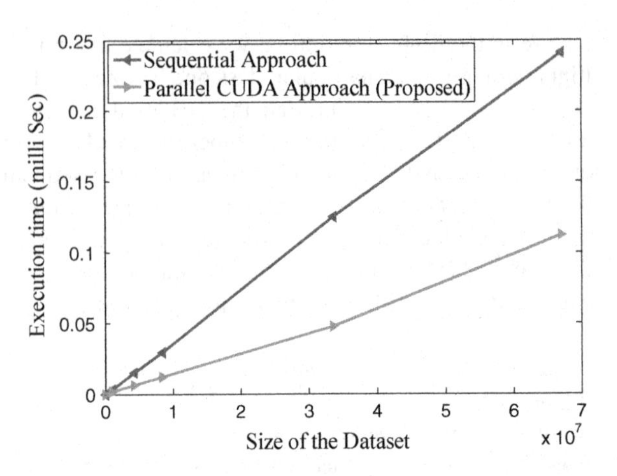

Fig. 6 Parallel CUDA implementation with the sequential method

4.2 *Experimental Analysis*

To perform the experiment-related approaches, we use the airline dataset of the American Statistical Association (ASA) [9]. The system on which we have conducted our tests has Windows 10 operating system running over Intel core i5-3337U processor 1.80 GHz with turbo boost up to 2.70 GHz and 6 GB of DDR3L internal memory. The GPU version is NVIDIA Geforce GT 740 M with dedicated 1 GB DDR3 memory. The coding and debugging of the CUDA project are done over Microsoft Visual Studio 2015 community edition with the latest CUDA toolkit 8.0 installed on the system. The compared results with the sequential algorithmic approach are depicted in Fig. 6.

From the results, we can conclude that the inclusion of the GPU support in training the model and testing the sample test data has resulted in a speedup which is nearly equal to 2.5 times the sequential methods. It also outperforms the related time execution sequences. However, this calculation may vary with different versions of NVIDIA GPU cards with different CUDA core configurations used for the purpose [18].

5 Conclusion and Future Work

This paper aims to describe how the enormous parallelizing power of CUDA platform can be utilized to accelerate the process of developing the ML model for prediction and the classification purpose. To make this scenario's performance better, we used the reduction sum technique to parallelize the task of calculation of the mean in the training phase and parallelized the testing process using the simple CUDA computation scheme. Both these techniques result in combined actual improvement over the sequential methodology to calculate the model and evaluate the result. Similar improvements can be made in developing other ML models by utilizing the resources offered through CUDA computing platform to improve the time and resources used by overall prediction and classification process without using costly machines to process on it.

Acknowledgements This work is partially supported by Indian Institute of Technology (ISM), Government of India. The authors wish to express their gratitude and thanks to the Department of Computer Science and Engineering, Indian Institute of Technology (ISM), Dhanbad, India, for providing their support in arranging necessary computing facilities.

References

1. Liao, S.H., Chu, P.H., Hsiao, P.Y.: Data mining techniques and application—a decade review from 2000 to 2011. Expert Syst. Appl. **39**(12), 11303–11311 (2012)
2. Carbonell, J.G., Michalski, R.S., Mitchell, T.M.: An overview of machine learning. In: Machine Learning, pp. 3–23. Springer, Berlin, Heidelberg (1983)
3. Che, S., Boyer, M., Meng, J., Tarjan, D., Sheaffer, J.W., Skadron, K.: A performance study of general-purpose applications on graphics processors using CUDA. J. Parallel Distrib. Comput. **68**(10), 1370–1380 (2008)
4. Wu, R., Zhang, B., Hsu, M.: GPU-accelerated large scale analytics. IACM UCHPC (2009)
5. Ghorpade, J., Parande, J., Kulkarni, M., Bawaskar, A.: GPGPU processing in CUDA architecture (2012). arXiv:1202.4347
6. Farber, R.: CUDA Application Design and Development. Elsevier (2011)
7. Yang, C.T., Huang, C.L., Lin, C.F.: Hybrid CUDA, OpenMP, and MPI parallel programming on multicore GPU clusters. Comput. Phys. Commun. **182**(1), 266–269 (2011)
8. Harris, M.: Optimizing CUDA. In: SC07: High Performance Computing with CUDA (2007)
9. Data for experimentation, American Statistical Association, Data Expo (2009). http://stat-computing.org/dataexpo/2009/the-data.html
10. Murphy, K.P.: Naive Bayes Classifiers. University of British Columbia (2006)
11. Rish, I.: An empirical study of the naive Bayes classifier. In: IJCAI 2001 Workshop on Empirical Methods in Artificial Intelligence, vol. 3, No. 22, pp. 41–46. IBM, New York (Aug 2001)
12. Jia, P.T., He, H.C., Lin, W.: Decision by maximum of posterior probability average with weights: a method of multiple classifiers combination. In: Proceedings of 2005 International Conference on Machine Learning and Cybernetics, vol. 4, pp. 1949–1954. IEEE (Aug 2005)
13. Jian, L., Wang, C., Liu, Y., Liang, S., Yi, W., Shi, Y.: Parallel data mining techniques on graphics processing unit with compute unified device architecture (CUDA). J. Supercomput. **64**(3), 942–967 (2013)
14. Zhou, L., Wang, H., Wang, W.: Parallel implementation of classification algorithms based on cloud computing environment. TELKOMNIKA Indones. J. Electr. Eng. **10**(5), 1087–1092 (2012)
15. Fang, W., Lau, K.K., Lu, M., Xiao, X., Lam, C.K., Yang, P.Y., Yang, K. et al.: Parallel data mining on graphics processors. In: Technical Report HKUST-CS08-07. Hong Kong University Science and Technology, Hong Kong, China (2008)
16. Chengpeng, Y., Zhanchun, G., Yanjun, J.A.: GPU-based Native Bayesian algorithm for document classification. http://www.paper.edu.cn/lwzx/en_releasepaper/content/4570429. Accessed 26 Nov 2013

17. Viegas, F., Andrade, G., Almeida, J., Ferreira, R., Gonçalves, M., Ramos, G., Rocha, L.: GPU-NB: a fast CUDA-based implementation of naive bayes. In: 2013 25th International Symposium on Computer Architecture and High Performance Computing (SBAC-PAD), pp. 168–175. IEEE (Oct 2013)
18. Zhou, L., Yu, Z., Lin, J., Zhu, S., Shi, W., Zhou, H., Zeng, X. et al.: Acceleration of Naive-Bayes algorithm on multicore processor for massive text classification. In: 2014 14th International Symposium on Integrated Circuits (ISIC), pp. 344–347. IEEE (Dec 2014)

IDPC-XML: Integrated Data Provenance Capture in XML

Dharavath Ramesh, Himangshu Biswas and Vijay Kumar Vallamdas

Abstract In the contemporary world, data provenance is an acute issue in the world of www due to its openness of the Web and the ease of copying and combining interlinked data from different database sources. The term data provenance is defined as lineage of data and movement between databases. Scientists and enterprises use their own analytical tools to process the data provenance. In the current scenario, workflow management systems are popular in scientific domains due to the level of standardization of data formats and analysis. Using graph visualizations, scientists can easily view the data provenance associated with a scientific workflow of any data to understand the methodology and to validate the results. In this paper, we emphasize on a tool-based PROV-DM for collecting provenance data in the XML file and visualizing it as directed graph. We also propose an approach named IDPC-XML for processing and managing the internal data using XML file. This tool collects data provenance obtrusively in a local system using self-generated log and also collects provenance data in XML format which can be visualized as a directed graph to understand the convergence. Relevant case studies of IDPC-XML are discussed and further research scope is pinpointed.

Keywords Data provenance · PROV-DM · IDPC-XML · Visualization XML

D. Ramesh (✉) · H. Biswas · V. K. Vallamdas
Department of Computer Science and Engineering,
Indian Institute of Technology (ISM), Dhanbad 826004, Jharkhand, India
e-mail: ramesh.d.in@ieee.org

H. Biswas
e-mail: himangshu4110biswas@gmail.com

V. K. Vallamdas
e-mail: vijay.vallamdas@gmail.com

D. Reddy Edla et al. (eds.), *Advances in Machine Learning and Data Science*,
Advances in Intelligent Systems and Computing 705,
https://doi.org/10.1007/978-981-10-8569-7_3

1 Introduction

Provenance is used to denote the documented history of artwork and the chain of ownership. It also helps to determine the authenticity and trustworthiness of the artwork tool [1]. With the revolutionized technologies, distributing computing provides substantial benefits to the enterprises and scientists to manage their information with appropriate features. Features like copying and combining the relevant data do not provide adequate guarantee, reliability, and accountability for that data [2]. Due to this, scientists, researchers, and private enterprises are not able to assure the quality, integrity, and authenticity of the data on the Web. If all the works and efforts of the researcher turn out to be based on substandard data, then the outcome of these works may damage the user's authenticity and trust. Such substandard data can be steered by finding its provenance, explaining its history, and other information which is helpful in verifying its quality and scientific measurements [3].

In other words, authenticated data reduce duplication and ensure the trust against data loss. Because, provenance ensures the genuineness of the result and can be used by the scientists and researchers to analyze their data which allows them for suspension and resumption of an experiment and provide detailed references to find the data collected for the experiment. In the context of e-Science, the provenance of workflow-generated data is a record of the workflow processing steps that contributed to the production of the data. Depending on the operating environment, provenance metadata may include (i) information, (ii) references of the workflow tasks, (iii) the computing resources involved, (iv) process the data which are running and dependencies between the data artifacts, (v) software and user interactions, and (vi) system environment settings, and many observable elements that may support the experiment of the scientist's claims [4].

Keeping the above constraints into an account, in this paper, we introduce an approach called Integrated Data Provenance Capture (IDPC) in XML format, which can be visualized to find provenance and the workflow graph of any application. In many applications, XML plays a crucial role and in the near future, it may become more crucial. In the present era of digitization, despite most databases are still relational, there are many online biological databases which are in XML format. XML-based message system and XML databases are also often used in distributed grid and workflow computing systems. The problem of tracking provenance information could be simplified by standardizing on XML data models and its related databases if it crosses the system boundaries [2]. An "*id*" attribute can be declared as a user-defined identifier in the XML document with a unique value and can be used to detect the element of the document. XPath [5] is a path description language for XML on which a key of XML-schema relies. Through the data and a key is a combination of two paths, the first path describes the target set. The target is a set of nodes on which a key constraint is imposed. The second path called *key path*, which uniquely identifies nodes in the target set where at the end of the key paths two distinct nodes in the target set must have distinct values. Unlike, well-organized databases and documents, complex language like XPath does not consider the hierarchical structure of keys [6]. With effective visualizations of these XML files formats, the complex hierarchical structure of system provenance data can be understood without much difficulty. Any provenance data mostly visualize as a node-link diagram, which helps to understand the local activity very effectively, but when it comes to interrelationships within the data and high-level abstract of activity, it is very unsuccessful [7]. But with our approach, provenance can be visualized as both node- and edge-link diagram with different two-dimensional layouts and visual styles like mapping node color, labeling node, border thickness.

2 Related Works

To meet the requirement of various applications and relevant domains, different approaches have been introduced by scientists and researchers to support data provenance. For example, scientists in the individual domain use the individual format of provenance for their work. But, for the experimental data and results, they follow a common representation policy which helps others to understand it [8]. Many approaches have been introduced to find data provenance from an application or system. Overviews of few approaches and techniques have been discussed in the literature. The Chimera [9] is a virtual data grid (VDG) based on a relational virtual data schema for tracking and managing provenance data derivations. A process-oriented model is used to record provenance and virtual data language is used to construct a workflow graph. For executing requests and querying database entries, a virtual data language interpreter is used. The VisTrails system [8] is a provenance management system which supports multiple view visualization and data exploration [10]. Among the views, the visualization trial pipeline is one of the key components of the system. It can also be connected to another visualization system, e.g., Kitware's Visualization Toolkit [11] for creating and maintaining visualization workflows.

In Python bytecode [12], the provenance tool computes dependencies between the input and the output of a given Python program. While execution, the input programs log their run-time data. The custom interpreters execute the input programs in a symbolic way and obtain dependencies between arguments and return the values. The MyGrid project [13] records provenance data for workflows in XScufl language that are executed using the *TAVERNA* workflow engine [14]. During the workflow execution, a provenance log is automatically recorded in a framework. The framework can be differentiated between four provenance levels: *process level, data level, organization level, and knowledge level.* Each level deals with invoked processes, derivation paths, metadata, linking with other provenance, respectively, but provenances are available in one of the two forms as *derivation path* and *annotations*. Derivation path is attached to process and produce the result and annotations are attached to a collection of objects. The Kepler [15] provenance framework is a scientific workflow system with provenance database and API in the core. It has three components: *recording API, query API, and management API.* The database is self-contained and stores all the provenance information using recording API. Query API is used to access the data and management API maintain the database.

The Karma provenance framework [16] records provenance data for scientific workflows, which are uniform and used with minimum performance overhead based on OPM [17]. In this work, the provenance is collected in two forms: process provenance, data provenance. Process provenance or workflow trace describes the workflow's execution and associated service invocations. It is also used to monitor the workflow progress. The data provenance provides metadata about the derivation history of data in the workflow. It also provides the services that create and use it, and the input data transformed to generate it. Komadu [18] is a standalone provenance collection tool, which meets W3C PROV [19] compliant and can be a conjunction to an existing cyberinfrastructure to collect and visualize provenance data. It is the successor of Karma provenance framework, which provides a Web services-based API for both provenance collection and querying collected data.

3 Data Provenance Approach

Provenance information is useful in estimating the standard of data, inspection of data, detecting repetition of data, accountability, and the discovery of substandard data. Our main objective through IDPC-XML provenance approach is to find the origin of the data along with some other factors like the owner of the data, the user of the data, with whom the data associated, from where it is derived, etc. We use IDPC-XML in our local system (PC, laptop/workstation) with UBUNTU basic configuration. However, its performance would be better with the greater RAM and may vary with bigger load and notification size. This approach can also used be to collect provenance from the processing tools, services, applications, and middleware systems. The generated provenance data are used to produce the workflow graph by visualization and to trace the link between nodes. The working of IDPC is depicted in Fig. 1.

From Fig. 1, we can observe that the client messenger gathers provenance data from generated log files and uses a message broker to send messages to IDPC. IDPC service captures provenance data from the system through log files by using message broker. Then, the provenance of the data is stored in a repository which is visualized through different tools.

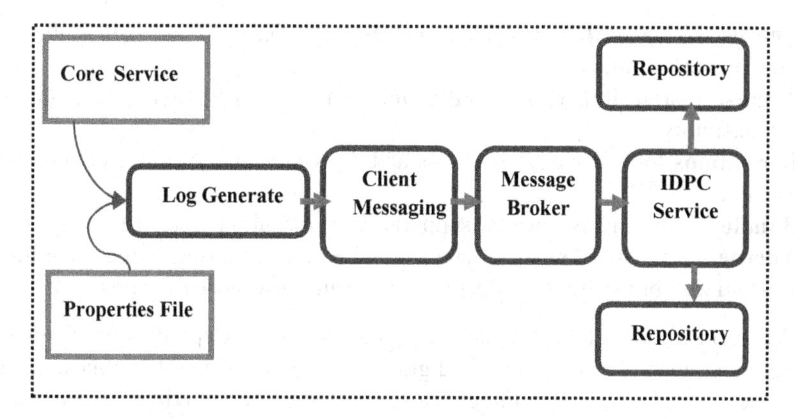

Fig. 1 Flowchart of IDPC

3.1 Data Provenance Model

To visualize the model, we use PROV-DM model which is a conceptual data model based on the W3C provenance (PROV) family of specifications [19, 20]. It is domain agnostic and application specific that forms the core structure of provenance from the extended structures serving for more specific uses of provenance. The core structure of PROV-DM is illustrated in Fig. 2.

It has six components where an entity, activity, and agent as PROV-DM types and rests are relations.

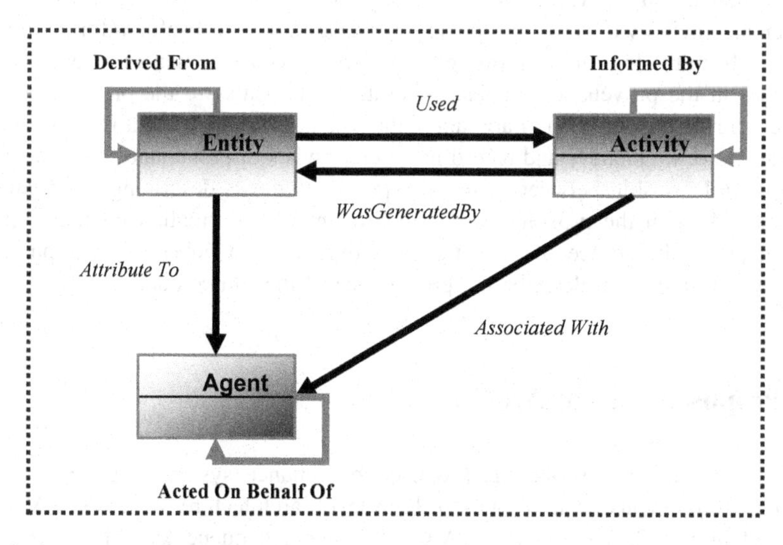

Fig. 2 Core structure of PROV-DM

- **Entities and activities** deal with entities, activities, and links. It is the only time-related component.
- **Agents, responsibility, and influence** deal with the agents regarding its responsibility.
- **Derivations** formed with derivations and derivation subtypes like entities from entities.
- **Bundle** is a mechanism which supports types of sub-provenance.
- **Alternate** consists of relations and linking entities referring to the same thing.
- **Collections** form a logical structure for its sub-provenance members.

The specification of this model is based on the concept of a world where "*things*" are physical, conceptual, or digital and "*activities*" involving these things. The concept of characterizing things fixes some aspects of activities involve things in multiple ways by consuming, processing, transforming, changing, modifying, relocating, using, generating, and controlling them. Apart from the three PROV-DM "*types,*" the relational concept of the PROV-DM deals with the generation, usage, communication, derivation, attribution, association, and delegation.

3.2 Data Provenance Analysis

Data provenance is extremely effective and efficient when data are created and used in distributed systems and applications. One of the critical tasks after capturing the provenance for users is to understand the provenance log and how to extract important and useful information from it. The physical form of the log file is represented in different formats. A tabular format is a coherent way to represent, where change entry or recorded events are presented in row wise [21]. But, users can create their own views of the log for extracting the useful information as required from the log. In our proposed IDPC, provenance elements are represented as nodes in the provenance graphs. Provenance elements are the piece of provenance information about any particular data or dataset, such as creator, user, of the dataset. Here, we identify and record in different types of provenance elements and trace the relationship between different types of elements describing the relationships as edges in the provenance graph. Provenance information contains information about the source of data. On the other hand, a provenance graph may contain subgraphs that describe the provenance of the source data.

4 Proposed System Architecture

In this section, we describe the proposed provenance system architecture with suitable components. The proposed high-level architecture of IDPC-XML is depicted in Fig. 3. We use RabbitMQ [22] message queue known as message

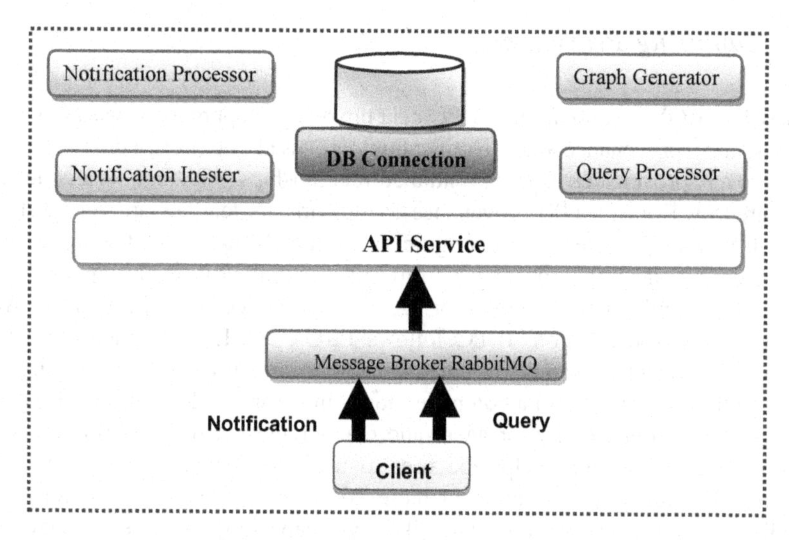

Fig. 3 System architecture of IDPC

broker which is an event system that implements the Advanced Message Queuing Protocol (AMQP) [23] and supports asynchronous communication. A client needs to be created to send queries and notification in IDPC. RabbitMQ is used to receive provenance notifications, to publish messages, and to send responses to the incoming messages queries. It also allocates persistent and reliable storage.

Provenance notification can be ingested into IDPC through tools, services, applications, and log processing scripts. After the execution of the application, log script translates execution log for provenance information in the form of metadata. To generate the provenance graphs and retrieve provenance information, queries can be issued to IDPC. Provenance information relates to elements like agents, entities, and activities of the data. For this, a configured channel of RabbitMQ setup is created.

There are two types of APIs used in IDPC: ingest and query API. These two are exposed through the RabbitMQ channel. During provenance collection, the ingested API sends provenance notifications. After that query API is issued once, the provenance is captured to collect information about activities or entities. To collect and store the information, MySQL database is used, where notification ingester ingests the XML notification and notification processor processes asynchronously. The processor also checks the unprocessed notification periodically in the database. The query processor processes all the incoming queries of the agents, activities, and entities and responds the queries by accessing the database. After this, the graph generator is invoked to create nodes from activities, entities, and agents. Finally, connects the nodes as per their relationship with each other to form a provenance graph. To perform this activity, the database connection pool helps to manage the database connection efficiently.

4.1 Capturing Provenance

As a W3C PROV compliant, the IDPC data provenance approach is used to collect and visualize provenance data. Notifications are used for provenance collection related to any provenance elements and their relationships. Although its architecture is inspired by Komadu [18] provenance framework, its features are very distinct, focused, user-friendly, easy to configure and setup. It also provides greater efficiency while collecting provenance from any system. Provenance graph details about any particular activity, entity, or agent can be collected by using query API. Due to PROV specification, IDPC follows PROV XML [24] schema. First, it generates provenance graphs by creating a node and uses its relationships for creating other nodes to form a complete graph. In other words, nodes of the graph represent the activity, entity, or agent and edges represent their relationship with other nodes. This feature helps the user to track the lineage of any particular elements. The related information of these nodes is captured in XML format to visualize through a visualization tool like Cytoscape [25] for different view and layouts.

4.2 Result: Visualization of Provenance Data

Through provenance visualization, the user can navigate the provenance. As a result, scientists are relying on visualization tools which help them understand the large and complex data generated from the complex virtual environment or cyberinfrastructure experiments. This also helps to interact and explore the provenance by providing different layouts of the configured map. For provenance visualization, we use Cytoscape [25] as it provides detailed visualizations with additional annotations. Cytoscape offers several default meaningful layouts for provenance visualization. It also allows users to encode any attribute of data as a visual property. These encoded or mapped attribute sets are known as visual style. The visualized provenance data are useful for manipulating provenance graphs and for displaying views. This helps the users of the system to navigate their experiment information by comparing it with other experiments those runs quantitatively. A sample view of IDPC captured log in XLS format is shown in Fig. 4.

The provenance information collected by IDPC is visualized using Cytoscape is shown in Fig. 5. From this, we can see that the activities, entities, and agents generated by IDPC are the nodes of the graph and edges representing the relationship between them. Here, we take Agent (Agent_91) as starting node which invokes two services (Activity_132) and (Activity_134). The (Activity_132) generates an output file (File_86). Agent (Agent_92) which acted on behalf of (Agent_91) invokes a service (Activity_131). (Activity_131) uses a file (File_85). This graph only shows the basic overview of the visualization graph for an understanding of elements and relationships. The other contained additional

	A	B	C	D	E
1	edgeType	source	dest	location	location.type
2	used	Activity_131	File_85	Location2	xsd:string
3	wasGeneratedBy	File_86	Activity_132	Location1	xsd:string
4	wasAssociatedWith	Activity_134	Agent_91		
5	wasAssociatedWith	Activity_131	Agent_92		
6	wasAssociatedWith	Activity_132	Agent_91		
7	actedOnBehalfOf	Agent_91	Agent_92		

Fig. 4 Filtered provenance log in XLS format

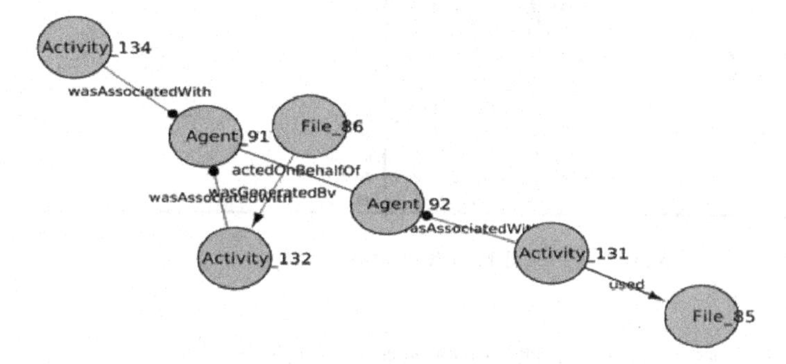

Fig. 5 Visualization of provenance data

attributes are displayed in a separate window, which are ignored while focusing on the specific part of the graph.

A complete visualization graph of IDPC in Cytoscape is depicted in Fig. 6. At the same time, a complete capture of IDPC log containing provenance information in the readable tabular form is shown in Fig. 7.

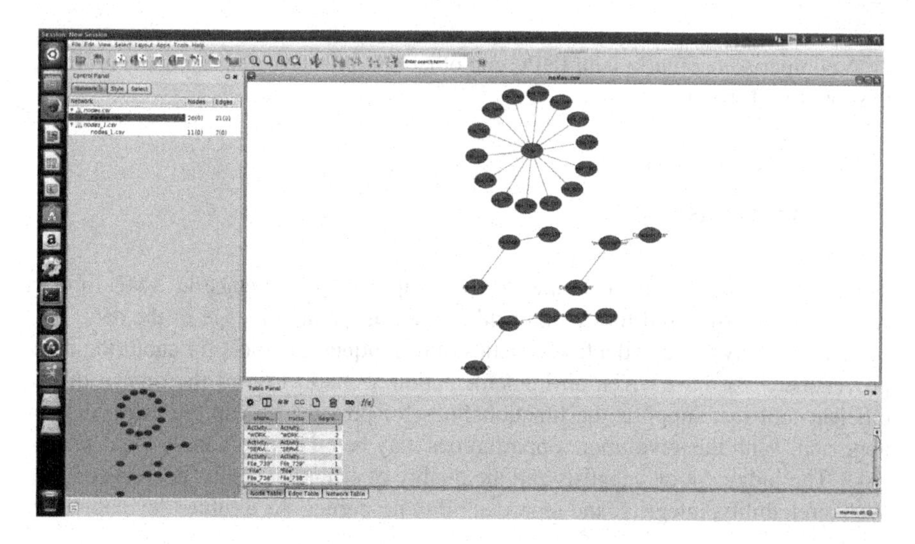

Fig. 6 Visualization of IDPC with Cytoscape

	A	B	C	D	E
1	edgeType	source	dest	location	location.type
2	used	Activity_335	Collection_726	Location2	xsd:string
3	used	Activity_334	File_725	Location2	xsd:string
4	wasGeneratedBy	Collection_730	Activity_336	Location1	xsd:string
5	wasGeneratedBy	File_728	Activity_336	Location1	xsd:string
6	wasGeneratedBy	File_732	Activity_334	Location1	xsd:string
7	wasDerivedFrom	File_734	File_733		
8	wasDerivedFrom	File_735	File_733		
9	wasDerivedFrom	File_736	File_733		
10	wasDerivedFrom	File_732	File_725		
11	wasDerivedFrom	File_733	File_728		
12	actedOnBehalfOf	Agent_199	Agent_200		
13	alternateOf	File_733	File_737		
14	wasAttributedTo	File_728	Agent_199		
15	specializationOf	File_733	File_738		
16	wasInformedBy	Activity_336	Activity_334		
17	wasInformedBy	Activity_336	Activity_335		
18	wasInvalidatedBy	File_728	Activity_334	Location4	xsd:string
19	wasStartedBy	Activity_335	File_729	Location2	xsd:string
20	wasEndedBy	Activity_335	File_731	Location3	xsd:string
21	hadMember	Collection_730	File_727		
22	hadMember	Collection_730	File_740		
23	hadMember	Collection_726	File_727		
24	hadMember	Collection_726	File_739		

Fig. 7 Example of provenance log data in XSL format

Table 1 Comparison of IDPC features with other tools/methods

	IDPC	VisTrails	Taverna2/Kepler	Karma
Record	Yes	Yes	Yes	Yes
API	Yes	No	No	Yes
Graph generation	Yes	Yes	Yes	No
W3C PROV	Yes	No	No	No
Visualization	Yes	Yes, workflow version tree	Yes, data flow graph	Yes

A comparison scenario with IDPC and other provenance capture models in XML is shown in Table 1.

5 Conclusions

This paper advances a better framework for capturing provenance in XML format which can be visualized for though study of the origin and lineage of the data. We have also briefly discussed a few varieties of techniques and tools for capturing data provenance. We have given an implementation scenario toward the design of an efficient tool that supports the functionalities for provenance capture in a suitable approach. With this revolution, opportunities may be created to handle the data log risks. The judgment of scientists and the quality of scientific results depends on the Web's reliability, integrity, and reproducibility properties. As a future development, we need to develop suitable scientific Web-based methodologies to provide a basis for performing these scientific functionalities online and retaining valuable results. This can be treated as an emerging technology development toward the scientific and research activities.

Acknowledgements This work is partially supported by Indian Institute of Technology (ISM), Govt. of India. The authors wish to express their gratitude and thanks to the Department of Computer Science and Engineering, Indian Institute of Technology (ISM), Dhanbad, India, for providing their support in arranging necessary computing facilities.

References

1. Moreau, L., Ludäscher, B., Altintas, I., Barga, R.S., Bowers, S., Callahan, S., Davidson, S.: Special issue: the first provenance challenge. Concur. Computat. Pract. Exp. **20**(5), 409–418 (2008)
2. Cheney, J.: Provenance, XML and the scientific web. In: ACM SIGPLAN Workshop on Programming Language Technology and XML (PLAN-X 2009) (2009)
3. Ram, S., Liu, J.: A new perspective on the semantics of data provenance. In: Proceedings of the First International Conference on Semantic Web in Provenance Management, vol. 526, pp. 35–40. (Oct 2009). CEUR-WS.org
4. Davidson, S.B., Freire, J.: Provenance and scientific workflows: challenges and opportunities. In: Proceedings of the 2008 ACM SIGMOD International Conference on Management of data, pp. 1345–1350. ACM (June 2008)
5. Buneman, P., Chapman, A., Cheney, J.: Provenance management in curated databases. In: Proceedings of the 2006 ACM SIGMOD International Conference on Management of data, pp. 539–550. ACM (June 2006)
6. Buneman, P., Khanna, S., Tan, W.C.: Data provenance: some basic issues. In: International Conference on Foundations of Software Technology and Theoretical Computer Science, pp. 87–93. Springer, Berlin, Heidelberg (Dec 2000)
7. Borkin, M.A., Yeh, C.S., Boyd, M., Macko, P., Gajos, K.Z., Seltzer, M., Pfister, H.: Evaluation of filesystem provenance visualization tools. IEEE Trans. Visual Comput. Graph. **19**(12), 2476–2485 (2013)
8. Simmhan, Y.L., Plale, B., Gannon, D.: A survey of data provenance in e-science. ACM SIGMOD Record **34**(3), 31–36 (2005)
9. Foster, I., Vöckler, J., Wilde, M., Zhao, Y.: Chimera: a virtual data system for representing, querying, and automating data derivation. In: International Conference on Scientific and Statistical Database Management, vol. 37
10. Callahan, S.P., Freire, J., Santos, E., Scheidegger, C.E., Silva, C.T., Vo, H.T.: VisTrails: visualization meets data management. In: Proceedings of the 2006 ACM SIGMOD International Conference on Management of Data, pp. 745–747. ACM (June 2006)
11. Simmhan, Y.L., Plale, B., Gannon, D.: Towards a quality model for effective data selection in collaboratories. In: 22nd International Conference on Data Engineering Workshops (ICDEW'06), pp. 72–72. IEEE (2006)
12. Yang, C., Yang, G., Gehani, A., Yegneswaran, V., Tariq, D., Gu, G.: Using provenance patterns to vet sensitive behaviors in Android apps. In: International Conference on Security and Privacy in Communication Systems, pp. 58–77. Springer International Publishing (Oct 2015)
13. Goble, C., Wroe, C., Stevens, R.: The myGrid project: services, architecture and demonstrator. In: Proceedings of the UK e-Science All Hands Meeting, pp. 595–602

14. Groth, P., Miles, S., Fang, W., Wong, S.C., Zauner, K.P., Moreau, L.: Recording and using provenance in a protein compressibility experiment. In: Proceedings of 14th IEEE International Symposium on High-Performance Distributed Computing, 2005 on HDPC-14, pp. 201–208. IEEE
15. Mouallem, P., Barreto, R., Klasky, S., Podhorszki, N., Vouk, M.: Tracking files in the Kepler provenance framework. In: International Conference on Scientific and Statistical Database Management, pp. 273–282. Springer, Berlin, Heidelberg (June 2009)
16. Simmhan, Y.L., Plale, B., Gannon, D.: A framework for collecting provenance in data-centric scientific workflows. In: 2006 IEEE International Conference on Web Services (ICWS'06) pp. 427–436. IEEE
17. Moreau, L., Clifford, B., Freire, J., Futrelle, J., Gil, Y., Groth, P., Plale, B.: The open provenance model core specification (v1. 1). Fut. Generat. Comp. Syst. **27**(6), 743–756 (2011)
18. Suriarachchi, I., Zhou, Q., Plale, B.: Komadu: a capture and visualization system for scientific data provenance. J. Open Res. Softw. 3.1 (2015)
19. Missier, P., Belhajjame, K., Cheney, J.: The W3C PROV family of specifications for modeling provenance metadata. In: Proceedings of the 16th International Conference on Extending Database Technology, pp. 773–776. ACM (Mar 2013)
20. Moreau, L., Missier, P.: Prov-dm: the proven data model (2013)
21. Asuncion, H.U.: Automated data provenance capture in spreadsheets, with case studies. Fut. Generat. Comput. Syst. **29**(8), 2169–2181 (2013)
22. Videla, A., Jason, J.W.: RabbitMQ in action. Manning (2012)
23. Vinoski, S.: Advanced message queuing protocol. IEEE Internet Comput. **10**(6), 87 (2006)
24. Hua, H., Curt, T., Stephan, Z.: Prov-xml: the prov xml schema (2013)
25. Smoot, M.E., Ono, K., Ruscheinski, J., Wang, P.L., Ideker, T.: Cytoscape 2.8: new features for data integration and network visualization. Bioinformatics **27**(3), 431–432 (2011)

Learning to Classify Marathi Questions and Identify Answer Type Using Machine Learning Technique

Sneha Kamble and S. Baskar

Abstract One of the budding fields of artificial intelligence is Question Answering (QA). QA is a type of information retrieval in which a set of documents is given, and a QA system attempts to search for the correct answer to the question posed in natural language. Question classification (QC), which is a part of QA system, helps to categorize each question. In QC, the entity type of the answering sentence for a given question in natural language is predicted. QC is a very crucial step in QA system as it helps to take the important decision. For example, QC helps to reduce the possible options of the answer, and thus the answers that match the question class are to be considered. This research takes the first step toward the development of QC system for English–Marathi QA system. This system analyzes the user's question and deduces the expected Answer Type (AType), for which a dataset of 1000 questions from Kaun Banega Crorepati (KBC) was scrapped and manually translated into Marathi. Right now, the result for translation approach for the coarse-grained class is 73.5% and the fine-grained class is 47.5%, and for the direct approach, it is 56.5 and 30.5% for coarse and fine, respectively. Experiments are going on to improve the results.

1 Introduction

Question Answering (QA) is an application of Natural Language Processing. A system which provides people with a convenient and natural interface for accessing information is Question Answering system. Nowadays, the need to develop accurate systems gains more importance due to available structured knowledge bases and

S. Kamble (✉) · S. Baskar
Goa University, Taleigao, Goa, India
e-mail: sneha311093kamble@gmail.com

S. Baskar
e-mail: baskar@unigoa.ac.in

D. Reddy Edla et al. (eds.), *Advances in Machine Learning and Data Science*,
Advances in Intelligent Systems and Computing 705,
https://doi.org/10.1007/978-981-10-8569-7_4

the continuous demand to access information rapidly and efficiently. QA system has applications in various domains like education, health care, and personal assistance. QA system retrieves the precise information from large documents according to the user question [1].

India is a multilingual society with 30 languages which are spoken by more than a million native speakers, and these languages are written in different scripts. However, users who do not use English as the first language are also significantly high in number. In India, most of the people speak in their regional language and hence find it difficult to express their queries in English and a huge amount of information on the Internet is available in English. Thus, this work might help to develop a search system in the Marathi language [2].

Question classification maps a question into a category that indicates which information should be present in the answer. QC is a very important step because the question category helps mainly in two cases. One is to reduce the number of possible answer options and other is to decide the strategy to search for answer depending on the question category.

Example:
Who is Mozart?
Is directly mapped into Human: Description
What is Australia's national flower?
The headword flower is identified and mapped into the category Entity: Plant.

The paper is organized as follows: Sect. 2 presents the background study; Sect. 3 discusses the data creation; Sect. 4 describes the QC system; Sect. 5 presents the results and experiments for various features used; Sect. 6 discusses error analysis; and Sect. 7 concludes the paper with the future work.

2 Background Research

2.1 "Learning Question Classifiers: The Role of Semantic Information"—Xin Li and Dan Roth

Li et al. [3] present a question classification system that is modeled with the machine learning approach with a multi-class classification task with 50 classes. A hierarchical classifier was used to classify questions in fine-grained classes. Results of [3] conclude that a learning approach is better as the problem can be solved with 90% of accuracy.

2.2 *"From Symbolic to Sub-symbolic Information in Question Classification"—Joao Silva, Luisa Coheur, Ana Cristina Mendes, and Andreas Wichert*

Silva et al. [4] present the evaluation of a question classifier which is rule based. Direct match to the question or identifying the question headword and then applying WordNet to map the question are the two scenarios that are followed [5]. Direct match uses a set of rules that are manually built for compilation. The second scenario also uses manually built rules, but these rules express a path in the parse tree for question headword. After headword extraction, the headword hypernyms are analyzed until it matches to the possible question category. Results obtained by [4] gave a precision of 99.9% for coarse-grained class and 98.1% for fine-grained class in case of a direct approach. Whereas a precision of 86.4% for coarse-grained class and 78.5% for fine-grained class was obtained for headword extraction approach with WordNet mapping.

2.3 *"Answer ka type kya he?" Learning to Classify Questions in Code-Mixed Language—Khyathi C. Raghvavi, Manoj Chinnakotla, and Manish Shrivastava*

Raghvavi et al. [6] present a QC system for English–Hindi code-mixed questions which is Support Vector Machine (SVM) based. The language resources like chunker, parser does not exist for code-mixed languages. [6] made use of word-level resources such as language identification, transliteration, and lexical translation to the features in a single common language and then trained the SVM model to predict the question classes. Translating into a resource-rich language such as English would be helpful since it has QC training data. [6] created a code-mixed question dataset for English–Hindi language pair from college students and evaluated the approach to it. [6] achieves a coarse-grained average accuracy of 63% and fine-grained accuracy of 45%

Table 1 summarizes the discussed papers.

3 Data Creation

Predicting the entity type of the answering sentence for a given question is performed in question classification. QC [7] is performed by classifying the question to a category from a set of predefined categories.

Table 1 Literature review

Paper	Author name and year	Experiments (Classifier-features)	Results
From symbolic to sub-symbolic information in question classification	Joo Silva Lusa Coheur Ana Cristina Mendes Andreas	SVM-Headword, categories, Unigram	Coarse: 95.00% Fine: 90.8%
Learning Question Classifiers: The Role of Semantic Information	Xin Li and Dan Roth (2009)	VM - POS, NER, Chunk Tags	Coarse: 92.5% Fine: 85.00%
Answer ka type kya he? Learning to Classify Questions in Code-Mixed Language	Khyathi C. Raghavi, Manoj Chinnakotla, Manish Shrivastava (2015)	SVM Unigram, Adjacent	Coarse: 63% Fine: 45.00%

English
ENTY:animal Kathiawari Marwari Zanskari and Bhutia are all breeds of what animal found in India ?
Marathi
ENTY:animal काठीयावाडी मारवाडी जन्सकाडी आणि भुतिया ह्या कोणत्या प्राण्यांच्या प्रजाती आहेत ?
Translation
ENTY:animal kathiyavadi Marwari jansakadi and Bhutia These what animals species are ?

Fig. 1 Dataset example

For research and development for QC system, data was obtained. The dataset was created through KBC questions from[1] and comprises a total of 1000 questions. For each question, the dataset includes the questions and their question class as the class for the classifier. These questions were tagged into their respective classes using the hierarchical question ontology defined by [3] which is shown in Table 2. All question have been annotated with syntactic features. Further, the complexity of the dataset was analyzed. Since the data is of KBC, it had English–Hindi code-mixed words in it, which were taken care separately.

For Marathi questions, the same questions were translated in Marathi manually and under linguist guidance. So, total parallel dataset of 1000 questions in English and Marathi has been generated. The dataset created is expected to contribute to others for their research and also simulate other research in the field of question classification. Example of the dataset is given Fig. 1.

[1] http://www.smsoye.in/quiz.php.

Table 2 Taxonomy for question classification

Coarse	Fine
Abbreviation	Abbreviation, Explanation
Entity	Animal, Body, Color, Creative, Currency, Disease, Event, Food, Instrument, Language, Letter, Other, Plant, Product, Religion, Sport, Substance, Symbol, Technique, Term, Vehicle, Word
Description	Definition, Description, Manner, Reason
Human	Group, Individual, Title, Description
Location	City, Country, Mountain, Other, State
Numeric	Code, Count, Date, Distance, Money, Order, Period, Percent, Speed, Temperature, Size and Weight

4 Question Classification System

This section describes the QC system in detail. For classification, the taxonomy of question classes proposed by [3] is used. This research mainly focuses on the classification of Marathi questions for which two approaches were used. One is translation and second is direct, i.e., using syntactic features on Marathi words as it is [8].

4.1 Translation Approach

The architecture of the system for translation approach is shown in Fig. 2. Given a question in the Marathi language, we initially perform word-to-word translation. The questions are translated into English. Given a translated question, we transform it into a feature vector and pass it through all the SVMs and output the SVM class which outputs the maximum score. We use Part Of Speech (POS), Named Entity Recognition (NER), and Chunk as the features for all the words in the question [9]. The classifier according to training data classifies the test questions.

4.2 Direct Approach

In this approach, the process starts from the second step of the architecture. Given a question in the Marathi language, we directly transform the Marathi question into

Table 3 Results

Dataset	SVC		SVM		Random forest	
	Coarse (%)	Fine (%)	Coarse (%)	Fine (%)	Coarse (%)	Fine (%)
English	72.5	47.5	73.5	41.5	69.0	42.5
Marathi	56.0	30.5	59.0	31.5	61.0	30.5
Translated	73.5	47.5	67.0	36.0	67.0	35.0

a feature vector, rather than translating it, and pass it through all the SVMs and output the SVM class which outputs the maximum score [10]. We use POS, NER, and, Chunk as the features for all the words in the question. Later, the classifier according to training data classifies the test questions.

To compare and study the results, the English questions were classified using the existing question classification system.

5 Results and Experiments

The overall results are presented in Table 3. It shows the results obtained for English and Marathi dataset for coarse and fine class.

Along with this, we would like to mention the challenges faced in translation. For translation, Goslate API, Google translator, and Shata Anuvaadak[2] were tried.

The results obtained by Goslate API and Google translator are almost same. But Goslate API is available only limited times a day. None of the above gave better results for an entire question at once; hence, word-level translation was experimented and better results were obtained.

From the example given in Fig. 3, we can see that word-level translation gives a better result. When the results were compared to the results of Shata Anuvaadak, it was observed that few words remained as it is which it could not translate. In addition, the coverage of the translation model was less.
On the other hand, Google translator translates most of the word; the words that were difficult to translate were transliterated.

6 Error Analysis

During the course of our experiments, a few observations stood out that highlight the errors made by the approach used to extract the features. We observed that when the features were passed on word level, the results were poor.

[2]http://www.cfilt.iitb.ac.in/~moses/shata_anuvaadak/resource.phptext.

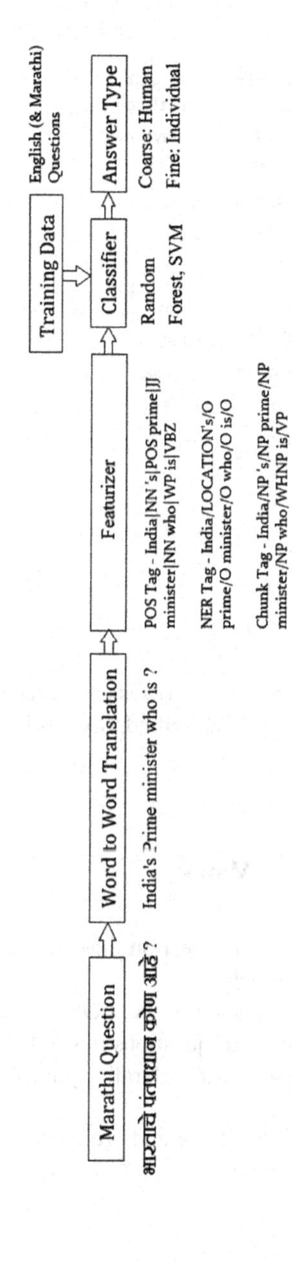

Fig. 2 Architecture

Fig. 3 Translation example

Sentence Level Translation:

Marathi Question: काठीयावाडी मारवाडी जन्सकाडी आणि भुतिया ह्या कोणत्या प्राण्यांच्या प्रजाती आहेत ?

Translated Question: What are these animal species and Bhutia kathiyavadi Marwari jansakadi ?

Word Level Translation:

Marathi	Translated
काठीयावाडी	kathiyavadi
मारवाडी	Marwari
जन्सकाडी	jansakadi
आणि	and
भुतिया	Bhutia
ह्या	These
कोणत्या	what
प्राण्यांच्या	animals
प्रजाती	species
आहेत	are

Example:
How far is Goa from Gujarat?
numeric : distance
How are you?
Entity: other

Here is an example in which the word how is categorized into two, and in the test example, the word might get wrongly classified. So, it is better to consider the whole sentence for better results.

7 Conclusion and Future Work

This paper presents our proposal of question classification for Marathi questions using translation and direct approach.

For this, we have used syntactic features like POS, NER, Chunk. The experimental results show that for a given set of questions in English and Marathi, the results are almost the same, whereas while using Marathi syntactic features directly gives a poor result.

Further, the features that are available and are useful to improve results can be experimented using better techniques.

References

1. Ferrucci, D., Brown, E., Chu-Carroll, J., Fan, J., Gondek, D., Kalyanpur, A.A., Lally, A., Murdock, J.W., Nyberg, E., Prager, J., et al.: Building watson: an overview of the DeepQA project. AI Mag. **31**(3), 59–79 (2010)
2. Zhang, D., Lee, S.W.: Question classification using support vector machines. In: Proceedings of the 26th Annual International ACM SIGIR Conference on Research and Development in Informaion Retrieval, pp. 26–32. ACM (2003)
3. Li, X., Roth, D.: Learning question classifiers. In: Proceedings of the 19th International Conference on Computational Linguistics, vol. 1, pp. 1–7. Association for Computational Linguistics (2002)
4. Silva, J., Coheur, L., Mendes, A.C., Wichert, A.: From symbolic to sub-symbolic information in question classification. Artifi. Intell. Rev. **35**(2), 137–154 (2011)
5. Zhiheng, H., Marcus, T., Zengchang Q.: Question classification using head words and their hypernyms. In: Proceedings of the Conference on Empirical Methods in Natural Language Processing, pp. 927–936. Association for Computational Linguistics (2008)
6. Raghavi, K.C., Chinnakotla, M.K., Shrivastava, M.: Answer ka type kya he? Learning to classify questions in code-mixed language. In: Proceedings of the 24th International Conference on World Wide Web, pp. 853–858. ACM (2015)
7. Metzler, D., Croft,W.B.: Analysis of statistical question classification for fact-based questions. Informat. Retr. **8**(3), 481–504 (2005)
8. Moschitti, A., Chu-Carroll, J., Patwardhan, S., Fan, J., Riccardi, G.: Using syntactic and semantic structural kernels for classifying definition questions in jeopardy! In: Proceedings of the Conference on Empirical Methods in Natural Language Processing, pp. 712–724. Association for Computational Linguistics (2011)
9. Vyas, Y., Gella, S., Sharma, J., Bali, K., Choudhury, M.: Monojit: Pos tagging of English-Hindi code-mixed social media content. EMNLP **14**, 974–979 (2014)
10. Jinzhong, X., Yanan, Z., Yuan, W.: A classification of questions using svm and semantic similarity analysis. In: 2012 Sixth International Conference on Internet Computing for Science and Engineering (ICICSE), pp. 31–34. IEEE (2012)

A Dynamic Clustering Algorithm for Context Change Detection in Sensor-Based Data Stream System

Nitesh Funde, Meera Dhabu and Umesh Balande

Abstract Sensor-based monitoring systems are growing enormously which lead to generation of real-time sensor data to a great extent. The classification and clustering of this data are a challenging task within the limited memory and time constraints. The overall distribution of data is changing over the time, which makes the task even more difficult. This paper proposes a dynamic clustering algorithm to find and detect the different contexts in a sensor-based system. It mines dynamically changing sensor streams for different contexts of the system. It can be used for detecting the current context as well as in predicting the coming context of a sensor-based system. The algorithm is able to find context states of different length in an online and unsupervised manner which plays a vital role in identifying the behavior of sensor-based system. The experiments results on real-world high-dimensional datasets justify the effectiveness of the proposed clustering algorithm. Further, discussion on how the proposed clustering algorithm works in sensor-based system is provided which will be helpful for domain experts.

Keywords Clustering · Context state · Data streams · Principal component analysis (PCA) · Sensor-based system

1 Introduction

Nowadays, infinite volume of data stream is generated by the industry processes equipped with sensors, real-time monitoring systems, online transactions, social networks, and Internet of Things (IoT). In contrast to the traditional data, sensor

N. Funde (✉) · M. Dhabu · U. Balande
Department of Computer Science and Engineering,
Visvesvaraya National Institute of Technology, Nagpur, Maharashtra, India
e-mail: nitesh.funde@gmail.com

D. Reddy Edla et al. (eds.), *Advances in Machine Learning and Data Science*,
Advances in Intelligent Systems and Computing 705,
https://doi.org/10.1007/978-981-10-8569-7_5

data streams are continuous, ordered, potentially infinite, and require fast real-time response. Almost all the real-time sensor-based monitoring systems produce a fast, huge amount of data which need to be analyzed for real-time decision making. Data stream mining is the process of analyzing continuous, fast records within limited storage and time constraints. For more efficient data stream mining technique, sensor data streams must be processed online in single-scan, and it should incorporate the ability to handle multidimensional data [1]. Consider an example of a satellite remote sensing system, in which a sensor constantly produces a data over a period of time. The statistical properties of data distribution are often changes in an unexpected ways in sensor data streams. This property is called as a concept drift [2].

The data stream clustering is a challenging problem in sensor-based system applications because of its operational constraints. In this paper, the data stream clustering for context change detection in sensor-based system is addressed. Context is an information which can be used to discover the behavior of a sensor-based system [3]. Although many data stream clustering algorithms are available, they are not suitable for detecting the current context state of sensor-based system. The context state of a sensor-based system is constantly changing which represents the different modes or states. This behavior of system can further modeled in order to develop some useful applications. For example, smartphone nowadays consists of in-built sensors such as accelerometer and gyrometer. In the near future, it is easily possible for the sensor-based system to provide more personalized services. The behavior analysis of smartphone user will be useful for developing personalized recommender system.

This paper is organized as follows. Section 2 discusses the related work of data stream clustering algorithm. Section 3 describes the proposed dynamic clustering algorithm. Section 4 provides experimental details, results, and discussion, and at last, Sect. 5 gives conclusion.

2 Related Work

Silva et al. presented a detailed survey of different existing data stream clustering algorithms such as Clustream, D-stream, Denstream [4–7]. The experimental methodology and temporal aspects of data stream clustering are discussed in this survey. There are large number of references provided regarding applications of data stream clustering in the different fields such as network intrusion detection, stock market analysis, and sensor networks. Authors also presented the information regarding various datasets and software packages. The most important part of the survey is the discussion of open challenges and issues regarding the design of algorithms with or without using ad hoc user-defined parameters such as number of clusters, window length. Zhang et al. proposed Strap clustering algorithm by combining affinity propagation (AP) algorithm with Page-Hinkley test [8]. The Strap is validated on a real-world application and performed well in data stream environments.

Qahtan et al. discussed the framework for change detection in high-dimensional data streams using PCA [9]. Here, change detection is performed by comparing data distribution in a test window with reference window which uses divergence metrics and density estimators. This framework has many benefits over the existing approaches of reduction in computational costs of density estimator. The threshold value is dynamically calculated by using Page-Hinkley test. However, the clustering procedure is yet to be experimented in sensor-based system. Mirsky et al. proposed the pcStream which is dynamically detecting the sequential temporal context of an entity from sensor-fused datasets [10].

In contrast to the above-mentioned algorithms, the proposed algorithm transforms the data into the suitable representation for handling high-dimensional data. It is capable of identifying different changing context states of various length in sensor-based system in an online and unsupervised manner within limited memory and time constraints. The context can be of any repeating nature or representing the current state of an operation of a system. The experiments are performed on two real-world dataset such as KDD'99 and home activity monitoring.

3 Proposed Dynamic Clustering Algorithm (Dclust)

The proposed dynamic clustering algorithm uses the principal component analysis (PCA) for modeling the data distribution in high-dimensional data streams. PCA is a popular technique for dimensionality reduction [11]. The proposed clustering algorithm exploits PCA to capture the set of uncorrelated information, i.e., principal components (eigenvectors) from the original data such that it better represents the transformed data. PCA on original data S gives the eigenvectors and eigenvalues. The advantage of using PCA is as follows: It allows the algorithm to detect changing context state in the form of mean, variance, and correlation. The original data distribution changes are reflected in the PC projections over a period of time.

Table 1 Notations

Notations	Descriptions
S	An unbounded data stream, $S = \{x_1, x_2, \ldots, x_n\}$ and $x_i \in R^d$
$\|Clust\|$	Set of clusters representing the different context states of system
δ	Threshold value
p	Number of stream instances
β	Minimum context state drift size (# drifting instances that represent new context state)
P_v	Percentage of variance required to select principal components
B_d	A buffer with at most β consecutive drifting instances
m	The maximum # records a system can process (window size)

It also reduces the computational cost by removing the principal components which is having smaller variance.

The notations used in the proposed algorithm are as shown in Table 1. The basic idea of this algorithm is to follow data stream distribution for identifying the different context states of the sensor-based system. The statistical similarity between a stream instance and known context is calculated using Mahalanobis distance method by taking only selected PCs of that data distribution which satisfies the criteria of variance P_v. As long as the data streams fit within a known context's data distribution, we assign it to that existing context state. Otherwise, new context state is defined for those data points. Each context state has window memory size m for concept drift. Finally, algorithm's different parameters such as β, δ, and P_v are adjusted accordingly to identify different context state in the system. The code of the algorithm is as shown below.

At first, all parameters are initialized: p is initialized with the number of rows which is equal to min context state drift size β, $|Clust| = \phi$, $m = 100$, and $P_v = 0.98$. Then, by applying function ClusterModel to first p instances of S for initial model building, the first context state is obtained. The ClusterModel function calculates the eigenvectors, eigenvalues and mean using P_v. The memory window of the context model is a maximum length m which forgets older observations. As the stream instance arrives, the statistical similarity is calculated by using selected eigenvectors to existing **clusters (contexts)**. If it fits within that known context distribution by checking with δ, then that instance is assigned to the best model in $|Clust|$ and B_d is emptied. If it does not fit, then we add stream instance to B_d and check if it is equal to β (full). If B_d is full, then new context state has been discovered in sensor-based system. In this case, new context model C_j is added to $|Clust|$. In this way, different contexts of the sensor-based system are discovered.

```
Algorithm: A Dynamic Clustering Algorithm
Input: An unbounded data streams S
Parameters: Threshold value δ, min context state drift
size β, variance P_v .
Output: Clusters in |Clust| representing different context
state of a system
Initialization: p=β ; |Clust|=ϕ ; m=100; countDrift=0
ClusterModel(S(1:p),P_v);
numClust = numClust+1;
AddModel(C_1,|Clust|);
for i=p+1 to size(S) do
   for j=1 to |Clust| do
     Calculate dist =[(S_i)-Centroid(j)×eigenvector(j)];
     scores(S_i) = dist × dist';
   end
   score =sqrt(dist);
   bestModel= min(scores);
   if min(scores) > δ
      countDrifters = countDrifters+1;
      B_d(countDrifters,:)=S(i,:);
      if countDrifters == β
         numClust=numClust+1;
         AddModel(C_j,|Clust|);
         ClusterModel(B_d, P_v);
      end
   else
      if countDrifters >0
         Add new instance to the S;
         Check the window size m of instances;
         ClusterModel(B_d , P_v);
         S_i =bestModel; // S_i belongs to bestModel
      end
   end
end

ClusterModel (S,P_v){
   [EigenVectors, EigenValues] =PCA(S);
   Centroid=Mean(S);
}
```

The contextual sensor stream can be illustrated as a sequence of attributes in geometric space. The clusters C_1, C_2, and C_3 with their principal components and mean values in geometric space are as shown on Fig. 1. The different clusters are considered to be as different context state in sensor-based system.

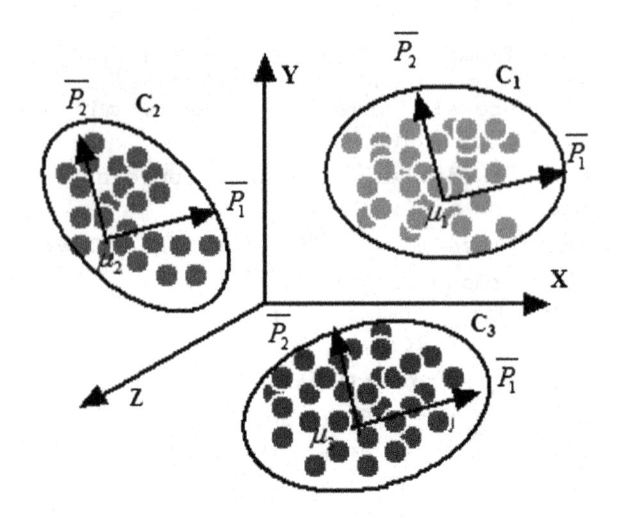

Fig. 1 Clusters with its principal components

4 Experiments

4.1 Datasets

We used two real-world datasets to demonstrate the clustering in sensor-based system. The first is the network intrusion dataset (KDD) and second is home activity monitoring dataset (HAM) [12, 13]. The aim of KDD'99 is to differentiate various attacks from normal connection. The sequence of TCP packets starting from source to target is considered as a connection. There are approximately 4,900,000 records and 41 numeric attributes. We normalized dataset and used 38 attributes for the proposed clustering algorithm. The objective of using this dataset is to evaluate the proposed algorithm on high-dimensional data. There are 23 classes for each connection for the normal and the particular type of attack such as ftp_write, neptune, and buffer_overflow. Overall, the dataset includes normal connection and specific attacks which are categorized in major type such as user to root attack (U2R), denial of service attack (DoS), probing attack, and remote to local attack (R2L). These major types can be considered as different contexts.

The HAM dataset has recordings of 10 sensors, i.e., 8 MOX gas sensors, 1 temperature, and 1 humidity sensor over a period of time. These sensors are exposed to different conditions in home such as normal background and two stimuli: wine and banana. The aim of HAM dataset is to build an application like electronic nose which differentiates among background, banana and wine.

4.2 Experimental Setting

The dynamic clustering algorithm is implemented in R which is a language and environment for statistical computing and graphics. We have used Massive Online Analysis (MOA) tool for comparison of our proposed clustering algorithm (Dclust) with other algorithms such as Clustream, Denstream, StreamKm, and Strap. MOA is an open-source framework for data stream mining which includes various machine learning algorithms and tool for evaluation. The setup for evaluation of clustering algorithms is shown in Fig. 2.

The MOA tool also provides visualization of clustering over a period of time and user can control the different parameters such as visualization speed and evaluation measures. The evolving clusters at T_1 and T_2 are as shown in Fig. 3a and Fig. 3b, respectively. The different colors represent different clusters.

4.3 Results and Discussion

The clustering performance of proposed algorithm is evaluated in terms of adjusted Rand index (ARI) because of overlapping criteria of clusters (context states) in the system. The performance of clustering algorithms is measured by calculating ARI between dataset's labels and algorithm's clustering assignment. First, the data

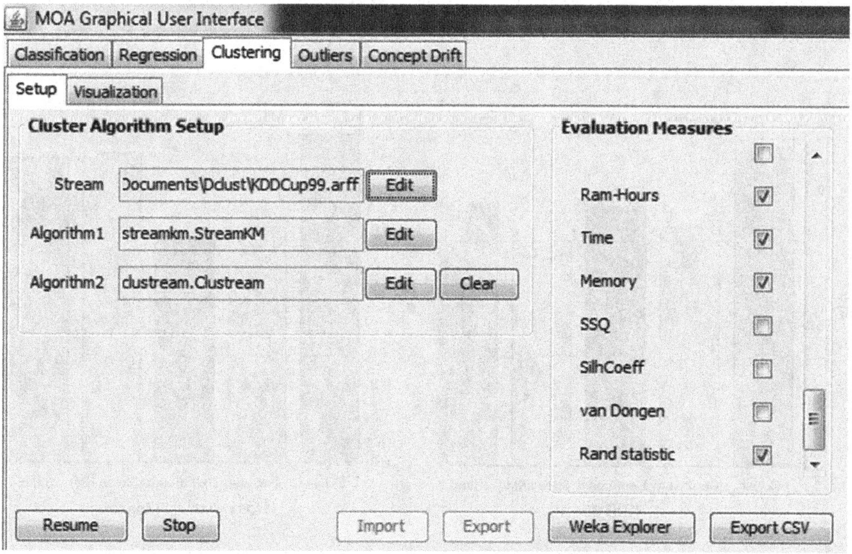

Fig. 2 MOA setup for clustering

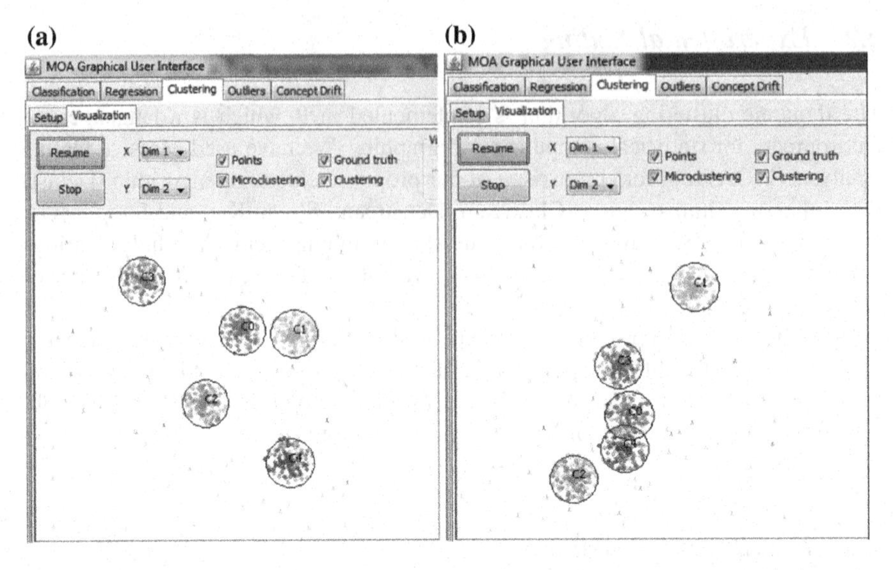

Fig. 3 **a** Evolving clusters at T_1, **b** evolving clusters at T_2

stream clustering is performed on a dataset, and then ARI is calculated using dataset's labels and respective clustering assignments.

Figure 4 shows the comparison of dynamic clustering (Dclust) algorithm with the other algorithms such as Clustream, Denstream, StreamKM, and Strap on two dataset. It shows that the dynamic clustering algorithm performs better than other clustering algorithms using all feasible input parameters of algorithms.

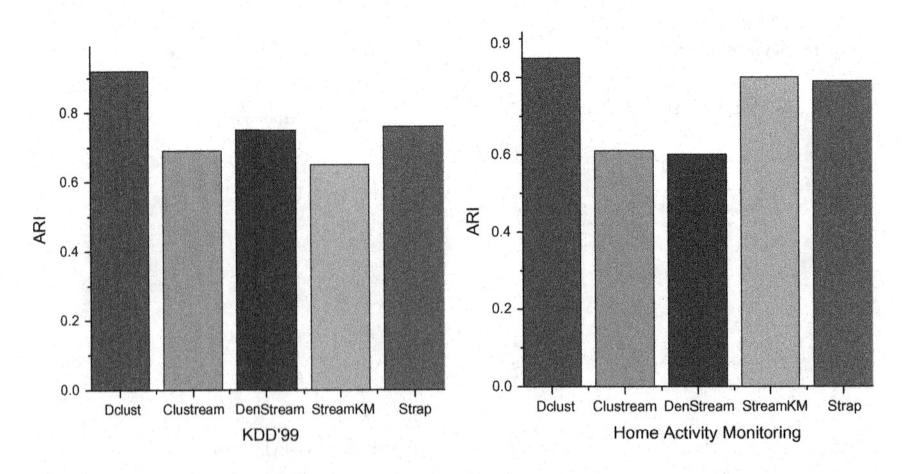

Fig. 4 Comparison of dynamic clustering (Dclust) algorithm with other clustering algorithms on two dataset, i.e., KDD'99 and home activity monitoring

5 Conclusion

This paper presented the dynamic clustering algorithm for detecting the different changing context states in sensor-based system. The algorithm is based on the concept of PCA. The different context states of the sensor-based system are on or off mode, specific working mode and not working. The context states can be of any length which represents the current property of the system. This dynamic clustering algorithm can be used for recognizing the current context and predicting the coming contexts of the system. The performance results on high-dimensional real-world dataset show that the algorithm outperforms the existing stream clustering algorithms. There are different user-defined parameters which control the proposed algorithm such as threshold value and window size. As the data distribution changes, threshold value should also be dynamic. We will consider the combining of dynamic threshold settings approach with this algorithm in real-time application as our future work.

References

1. Han, J., Kamber, M., Pei, J.: Data Mining: Concepts and Techniques. Elsevier (2011)
2. Gama, J., Žliobaitė, I., Bifet, A., Pechenizkiy, M., Bouchachia, A.: A survey on concept drift adaptation. ACM Comput. Surv. (CSUR) **46**(4), 44 (2014)
3. Dey, A.K.: Providing architectural support for building context-aware applications. Doctoral Dissertation, Georgia Institute of Technology (2000)
4. Aggarwal, C.C., Han, J., Wang, J., Yu, P.S.: A framework for clustering evolving data streams. In: Proceedings of the 29th International Conference on Very Large Data Bases, vol. 29, pp. 81–92. VLDB Endowment (2003)
5. Cao, F., Estert, M., Qian, W., Zhou, A.: Density-based clustering over an evolving data stream with noise. In: Proceedings of the 2006 SIAM International Conference on Data Mining, pp. 328–339. Society for Industrial and Applied Mathematics (2006)
6. Chen, Y., Tu, L.: Density-based clustering for real-time stream data. In: Proceedings of the 13th ACM SIGKDD International Conference on Knowledge Discovery and Data Mining, pp. 133–142. ACM (2007)
7. Silva, J.A., Faria, E.R., Barros, R.C., Hruschka, E.R., de Carvalho, A.C., Gama, J.: Data stream clustering: a survey. ACM Comput. Surv. (CSUR) 46–53 (2013)
8. Zhang, X., Furtlehner, C., Germain-Renaud, C., Sebag, M.: Data stream clustering with affinity propagation. IEEE Trans. Knowl. Data Eng. **26**(7), 1644–1656 (2014)
9. Qahtan, A.A., Alharbi, B., Wang, S., Zhang, X.: A pca-based change detection framework for multidimensional data streams: change detection in multidimensional data streams. In: Proceedings of the 21th ACM SIGKDD International Conference on Knowledge Discovery and Data Mining. ACM (2015)
10. Mirsky, Y., Shapira, B., Rokach, L., Elovici, Y.: pcstream: a stream clustering algorithm for dynamically detecting and managing temporal contexts. In: Advances in Knowledge Discovery and Data Mining, vol. 2015, pp. 119–133. Springer (2015)
11. Jolliffe, I.T., Cadima, J.: Principal component analysis: a review and recent developments. Philos. Trans. R. Soc. A **374**(2065), 20150202 (2016)
12. Bache, K., Lichman, M.: UCI Machine Learning Repository. http://archive.ics.uci.edu/ml (2013)
13. Huerta, R., Mosqueiro, T., Fonollosa, J., Rulkov, N., Rodriguez-Lujan, I.: Online decorrelation of humidity and temperature in chemical sensors for continuous monitoring. Chemom. Intell. Lab. Syst. **157**, 169–176 (2016)

Predicting High Blood Pressure Using Decision Tree-Based Algorithm

Satyanarayana Nimmala, Y. Ramadevi, Srinivas Naik Nenavath and Ramalingaswamy Cheruku

Abstract High blood pressure, also called as hypertension, is a state developed in biological system of human beings by knowingly or unknowingly. It may occur due to varied biological and psychological reasons. If high blood pressure state is sustained for a longer cycle, then the person may be the victim of heart attack or brain stroke or kidney disease. This paper uses a decision tree-based J48 algorithm, to predict whether a person is prone to high blood pressure (HBP). In our experimental analysis, we have taken certain biological parameters such as age, obesity level, and total blood cholesterol level. We have taken the real-time data set of 1045 diagnostic records of patients in the age between 18 and 65. These are collected from a medical diagnosis center Doctor C, Hyderabad. Records (66%) are used to train the model, and remaining 34% records are used to test the model. Our results showed 88.45% accuracy.

Keywords Classification · Decision tree · Blood pressure monitoring

---

S. Nimmala (✉)
Department of Computer Science and Engineering, Osmania University,
Hyderabad 500007, India
e-mail: satyauce234@gmail.com

Y. Ramadevi
Department of Computer Science and Engineering, CBIT, Hyderabad 500075, India
e-mail: yrdcse.cbit@gmail.com

S. N. Nenavath
School of Computer and Information Sciences, University of Hyderabad,
Hyderabad 500046, India
e-mail: srinuphdcs@gmail.com

R. Cheruku
Department of Computer Science and Engineering, National Institute of Technology Goa,
Ponda 403401, Goa, India
e-mail: rmlswamygoud@gmail.com

D. Reddy Edla et al. (eds.), *Advances in Machine Learning and Data Science*,
Advances in Intelligent Systems and Computing 705,
https://doi.org/10.1007/978-981-10-8569-7_6

1 Introduction

When the heart beats, it pushes the blood against arteries with some force, which creates some pressure called as systolic blood pressure. The heart takes rest during the beats, while the pressure inside the arteries is called diastolic blood pressure. Most of the literature says that hypertension (HBP) or heart rate change leads to heart failure or stroke [1]. Recent health statistics show that men and women of age above 25 are prone to hypertension [3]. We consider blood pressure level, obesity condition, total cholesterol level as reported by laboratories and scientific studies as shown in Table 1.

In this research work, we address the classification methodology to predict whether a person is a victim of HBP or not. We have taken parameters like age, obesity level, and complete blood cholesterol level of a person.

The rest of the paper is organized as follows. A background of the factors influencing HBP is given in Sect. 2. Procedure for identifying a specific classifier based on the performance measures is given in Sect. 2.1, and Sect. 4 evaluates the performance of a classifier. Section 5 concludes and gives future outline of the paper.

2 Background Work

This section describes overview of the factors influencing HBP.

2.1 Factors Effecting BP

Blood pressure (BP) is mainly effected by the cardiac output (CO) and total peripheral resistance (TPR), which is calculated using (1).

$$BP = CO * TPR \tag{1}$$

Table 1 Ranges for BP, obesity, and cholesterol

BP	Low range	Normal range	Borderline	High range
Systolic	<90	90–130	131–140	>140
Diastolic	<60	60–80	81–90	>90
Total cholesterol level	<125	125–200	200–239	>240
Obesity	<18 (underweight)	18–24.9 (normal)	25–29.9 (overweight)	>30 (obese)

Here, CO is effected by increased venous return or stroke volume or heart rate and sympathetic activity. TPR is effected by the resistance that acts against the blood flow in the arteries. It may be due to a blood clot or fat in the blood vessels. CO effects the systolic blood pressure, whereas TPR effects the diastolic blood pressure [4]. So, CO is proportional to a number of heart beats per minute (HBM) and volume of blood (BV) pushed in each beat.

$$CO = HBM * BV \tag{2}$$

$$BP = Systolic/Diastolic \tag{3}$$

$$Systolic/Diastolic = CO/TPR \tag{4}$$

Sometimes, we may consider mean arterial blood pressure (MABP) which can be calculated as

$$MABP = (2 * (Diastolic + Systolic))/3 \tag{5}$$

If MABP [2] is within the range, then all organs and tissues will get enough blood, oxygen, and nutrients. Our paper focuses on age, obesity, and cholesterol levels of a person in elevating the blood pressure.

2.2 Impact of Age, Obesity, and Cholesterol on Hypertension

Aging is inevitable although a person has a healthy diet and exercise regularly. As we age, arteries may become narrow and harden and the ability of body to process sodium in the diet decreases [6]. The person is said to be obese if his body mass index (BMI) value is more than 30. In obese people, there is increased fatty tissue which needs more blood to live. High blood cholesterol is one of the main reasons for fat deposits in arteries which may harden the arteries also. However, age, obesity, and high blood cholesterol are playing their role to elevate the blood pressure of a person.

2.3 Nervous System

It will monitor changes in the state of a human body and transmits signals to and from different parts of the body. It is of two types [5]: voluntary system (VS) and automatic nervous system (ANS).

2.3.1 Voluntary System

It deals with movement and sense. The reaction of VS happens, when we move our hand in any direction or close our eye. It is under the control of human sense.

2.3.2 Automatic Nervous System

It controls the heartbeat, digestive system, BP, etc. It is not under the control of human sense. It is of two types, namely sympathetic nervous system (SNS) and parasympathetic nervous system (PNS).

Sympathetic Nervous System:

When we experience a sudden fear or high anxiety, SNS helps to prepare the body to defend itself by sending more blood to brain, muscles. This may increase BP. If SNS is active for a longer period, it is not good for human body (Fig. 1).

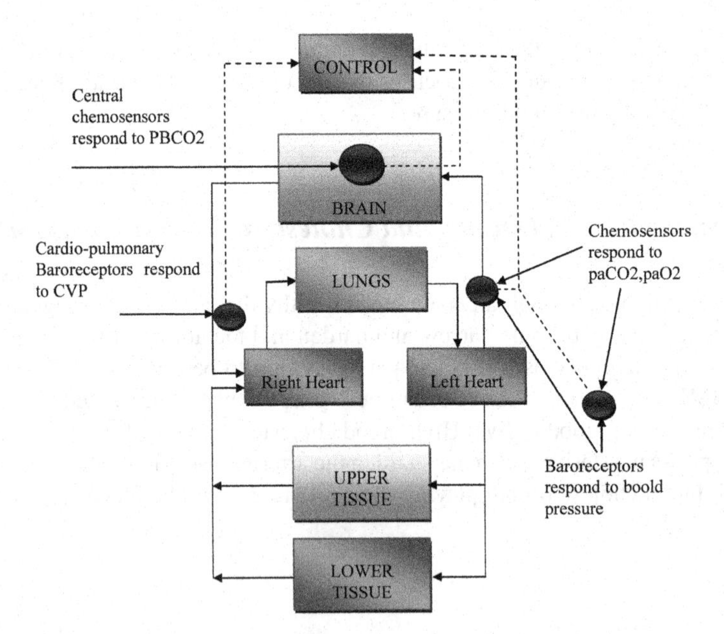

Fig. 1 Sympathetic action and its effects on circulation

Parasympathetic Nervous System:

PNS looks after the immune system to work effectively. PNS is active in the states like being calm, relaxed, and positive. It mostly works against the HBP.

2.4 *How Brain Monitors BP*

The BP management refers to heartbeat, arterial blood pressure, total blood volume, O_2, and CO_2 levels in the blood [7]. It uses the sensory system to know the state of the human system and update the same to the brain. Then brain converts this information to response signals via the peripheral nervous system, where the control listeners are waiting to take action on the received information. They may act to increase or decrease blood flow to various tissues by a change in systemic resistance. The response of the control is designed to maintain overall stable blood flow in the human body.

3 Proposed Methodology

In our work, we use a data mining classification technique. It borrows concepts from probability, statistics, fuzzy logic, neural networks to predict class labels of an object. The classification model will have two stages [8]: 1. training stage, where the model is trained with a set of records whose class labels are already known and 2. testing stage, where the model is going to predict class labels of a set of records whose class labels are unknown. There are various classifiers; we used a decision tree-based classifier named J48.

The classifier evaluation is most often based on prediction accuracy (the percentage of correct prediction divided by the total number of predictions). If accuracy is unsatisfactory, then to find the reason different factors must be observed like: 1. Relevant attributes of the problem are considered, 2. it may need more training records, 3. the dimensionality of the problem is too high, 4. algorithm that we select may not be suitable, and 5. parameter tuning is needed.

4 Experimental Result and Analysis

In our experiments, we use a classifier called J48-C 0.25-M2 (weka.classifiers.trees) to classify the tuples. The input data set is doctorcwekacsv-weka.filters. unsupervised.attribute.Remove-R2-weka.filters.unsupervised.attribute.Remove-R4. We have used 1045 number of records for our analysis and considered four attributes like age, obesity level, cholesterol, and BP for evaluation (Table 2).

Table 2 Performance measures of a classifier

Measure	Value
Classified instances (correctly)	314
Kappa statistic	0.7586
Mean absolute error	0.1506
Root mean squared error	0.3026
Relative absolute error	31.9906%
Root relative squared error	62.1366%
Total number of instances	355
Accuracy	88.4507%
Error rate	11.5493%

TP rate, FP rate, precision, recall, F-measure, Matthews correlation coefficient (MCC), receiver operating characteristic curve (ROC), precision–recall curve (PRC) are calculated for each class for the test data as presented in Table 3.

We trained the algorithm using 66% of input records and used remaining 34% of records to test the algorithm. The time taken to build the model is 0.05 s and time taken to test the model is 0.05 s. Some instances predicted correctly, some are predicted incorrectly, and the accuracy of the classifier, values of different errors and error rate of a classifier are as presented in Table 2.

The matrix in Table 4 represents 120 records, whose class labels are YES and predicted as YES, but 17 records whose class labels are YES and predicted as NO. For class label NO, 24 records are predicted as YES, and 194 records are predicted as NO.

The total number of right and wrong predictions is drawn using decision tree-based J48 algorithm. Figure 2a represents the distribution of test samples predicted right, wrong for selected attribute obesity. Figure 2b represents the distribution of test samples predicted right, wrong for selected attribute age. Figure 3a represents the

Table 3 Detailed accuracy by class

TP rate	FP rate	Precision	Recall	F-measure	MCC	ROC area	PRC area	Class
0.876	0.110	0.833	0.876	0.854	0.759	0.932	0.871	Yes
0.890	0.124	0.919	0.890	0.904	0.759	0.931	0.942	No

Table 4 Confusion matrix

Actual class versus predicted class	a = YES	b = NO
a = YES	120	17
b = NO	24	194

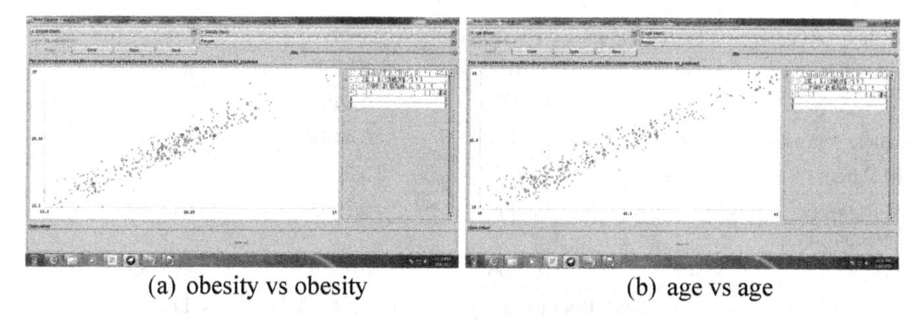

(a) obesity vs obesity (b) age vs age

Fig. 2 Comparison of error rate

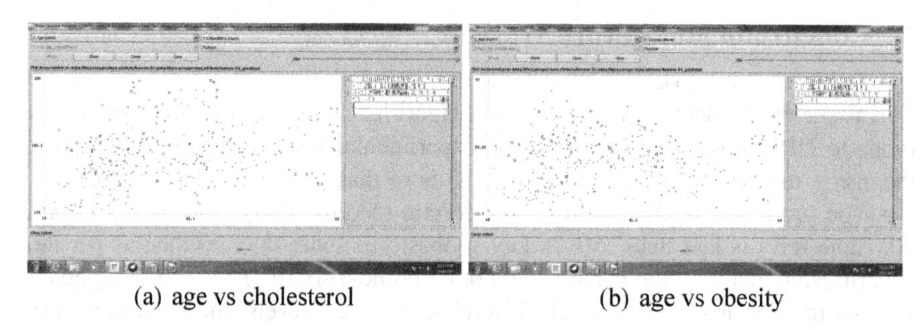

(a) age vs cholesterol (b) age vs obesity

Fig. 3 Comparison of error rate

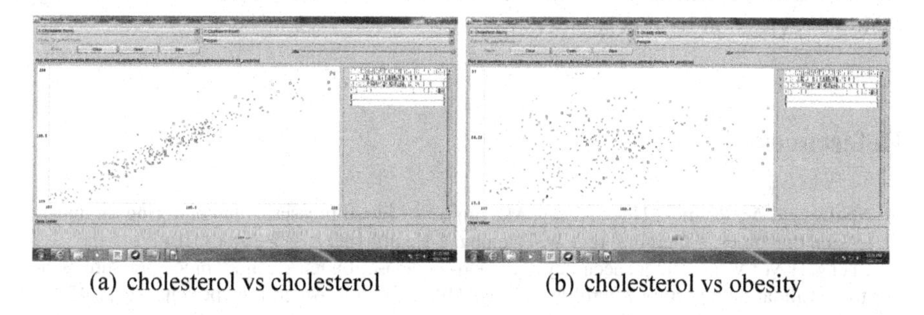

(a) cholesterol vs cholesterol (b) cholesterol vs obesity

Fig. 4 Comparison of error rate

distribution of test samples predicted right, wrong for selected attributes age, cholesterol. Figure 3b represents the distribution of test samples predicted right, wrong for selected attributes age, obesity. Figure 4a represents the distribution of test samples predicted right, wrong for selected attribute cholesterol. Figure 4b represents the distribution of test samples predicted right, predicted wrong for selected attributes cholesterol, obesity. Accuracy and error comparison of different classifiers are shown in Table 5.

Table 5 Accuracy and error comparison of different classifiers

Classifier	Accuracy	MAE	RMSE	RAE
Naive Bayes	70.7042	0.3714	0.4483	78.9052
Linear regression	70.9859	0.3859	0.4481	81.9946
REP tree	87.3239	0.1659	0.3093	35.2512
Multilayer perceptron	75.2113	0.3009	0.3993	63.9322
J48	88.4507	0.1506	0.3026	31.9906

MAE Mean absolute error; *RMSE* Root mean squared error; *RAE* Relative absolute error

5 Conclusion and Future Work

This paper used a decision tree-based J48 algorithm to predict whether a person is prone to HBP or not. By analyzing the experimental results, we conclude that if the age is more than 30 and cholesterol is more than 205, then the possibility of a person prone to the HBP is high. For certain records, if age is between 35 and 50, cholesterol is less than 150, and even obesity is more than 30 then as per the experimental analysis, the person may not be prone to HBP. If the age of a person is more than 55 irrespective of cholesterol and obesity levels, then that person is prone to HBP. As part of the future work, we will consider parameters like anxiety, depression along with age, obesity, and cholesterol, to predict HBP of a person using multiple regression-based classifiers.

References

1. AbuDagga, A., Resnick, H.E., Alwan, M.: Impact of blood pressure telemonitoring on hypertension outcomes: a literature review. Telemed. e-Health **16**(7), 830–838 (2010)
2. Cheng, H.M.: Measurement accuracy of a stand-alone oscillometric central blood pressure monitor: a validation report for MicrolifeWatchBP Office Central. Amer. J. Hypertens. **26**(1), 42–50 (2013)
3. Ding, X.R.: Continuous cuffless blood pressure estimation using pulse transit time and photoplethysmogram intensity ratio. IEEE Trans. Biomed. Eng. **63**(5), 964–972 (2015)
4. Forouzanfar, M.: Oscillometric blood pressure estimation: past, present, and future. IEEE Rev. Biomed. Eng. **8**, 44–63 (2015)
5. Forouzanfar, M.: Ratio-independent blood pressure estimation by modeling the oscillometric waveform envelope. IEEE Trans. Instrum. Measure. **63**, 2501–2503 (2014)
6. Liu, J.: Model-based oscillometric blood pressure measurement: preliminary validation in humans. In: Proceedings of IEEE Conference on Engineering, Medical Biological Soc, pp. 1961–1964 (2014)
7. Mukkamala, R.: Toward ubiquitous blood pressure monitoring via pulse transit time: theory and practice. IEEE Trans. Biomed. Eng. **62**, 1879–1901 (2015)
8. Satyanarayana, N., Ramalingaswamy, C.H., Ramadevi, Y.: Survey of classification techniques in data mining. Int. J. Innov. Sci. Eng. Technol. **1**, 268–278 (2014)

Design of Low-Power Area-Efficient Shift Register Using Transmission Gate

Akash S. Band and Vishal D. Jaiswal

Abstract The shift register is sequential logic circuit to store the digital data, also basic structure block in VLSI circuit. This proposes a low-power and area-efficient shift register using transmission gate. Shift register is used small number of pulse clock signal by alignment latches to several shift register and additional temporary latches. The non-stop flow of pulse signal from input side due to this unnecessary of signal flow so the power and delay will essential more, as to overcome this Clock Gating technique is used. The area and power consumption are compact by replacing latches with transmission gate. The area, power, and transistor count have been compared and designed using several latches and flip-flop stages. This technique solves the timing problem between pulsed latches through the use of multiple non-overlap delayed pulsed clock signals instead of the conventional single pulsed clock signal. Clock Gating technique is used for power consumption also delay factor as much as low and latches should be replaced by transmission gate. The static sense amp shared pulse latch (SSASPL) which is the smallest latch has been selected and also power PC Style flip-flop (PPCFF) which is used for calculating power, area, and delay. The shift register uses a small number of the pulsed clock signals by grouping the latches to several sub-shift registers and using additional temporary storage latches. The analysis is carried out using Tanner EDA–Industry Standard EDA design environment using 180 nm technologies. The simulation results are shown. A four-bit shift register using transmission gate in tanner tools used $V_{DD} = 3.3$ V and power consumption is 1.063 μW.

Keywords Shift register · Clock Gating technique · Transmission gate SSASPL · PPCFF

A. S. Band (✉) · V. D. Jaiswal
Electronics and Telecommunication, DMIETR, Wardha, India
e-mail: akashband7@gmail.com

V. D. Jaiswal
e-mail: vishaljaiswal1987@gmail.com

D. Reddy Edla et al. (eds.), *Advances in Machine Learning and Data Science*,
Advances in Intelligent Systems and Computing 705,
https://doi.org/10.1007/978-981-10-8569-7_7

1 Introduction

In a digital circuit, a shift register is a cascade of flip-flops sharing the clock, in which output of each flip-flop is connected to data input of next flip-flop in a chain. Shift registers are normally used in various applications, such as digital filters, communication receivers, and image processing ICs. The size of the image data continues to rise due to the high request for high superiority image materials, and the word length of the shift register increases to process large image data in image processing ICs [1]. During the switching process, adiabatic technology decreases the power or energy dissipation and reuses some part of the energy by recycling it from the load capacitance. Clock Gating is a standard method used in many synchronous paths for tumbling dynamic power dissipation. This clock gating method can also save significant area as well as power, since it removes large numbers of mux and replaces them with clock gating logic [2, 3].

2 Transmission Gate

Transmission gates represent another class of logic circuits, which use transmission gates as basic building block. A transmission gate consists of a PMOS and NMOS connected in parallel. Gate voltage applied to these gates is harmonizing of each other (C and C bar shown in figure). Transmission gate, or analog switch, is defined as an electronic element that will selectively block or pass a signal level from the input to the output. This solid-state switch is comprised of a PMOS transistor and NMOS transistor. Transmission gates act as bidirectional, switch among two nodes A and B controlled by signal C gate of NMOS is connected to C and gate of PMOS is connected to C bar (invert of C). When control signal C is high, i.e., VDD, both transistors are on and provides a low-resistance path between A and B. On the other hand, when C is low, both transistors are turned off and provide high-impedance path between A and B [4] (Fig. 1).

Fig. 1 Transmission gate

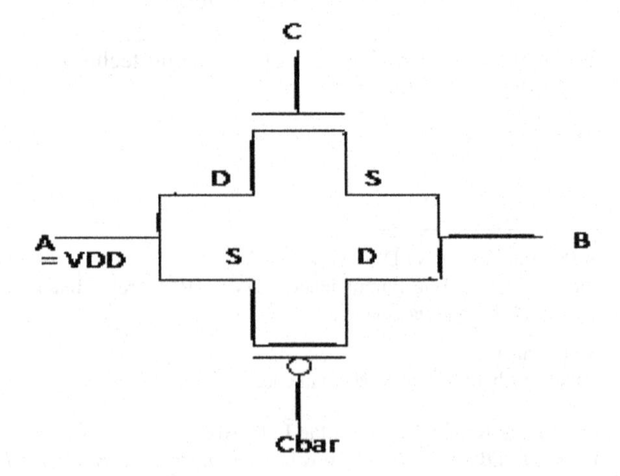

When input node A is associated to VDD and at the top shown in figure control logic C is also high, C = 1. The output node B may be related to capacitor. Let us say, voltage at output node is Vout. For PMOS, source of it is at higher voltage than drain. For NMOS, drain is at advanced voltage than source terminal. Hence, node A will act as source terminal for PMOS and as drain terminal for NMOS.

3 Static Differential Sense Amplifier Shared Pulsed Latch

The original static differential sense amplifier shared pulse latches (SSASPLs) with nine transistors is modified to the SSASPL with seven transistors [5] by removing an inverter to generate the complementary data input (Db) from the data input (D). In the proposed shift register, the difference data inputs (D and Db) of the latch come from the differential data outputs (Q and Qb) of the previous latch. The SSASPL uses the smallest number of transistors and it chomps the lowest clock power because it has a single transistor driven by the pulsed clock signal [5]. The SSASPL apprises the data with three NMOS transistors and it holds the data with four transistors in two cross-coupled inverters [6] (Figs. 2 and 3).

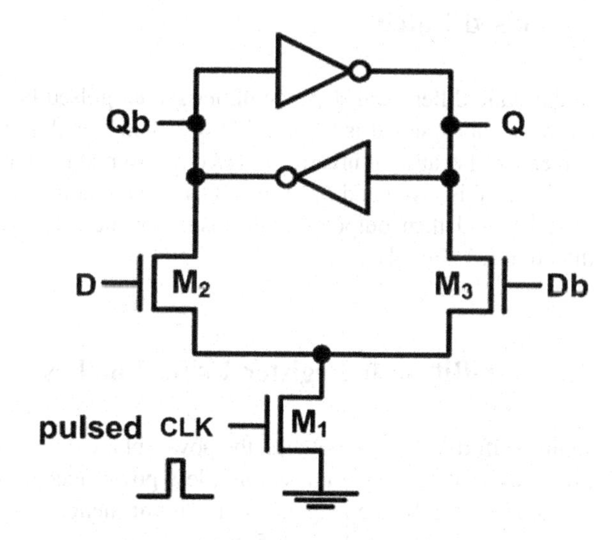

Fig. 2 Schematic of SSASPL

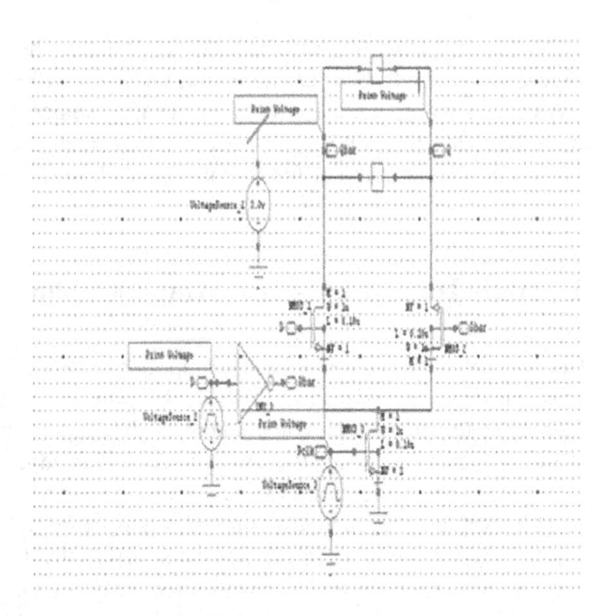

Fig. 3 Static differential sense amplifier shared pulse latches

4 Output of Static Differential Sense Amplifier Shared Pulsed Latch

In the static differential sense amplifier shared pulsed latch, waveforms show when the pulse clock signal is high and D = 1 then signal at the output side shows high waveforms because of inverter works on it from D to Q. [7] Now at next level when clock signal is given and M3 gate is ON, then at negative cycle the clock is at down direction and then output waveforms show the negative pulse, i.e., down world direction [5] (Fig. 4).

5 Four-Bit Shift Register Using Latches

In this shift register with latches, the power phase clock flip-flop is used to calculate power as well as the circuit required less-power technique because of clock signal we applied step-by-step signal to the circuit; hence, the circuit required less technique power is used [5] (Figs. 5 and 6).

When input is given to the circuit as the 1 and 0 as the D1 became also the circuit of the signal clock is given to the circuit as we compulsory to the Q as output will be shifted to the next level of the circuit of the system next level toward the next level of the fundamental of the continue shift register to the circuit of the level output will be shifted to the next bit level as shown in output figure.

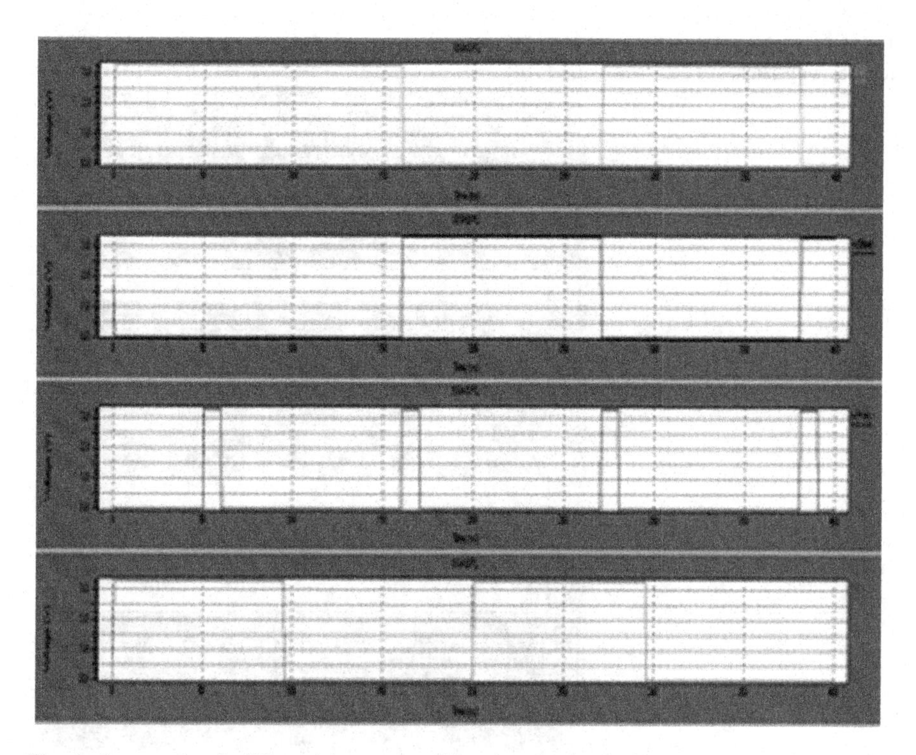

Fig. 4 Output of static differential sense amplifier shared pulse latches

Fig. 5 Four-bit shift register with latches

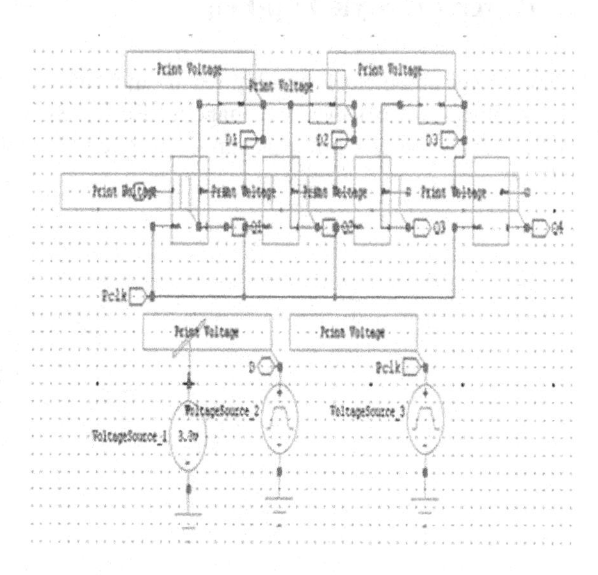

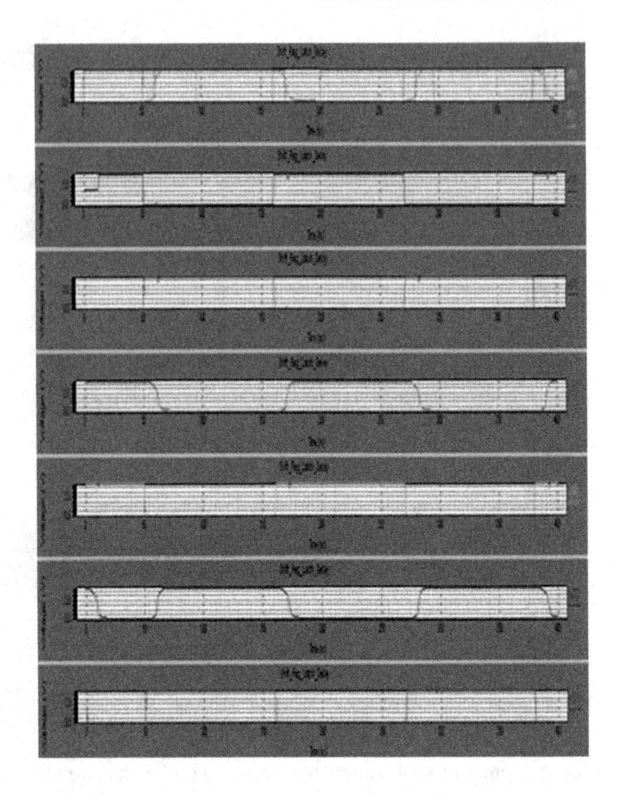

Fig. 6 Waveforms of shift register with latches

6 Power PC Style Flip-Flop

In this power phase clock flip-flop when input functional to D flip-flop as 1 then as well as clock is also 1 then as voltage is also on [8]. When D is 1 is overturned on output side as the D flip-flop is applied to circuit as output side to other circuit as

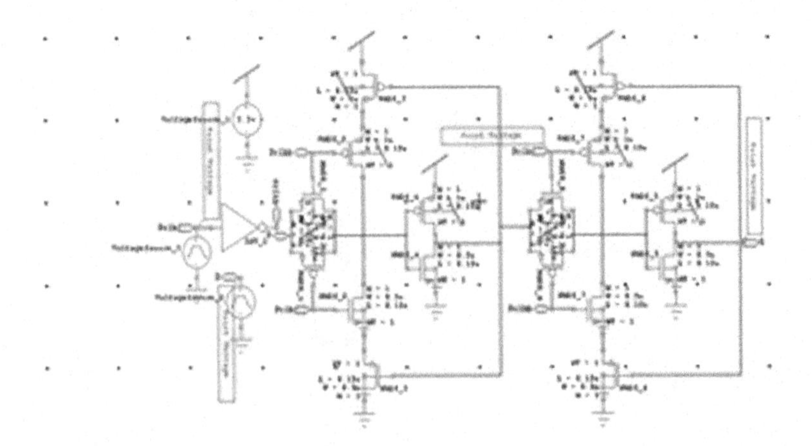

Fig. 7 Power PC Style flip-flop

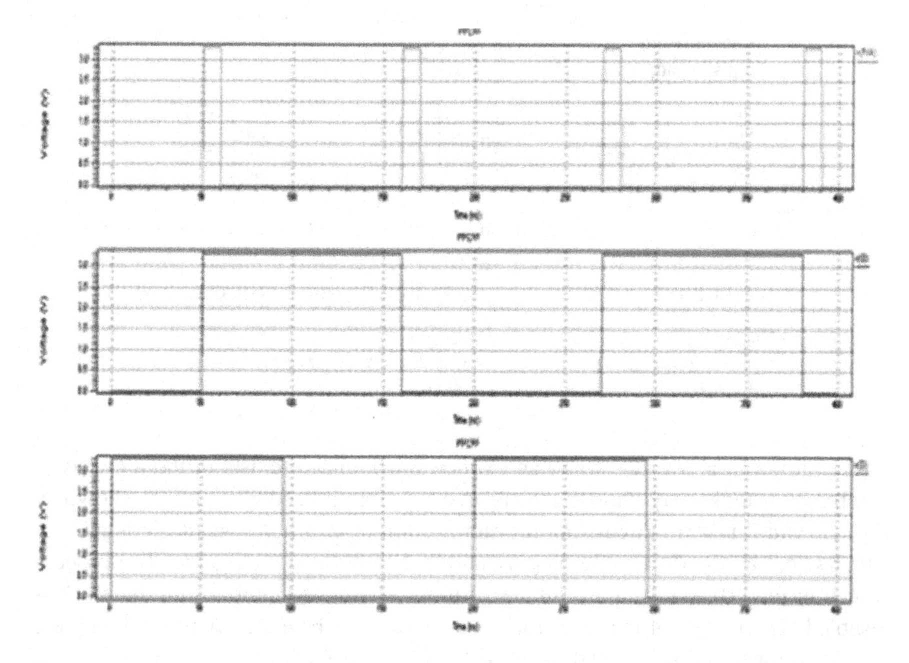

Fig. 8 Output waveforms of PPCFF

used as a input side to the circuit and the output side we getting output of D as we input is given to circuit [8] (Figs. 7 and 8).

7 Clock Gating Technique

The clock gating is a popular procedure used in many circuits for reducing dynamic power overindulgence. Clock gating saves power by adding more logic to a circuit to clip the clock tree. The clock gating logic is commonly in the form of "integrated clock gating" (ICG) cells. Though, the clock gating technique will be varied with the clock tree structure, and clock gating logic can be added into a proposal in several of ways. The dropping clock power is very important [7, 9]. Clock gating is a key power-saving technique used by many designers and is naturally implemented by gate-level power mixture tools. While there is no achievement at a register "data" input, there is no central to clock the register, and hence, the "clock" can be gated to change it off. At the clock side, an "enable" signal can be used to gate the clock, that is called as "clock gating enable."

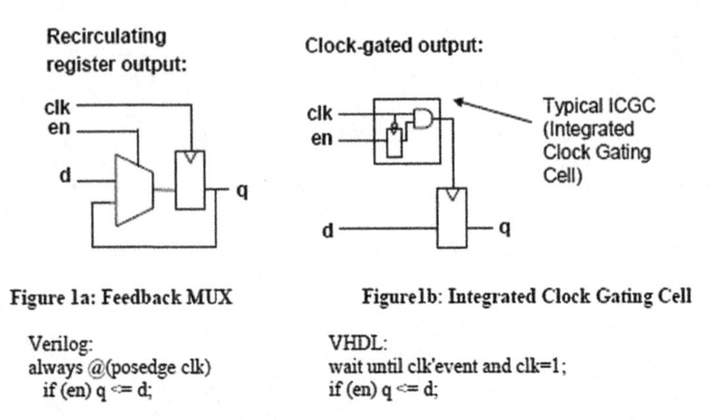

Figure 1a: Feedback MUX **Figure1b: Integrated Clock Gating Cell**

```
Verilog:
always @(posedge clk)
  if (en) q <= d;
```

```
VHDL:
wait until clk'event and clk=1;
if (en) q <= d;
```

As shown in above fig, when an explicit clock enable exists in the RTL code, combination tools may choose among two possible procedures such as shown in Fig. 1a, and it is a "re-circulating register" submission, where the enable is used to either select a new data value or re-circulate the previous data value. The process shown in Fig. 1b is a "gated clock" execution. When the enable is off, the clock is disabled. The output of the two processes will always be equal, but the timing and power performance will be different [10] (Figs. 9 and 10).

In this power PC Style flip-flop, we used Clock Gating technique at input side of the pulse at the flip-flop. In this circuit the pulse clock signal step by step fashioned at input pulse that reason the excessive power we can saved in the circuit. The delay element used to produce the delay in the circuit for shift waveforms in the circuit which will be used in the flip-flop. This is 4-bit shift register which will Clock Gating technique used for power reduction in the circuit [8].

In the waveforms of the power PC Style flip-flop, the waveforms show that the D is input which will be given to the circuit now at next level side q1, q2, q3, q4 are

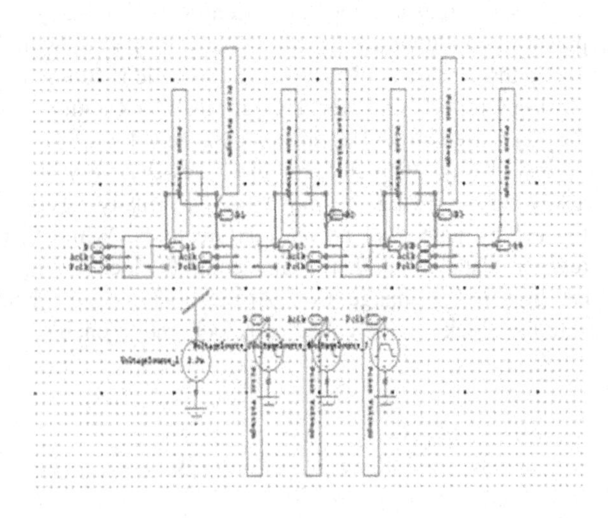

Fig. 9 PPCFF with clock gating technique

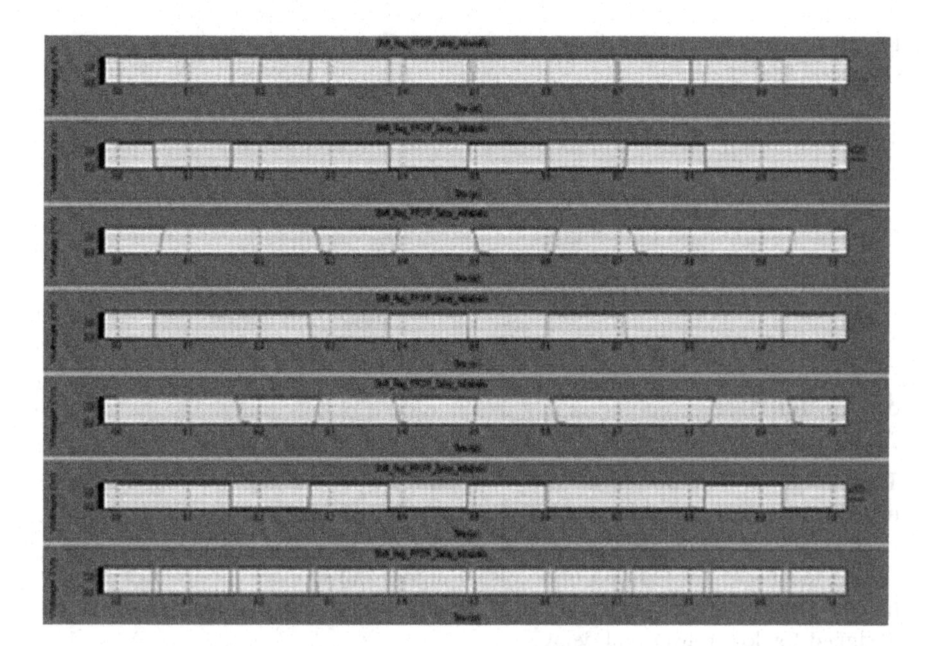

Fig. 10 Output of power PC Style flip-flop

the output waveforms of the shift register which will be shifted shown in the waveforms. Now, the delay element which will also be shifted toward right side of the shift register is shown in d1, d2, d3 in the waveforms.

Output: Performance comparison of shift register.

	SSASPL	PPCFF with adiabatic clock
Voltage (V)	3.3	3.3
Power	10.49 MW	1.06 μW
Delay (ns)	10.35	8.44
Current	3.457 mA	322.86 nA
No. of transistor	54	95

The graphical representation shows that power required for PPCFF with transmission gate as low as SSASPL, also the minimum delay optimized than SSASPL at the voltage source 3.3 v the result shows that the PPCFF with transmission gate better result than SSASPL latches (Fig. 11).

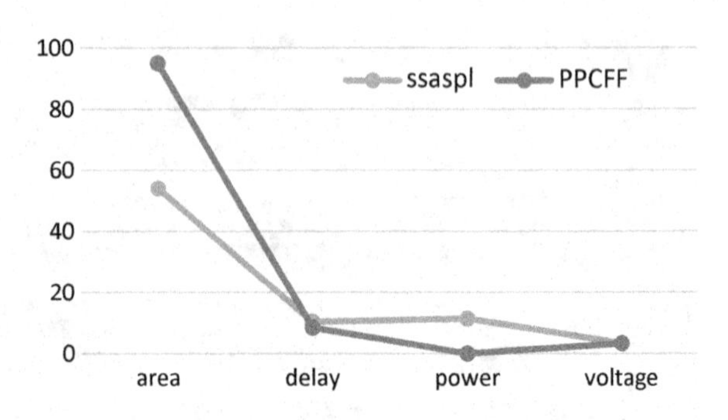

Fig. 11 Comparison of latches and flip-flop

8 Conclusion

The design of low-power area-efficient shift register using transmission gate understands the regulation problem between pulsed latches through the use of binary non-overlap delayed periodic clock signals. This work also includes Clock Gating technique for reduced unnecessary signal and reduces the leakage power; hence, low-power area-efficient shift register using transmission gate can be designed for low power and delay.

References

1. Yang, B.-D.: Low-power and area-efficient shift register using pulsed latches. IEEE Trans. Circuits Syst.—I: Regul. Pap. **62**(6) (2015)
2. Goyal, S., Singh, G., Sharma, P.: Variation of power dissipation for adiabatic CMOS and conventional CMOS digital circuits. In: IEEE Sponsores 2nd International Conference on Electronics and Communication System (ICECS 2015)
3. Rajesh, P., Chandra, D.S., Kumar, L.S., Kaushik, G.: Comparative analysis of pulsed latch and flip-flop based shift registers for high-performance and low-power systems. IJECT **7** (2016)
4. Mohanty, B.K.: Area–delay–power efficient carry-select adder. IEEE Trans. Circuits Syst.—II: Express Briefs **61**(6) (2014)
5. Heo, S., Krashinsky, R., Asanovic, K.: Activity-sensitive flip-flop and latch selection for reduced energy. IEEE Trans. Very Large Scale Integr. (VLSI) Syst. **15**(9), 1060–1064 (2007)
6. Band, A., Jaiswal, V.: Analysis of D flip-flop and latches for reduced power. IJIRCCE **5** (6) (2017)
7. Baker, R.J.: Circuit Design Layout and Stimulation, 2nd edn. (2013)
8. Stojanovic, V., Oklobdzija, V.: Comparative analysis of master slave latches and flip-flops for high-performance and low-power systems. IEEE J. Solid-State Circuits **34**(4), 536–548 (1999)
9. Uyemura, J.P.: Introduction to VLSI Circuits and System
10. Razavi, B.: Design of Analog CMOS Integrated Circuit, 13th edn.

Prediction and Analysis of Liver Patient Data Using Linear Regression Technique

Deepankar Garg and Akhilesh Kumar Sharma

Abstract In the current scenario, it is very difficult for the doctors to diagnose liver patient and there should be some kind of automated support based on machine intelligence that can help to diagnose in advance so that doctors start the treatment faster and save time. The machine intelligence is a way to predict the liver-related problems; in this study, the linear regression is used to predict the same, more accurately. The albumin levels are highly related in diagnosing these kinds of liver problems. The proposed model worked efficiently on 583 observations provided as well as on new datasets. The total average accuracy achieved in this proposed model was 89.34% which is much more than the previously identified research work of Wold et al. (SIAM J Sci Stat Comput, 5(3), 735–743, 1984, [1]) of 84.22%.

Keywords Liver patient · Albumin · Regression data mining model R-Miner

1 Introduction

India is a country with a population of 1.2 billion. There are just around 1 million certified doctors to attend to these people. Hence, we tried implementing a multiple linear regression algorithm, to predict the 'Albumin' level of a patient based on some significant variables. The original dataset provided consisted of 583 observations with 11 variables. Prediction of albumin level will help the doctors save time, and they need not run test, and the patients would save money as well. It would also enable you to diagnose the patient faster and with much more appropriate measures.

D. Garg (✉)
Department of Computer Science & Engineering, Manipal University, Jaipur, India
e-mail: deepankar.garg5@gmail.com

A. K. Sharma
Department of Information Technology, Manipal University, Jaipur, India
e-mail: akhileshshm@gmail.com

D. Reddy Edla et al. (eds.), *Advances in Machine Learning and Data Science*,
Advances in Intelligent Systems and Computing 705,
https://doi.org/10.1007/978-981-10-8569-7_8

In the model designed, 'Albumin' is considered as the dependent/outcome variable and 'Age', 'Gender', 'Total Bilirubin', and 'A/G Ratio' as the independent/observational variables. The independent variables are highly significant and are independent of each other. They are not correlated among themselves. Finally, a test model has been designed using multiple linear regression which predicts the albumin level with 90% accuracy, based on test dataset as well.

2 Related Work

As proposed by Tomohiro et al., the hypoalbuminemia separately predicts cardiac morbidity, and also mortality, for various kinds of chronic kidney disease-related patients by CRT. These kinds of observations are helpful in identifying the albumin levels, by the information of long-term prognosis in chronic kidney disease-related patients who willingly undergo CR [2]. Multiple linear regression was used [3] to implement our algorithm. According to Mark Tranmer et al., a multiple linear regression analysis is used to estimate the values of outcome variable, Y, provided with a set of p explanatory variables (x1, x2, …, xp). Mu-Jung Huang et al. adopts data cleaning and analysis methods [4] to explore meaningful guidelines from health-related data and employs case-based reasoning that always favors the highly severe diseases diagnosis and their treatments and working on these processes for more efficient working system. Swati Gupta et al. provided with introduction to linear regression algorithm [5] and explained its implementation. The researcher concentrated on the formulation and test data for linear regression. According to Dimitris Bertsimas et al., linear regression model is considered with response vector, model matrix, regression coefficients, and errors [6]. The linear regression models the relationship between a dependent variable and explanatory variables. Linear discriminant analysis and logistic regression are the most widely used statistical methods [7] for analyzing the numerical (or categorical) outcome variables. Logistic regression is useful when the dependent variable is of binary outcome.

According to David Broadhurst et al., variables in a linear regression can be selected [8] using backward or forward selection methods. These methods of elimination will help in building a linear model. Various logics were used to differentiate diffuse liver disorders, to categorize liver disorders under healthy and unhealthy liver patients, to diagnose hepatitis, and to perform the necessary operation required [9–15].

3 Methodology

The multiple linear regression algorithm is used, on the basis of which this model has been developed. In this study, the dependent/outcome variable used is 'Albumin'. And independent/observation variables are 'Age', 'Gender', 'Total Bilirubin', 'Direct Bilirubin', 'Alkphos', 'SGOT', 'SGPT', 'Total Proteins', and 'A/G Ratio'.

The correlation between independent variables was found out. It has been observed that 'Direct Bilirubin' and 'Total Bilirubin' were highly correlated. Hence, 'Direct Bilirubin' was eliminated from the model (in preprocessing phase). In this study, the first regression model was formed including uncorrelated independent variables. This model helped to classify the variables as 'significant' and 'nonsignificant'. All the significant variables were included in the next regression model, and the nonsignificant variables were removed. The final regression model consisted of just the significant and uncorrelated independent variables (Fig. 1).

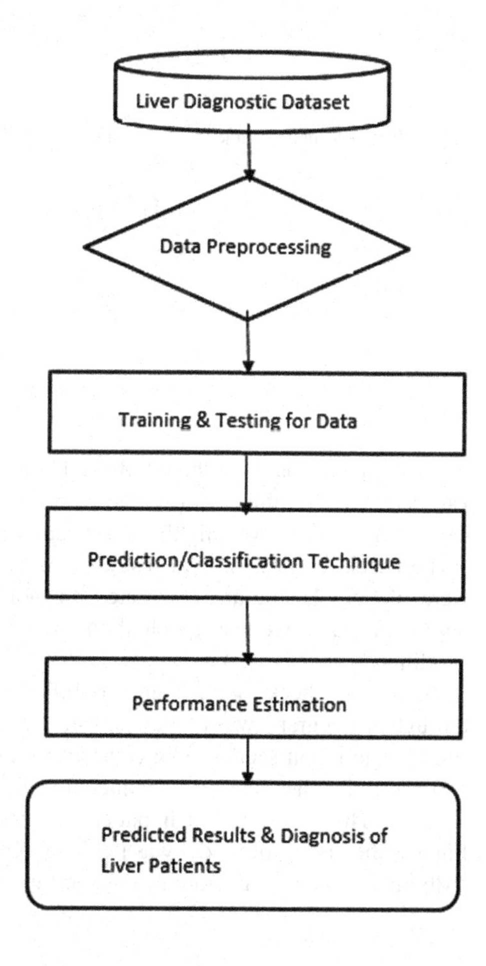

Fig. 1 Proposed model

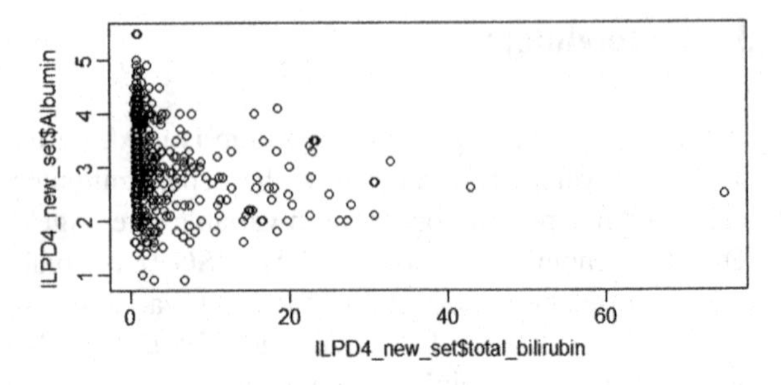

Fig. 2 Plot bet. albumin and total bilirubin

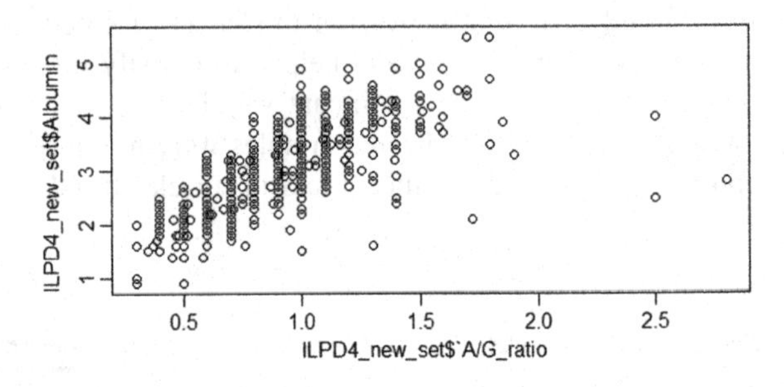

Fig. 3 Plot bet. albumin and A/G ratio

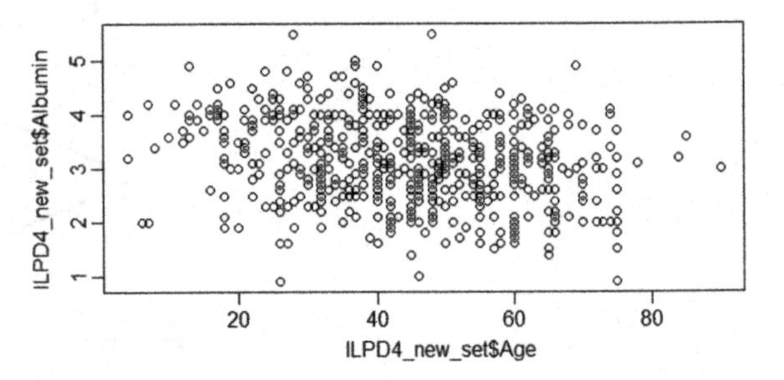

Fig. 4 Plot bet. albumin and age

In the next phase of this study, relation between dependent and independent variables was established (as shown in Figs. 2, 3, and 4). It was found that dependent variable was highly correlated with each of the independent variable.

To check for near normal residuals, a histogram (Fig. 5) and a qqplot (Fig. 6) were plotted. Histogram elaborates the normal residual values about 0 and the qqplot displays a linear graph, both of which satisfy the near normal residual condition.

In order to check for constant variability (i.e., whether the errors remain constant throughout), a graph was plotted between residuals and the fitted values as shown in Fig. 7. The graph seems to be constant in a particular region.

Lastly, the checking was conducted to find the existence of a pattern on x-axis (Fig. 8). There was no such pattern observed. Data was uniformly distributed. Finally, the testing of model was performed. It has been tried to predict the value of 'Albumin' on existing (training) as well as new (test) dataset. The model predicts

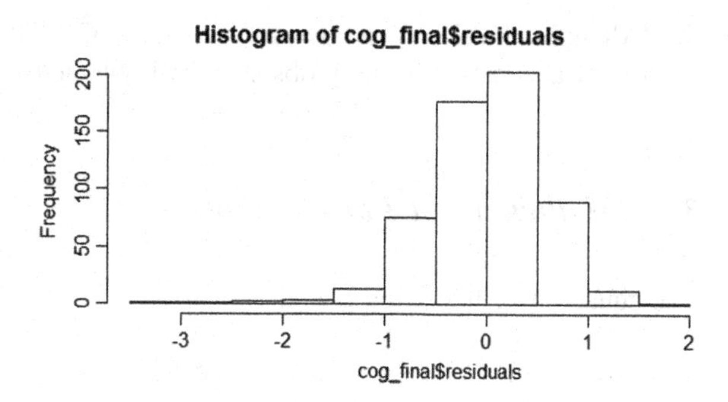

Fig. 5 Histogram plot of residuals

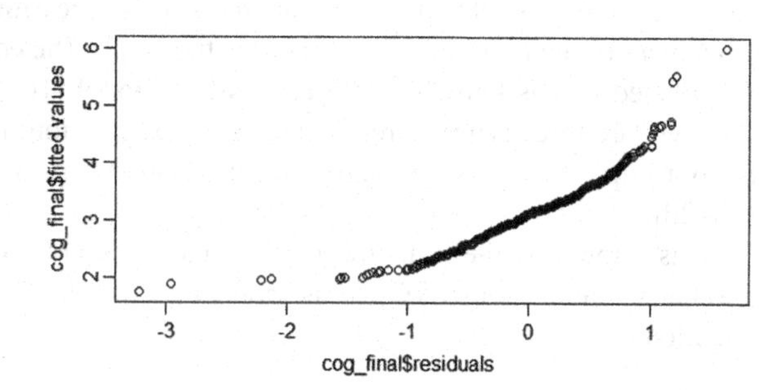

Fig. 6 qqplot between residuals and fitted values

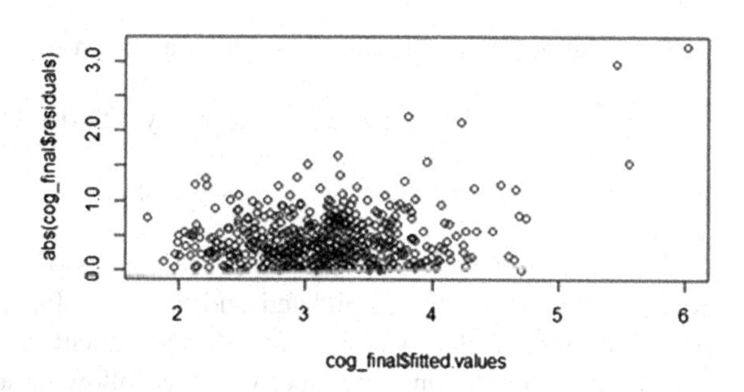

Fig. 7 Plot bet. absolute residuals and fitted values

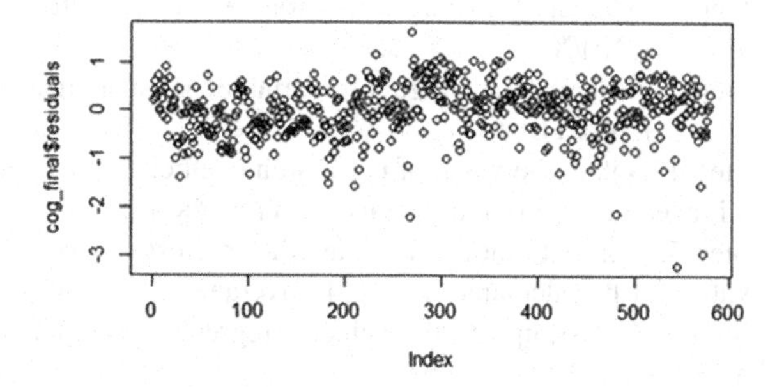

Fig. 8 Plot of residual values

the 'Albumin' level with an average accuracy of 89.353% which is a significant amount of accuracy this study observed and obtained.

3.1 *Mathematical Formulation*

The equation for linear regression is:

$$y = \alpha + \beta X$$

X and y are the two variables involved in the equation. The equation that describes how y is related to x is known as the **regression model** [13, 14]. α is the y intercept of the regression line, and β is the slope. The equation that describes how y is related to x is known as the **regression model** [16, 17].

There are three types of relationships in a regression line—first is positive linear relationship, second is a negative linear relationship, and third is no relationship [18–20].

Y is known as the outcome or dependent variable as a linear function [5] of another variable X also known as independent variable that is represented by the equation

$$Y = \alpha + \beta X \quad (1)$$

Here, the regression coefficients which are represented as α and β are given by

$$\beta = \Sigma(xi - x)(yi - y)/\Sigma(xi - x) \quad (2)$$

$$\alpha = y - \beta x \quad (3)$$

The mentioned values of regression coefficients of α and β can be computed in Eqs. (2) and (3) which are updated and putted in Eq. (1) to observe relationship among the independent variables and the dependent variable.

The linear regression technique works as following algorithm:

Step 1: Take the values of variable Xi and Yi
Step 2: Calculate average of variable Xi such that average is x = (X1 + X2 +· · · + Xi)/Xi
Step 3: Calculate average for variable Yi such that average is y = (Y1 + Y2 +· · · + Yi)/Yi
Step 4: Obtain the value of regression coefficient β by substituting the values of Xi, Yi, average of Xi, and average of Yi in the Eq. 2
Step 5: Compute the value of another regression coefficients α by substituting the values of β (calculated in step 4), average of Xi, and average of Yi in the Eq. (3).
Step 6: At last, update the value of regression coefficients α and β in the equation $Y = \alpha + \beta X$

1. **Relationship Between Dependent and Independent Variables: Linear Relation**

Figure 2 shows plot between 'Albumin' at y-axis and 'Total Bilirubin' at x-axis. It shows a linear relation between the two variables.

Figure 3 shows plot between 'Albumin' at y-axis and 'A/G Ratio' at x-axis. It shows a linear relation between the two variables.

2. **Nearly Normal Residuals: Condition Met**

The histogram of the residual as shown in Fig. 5 is taken to prove that variance is normally distributed. The histogram is evenly distributed about zero which indicates that it is normally distributed. If it were not normally distributed, it would mean that the model's assumptions may have been contradicted.

Relationship Between Dependent and Independent Variables: Linear Relation.

3.2 Database Terminology and Their Description

For dataset, the bilirubin test conducted to find any increased level in the blood. It can be used to determine the cause of jaundice or diagnose other liver diseases, hemolytic type anemia, the blockage of bile duct, etc. Direct bilirubin that moves much freely in the blood is known as conjugated bilirubin. Alkphos estimates the quantity of alkaline phosphate enzyme in anybody's bloodstream.

Increased level SGPT suggests medical problems such as viral hepatitis, diabetes, congestive heart failure, liver damage. This is a common way of screening for liver problems.

AST or aspartate aminotransferase is single or two liver enzymes. It is always found as serum glutamic oxaloacetic transaminase type or SGOT. The AST is a protein, i.e., made by liver cells. And when liver cells are damaged, AST leaks into the bloodstream and the level of the AST in the blood becomes more than basic/normal. An estimated serum protein test measures the total amount of protein in blood. It measures the amounts for two major groups of proteins in the blood: albumin and the globulin.

A/G ratio is the comparison between the amounts of albumin with those of globulin. This test is useful when your healthcare provider suspects you have liver disease. Albumin is a plasma (blood) protein. If it is low, it could be due to decreased synthesis by liver indicating liver problem.

Table 1 Table for accuracy and predicted albumin level

Levels	Total bilirubin	A/G ratio	Given albumin	Predicted albumin	Accuracy (%)
Level 1	0.8	0.8	3.7	3.03	81.89
Level 2	0.6	1.1	2.6	3.34	77.84
Level 3	0.8	1.1	3.7	3.46	93.51
Level 4	0.9	0.9	3.9	3.15	80.76
Level 5	9.4	0.8	2.8	2.79	99.64

4 Implementation and Result

Dataset collected in this work shows the different parameters that could affect a person's health, particularly the liver. All the variables are of numeric form, and a model is built with one independent/outcome variable and four dependent/observant variables. A multiple linear regression model is built which has an average accuracy of 89.34%. The model works well on new dataset as well.

Creating regression models # First regression model

model1 = lm(Albumin ~ Age + Gender + total_bilirubin + Alkphos + SGPT + SGOT + 'A/G_ratio', data = ILPD4_new_set summary (model1)

Removing insignificant variables from model1 # Final regression model created

model_final = lm(Albumin ~ Age + Gender + total_bilirubin + 'A/G_ratio', data = ILPD4_new_set)

summary(model_final) **# Testing the model**

Age = c(43), Gender = ("Male"), total_bilirubin = c(0.1), 'A/G_ratio' = c(1.2), new_patient_data <- data.frame(Age, Gender, total_bilirubin, 'A/G_ratio')

model_test <- lm(Albumin ~ Age + Gender + total_bilirubin + 'A/G_ratio', data = ILPD4_new_set) predict(model_test, newdata = new_patient_data)

Model Accuracy: Training and Test Datasets

The main parameters include 'Total Bilirubin' and 'A/G Ratio'. From these parameters, albumin level can be predicted, which will help in diagnosing the patient faster. The prediction has an average accuracy of 89.353%. Age and gender also play a significant role in determining the albumin level (Table 1).

Table 1 consists of total_bilirubin, a/g_ratio, and albumin, which were already given. The fifth column predicts the albumin level from the model built with an average accuracy of 86.72%.

5 Conclusion

A multiple linear regression model has been prepared, which has one dependent variable and four independent variables. Initial model was prepared using the backward elimination method. The variables were eliminated based on their significance level. Least significant variables were removed from the model. Highly correlated independent variables were excluded as well. A final linear model was prepared where the dependent variable was linearly related to each of the independent variable and the model consisted of just the significant, nonlinear independent variables.

The model prepared predicts with an average accuracy of 89.34%. It predicts the level of albumin in a patient, which is highly dependent on the four independent variables defined in the model. Albumin level will help the doctor think in the right direction and help diagnose the patient faster. Usually, it is assumed that significant changes in the albumin level indicate that the patient is probably having problems in his lever. The model made works well on new dataset as well.

References

1. Wold, S., et al.: The collinearity problem in linear regression. The partial least squares (PLS) approach to generalized inverses. SIAM J. Sci. Stat. Comput. **5**(3), 735–743 (1984)
2. Uchikawa, T., et al.: Serum albumin levels predict clinical outcomes in chronic kidney disease (CKD) patients undergoing cardiac resynchronization therapy. Intern. Med. **53**(6), 555–561 (2014)
3. Aiken, L.S., Stephen G.W., Steven, C.P.: Multiple linear regression. Handbook of psychology (2003)
4. Huang, M.-J., Chen, M.-Y., Lee, S.-C.: Integrating data mining with case-based reasoning for chronic diseases prognosis and diagnosis. Expert Syst. Appl. **32**(3), 856–867 (2007)
5. Gupta, S.: Int. J. Comput. Appl. **116**(9), 0975–8887 (2015)
6. Bertsimas, D., King, A.: OR forum—an algorithmic approach to linear regression. Oper. Res. **64**(1), 2–16 (2016). https://doi.org/10.1287/opre.2015.1436
7. Metodološki zvezki, vol. 1, No. 1, pp. 143–161 (2004)
8. Broadhurst, D., et al.: Genetic algorithms as a method for variable selection in multiple linear regression and partial least squares regression, with applications to pyrolysis mass spectrometry. Anal. Chim. Acta **348**(1–3), 71–86 (1997)
9. Badawi, A.M., Derbala, A.S., Youssef, A.B.M.: Fuzzy logic algorithm for quantitative tissue characterization of diffuse liver diseases from ultrasound images. Int. J. Med. Inform. **55**, 135–147 (1999)
10. Gadaras, I., Mikhailov, L.: An interpretable fuzzy rule-based classification methodology for medical diagnosis. Artif. Intell. Med. **47**, 25–41 (2009). Luukka, P.: Fuzzy beans in classification. Expert Syst. Appl. **38**, 4798–4801 (2011)
11. Ming, L.K., Kiong, L.C., Soong, L.W.: Autonomous and deterministic supervised fuzzy clustering with data imputation capabilities. Appl. Soft Comput. **11**, 1117–1125 (2011)
12. Neshat, M., Yaghobi, M., Naghibi, M.B., Esmaelzadeh, A.: Fuzzy expert system design for diagnosis of liver disorders. In: Proceedings of the 2008 International Symposium on Knowledge Acquisition and Modeling KAM 2008, pp. 252–256 (2008)

13. Breese, J.S., Heckerman, D., Kadie, C. (1998). Empirical analysis of predictive algorithms for collaborative filtering. In Proceedings of UAI-1998: The Fourteenth Conference on Uncertainty in Artificial Intelligence
14. Burges, C.: A tutorial on support vector machines for pattern recognition. Data Min. Knowl. Discov. **2**(2), 955–974 (1998)
15. CiteSeer: CiteSeer Scientific Digital Library (2002). http://www.citeseer.com
16. Singh A., et al.: Liver disorder diagnosis using linear, nonlinear and decision tree classification algorithms. IJET **8**(5) 2059–2069 (2016)
17. Qual Quant: Linear versus Logistic Regression vol. 43, pp. 59–74 (2009). https://doi.org/10.1007/s11135-007-9077-3
18. Fox, J.: Applied Regression Analysis, Linear Models, and Related Methods. Sage Publications, Thou-sand Oaks, CA (1997)
19. Robbins and Cotran's Pathologic Basis of Disease, 9th edn.
20. Hocking, Ronald R.: A biometrics invited paper. The analysis and selection of variables in linear regression. Biometrics **32**(1), 1–49 (1976)
21. Duda, R.O., Hart, P.E.: Pattern Classification and Scene Analysis. Wiley (1973)

Image Manipulation Detection Using Harris Corner and ANMS

Choudhary Shyam Prakash, Sushila Maheshkar and Vikas Maheshkar

Abstract Due to the availability of media editing software, the authenticity and reliability of digital images are important. Region manipulation is a simple and effective method for digital image forgeries. Hence, the potential to identify the image manipulation is current research issue these days and copy–move forgery detection (CMFD) is a main domain in image authentication. In copy–move forgery, one region is simply copied and pasted over other region in the same image for manipulating the image. In this paper, we have proposed a method based on Harris corner and adaptive non-maximal Suppression (ANMS). Initially, the input image is taken, and then Harris corner detection algorithm is used to detect the interest points, and ANMS is adopted to control the number of Harris points in an image. This gives an appropriate number of interest points for different size of images and gives the assurance for finding the manipulated region in manageable time. For each extracted interest point, SIFT is used for calculating the descriptors. Now obtained descriptors are matched using the outlier rejection with nearest neighbour. Here RANSAC is used to find the best set of matches to identify the manipulated regions. Experimental results show the robustness against different transformation and post-processing operations.

Keywords Copy–move forgery · Image forensics · SIFT descriptor · ANMS
Duplicate region detection

1 Introduction

In today's scenario, manipulating an image is an easy task due to the availability of sophisticated image editing tools like Photoshop, GIMP (GNU image manipulation program), the NIK collection (offered by Google) and many more software which

---

C. Shyam Prakash (✉) · S. Maheshkar
Indian Institute of Technology (Indian School of Mines), Dhanbad, India
e-mail: shyamprakash2008@yahoo.com

S. Maheshkar
e-mail: sushila_maheshkar@yahoo.com

V. Maheshkar
Netaji Subhas Institute of Technology, Delhi, India

D. Reddy Edla et al. (eds.), *Advances in Machine Learning and Data Science*,
Advances in Intelligent Systems and Computing 705,
https://doi.org/10.1007/978-981-10-8569-7_9

Fig. 1 Example of copy–move manipulation **a** original image and **b** manipulated image

make image editing easy. Due to this fact, relying on any image is hard to believe. Sometimes the perfection of manipulation is very high so that it cannot be detected by naked eyes. In addition, the images are accepted as the popular source of information and it plays a vital role when it is presented as the courtroom witness, insurance claims, scientific scams, etc. Hence, the authenticity of an image is a big question that can be answered through different image manipulation detection techniques.

Digital image forensics can be categorized as intrusive (active) and non-intrusive (passive) methods. In intrusive methods, digital watermark or digital signature is embedded into the images to authenticate the originality of an image. In non-intrusive methods, there is not such kind of external media presented in the image. We have to rely only on the intrinsic properties of image to detect the authenticity of an image.

In all manipulation techniques, copy–move (region duplication) is a common manipulation technique which is used to conceal some part of the image by copy–pasting in the same image. An example of copy–move manipulation can be seen from Fig. 1. The main intention of copy–move manipulation is to suppress some important object in the image by copy–pasting a set of pixels from one part to another part of an image, and it is hard to find the manipulation by naked eyes.

In this paper, proposed method is used to locate the manipulated region accurately with the help of Harris corner detection and ANMS. As it is known that the computational cost of matching is directly proportional to the number of interest points extracted from an image. Here, ANMS is adopted for better distribution of keypoints throughout the image which helps to detect the duplicated regions even if the image is post-processed by different operations.

The rest of the paper is organized as follows. In Sect. 2, we discuss the related work which has been proposed yet with their merits and demerits. In Sect. 3, proposed method is discussed in detail. Then we present the detailed experimentation in Sect. 4. In Sect. 5, we draw the conclusion of the paper.

2 Related Works

As far as the literature has been concerned, it is found that the detection techniques for copy–move manipulation deal with the additional post-processing operations like rotation, flipping, resizing, edge blurring, white Gaussian noise and JPEG compression. These post-processing operations are applied to conceal the originality of the image, and when it once applied, it is hard to detect the manipulation. Hence, to tackle these problems, there are two different techniques adopted to spot the manipulations: block-based techniques and keypoint-based techniques.

In block-based techniques, block of pixels is used to find the duplicated regions. Most of the techniques fall in this category. Fridrich et al. [7] proposed a technique for exact match, and it is considered as the basic algorithm for duplicate region detection. Popescue and Farid [18] proposed a method to reduce the dimension of the block using principal component analysis (PCA). Their method shows better results under Gaussian noise addition and JPEG compression. Luo et al. [12] used RGB channel and extract seven features which are based on the average of the pixel intensity. Their method achieved higher accuracy rate against JPEG compression, Gaussian blurring and additive Gaussian noise but it still fails in rotation, resizing and flipping operations performed on image. Mahadian and Saic [13] used blur moment invariants on each block for generating a feature vector. They have calculated 24 blur moment invariants; hence, the dimension of feature vector is 72 as the blur moment invariant calculated separately on each red, green, blue (RGB) channel. Zhang et al. [22] used discrete wavelet transform (DWT) and phase correlation for extracting the feature of blocks. Their method shows robustness under JPEG compression and blurring only. Li et al. [11] used DWT and singular value decomposition (SVD) to reduce the dimension and found a robust block representation. Their approach only works for JPEG compression with quality factor greater than 70%. Kang and Wei [10] proposed a method in which SVD is applied on overlapping blocks to extract the feature vectors.

As an alternative of block-based techniques, the methods based on keypoints are proposed for forgery detection by many authors. In keypoint-based techniques, image keypoints are detected and matched in the whole image for detection of duplicated region. Huang et al. [9] proposed a method in which scale-invariant feature transform (SIFT) descriptor is used for keypoint extraction. By comparing the descriptor, duplicated region is found but it fails for small duplicated regions. Pan and Lyu [16, 17] applied SIFT descriptor for forgery detection and their method detects forged region successfully even the images are post-processed using Gaussian noise and JPEG compression. Amerini et al. [1] carried out detailed examination on SIFT descriptor to locate the forged region. They have also considered geometrical transformation using hierarchical clustering procedure for the forgery detection. Xu et al. [2] considered speeded-up robust feature (SURF) for forgery detection instead of SIFT. Shivkumar and Baboo [19] used SURF for keypoints extraction and then stored those keypoints in a Kd-tree. The author has presented only visual results of two cases of resizing, rotation and Gaussian noise.

The block-based techniques are most popular in copy–move manipulation detection technique due to the significance with various feature extraction techniques and ability to obtain high matching performance. But it also has some limitations like it is not invariant to geometrical transformation (rotation and scaling). Apart from block-based techniques, keypoint-based technique is also becoming popular in copy–move forgery detection. In this category, SIFT feature is most reliable and popular detection technique because of its good performance in geometrical transformation. Hence, the main focus in keypoint-based detection techniques is to improve its accuracy while maintaining the robustness in geometrical transformation operations and it is the main advantage of using keypoint-based techniques over block-based techniques.

3 Proposed Method

The major goal of our proposed method is to perceive a method to recognize copy–move manipulated images and spot the duplicated region accurately. Since in copy–move manipulation, a part of image is copied and pasted on other part in the same image, so getting the accurate similar region is the key idea presented here as well as in other developed method. This can be achieved through extracting and matching the local features of the image to get the similar regions.

Here, we have proposed a manipulation detection method based on keypoint detection technique. There are many keypoints detection technique available like Harris corners, SIFT, MSER. In the proposed work, the Harris corner method is used to detect the keypoints.

3.1 Harris Corner Detection

In an image, interest point is a point having well-defined position and which can be detected robustly [15]. It means it does not change its characteristics even if the image gone through illumination adjustment or geometrical transformation. It can be an isolated point of local maximum or minimum intensity, a corner, a line ending or a point on a curve where the curvature is locally maximal.

In Harris corner detection method [8], it uses the fact that, the intensity of image changes a lot in various directions at a corner, but it changes enormously in a certain direction at an edge. To formulate this anomaly of change in intensity, a local window is used to get the result from shifting the local window. The image intensity changes enormously when the window is shifted in an arbitrary direction near a corner point. Around the edge point, it will change enormously when the window is shifted perpendicularly. By using this perception, in Harris corner detection method, a second-order moment matrix is used for the corner decisions.

Autocorrelation matrix M for a given image I at point (x, y) can be given by

$$M(i,j) = \sum_{i,j} w(i,j) \begin{bmatrix} I_i^2(i,j) & I_iI_j(i,j) \\ I_iI_j(i,j) & I_i^2(i,j) \end{bmatrix} \quad (1)$$

where I_i and I_j are the pixel intensity derivatives respectively in the x and y direction at point (i, j), i.e.

$$I_i = I \otimes [-1, 0, 1] \approx \partial I / \partial i \quad (2)$$

$$I_j = I \otimes [-1, 0, 1] \approx \partial I / \partial j \quad (3)$$

where operator ⊗ denotes convolution. The weighting function $w(i,j)$ can be given by

$$w(i,j) = g(i,j,\sigma) = \frac{1}{2\Pi\sigma^2} exp(-\frac{i^2 + j^2}{2\sigma^2}) \quad (4)$$

It assigns more weight to those values near the centre of a local region. It can be uniform, but it is more typically an isotropic, circular Gaussian.

To find the interest point, we first compute the corner strength function which is harmonic mean of the matrix M.

$$f_M(i,j) = \frac{detM_l(i,j)}{trM_l(i,j)} = \frac{\lambda_1 \lambda_2}{\lambda_1 + \lambda_2} \quad (5)$$

where λ_1 and λ_2 are the eigenvalues of M. These points are selected as interest point where $f_M(i,j)$ is a local maxima in window of 3×3 neighbourhood and above a threshold $t = 10$. When the local maxima is detected, its position is refined to sub-pixel accuracy by fitting a 2D quadratic to the corner strength function in the local window of 3×3 neighbourhood and finding its maxima. The orientation can be detected by two methods: the first method is to find the eigenvectors which denote the largest eigenvalue of the Harris matrix M and the second method is simply using the gradient at that point as its orientation. Here, we have used second method.

3.2 Adaptive Non-maximal Suppression

It is a novel approach which is used to control the density of features in image and provide an improved spatial distribution of features than earlier methods. The computational cost of matching is directly proportional to the number of interest points extracted from an image, and it should be restricted to minimize the computational cost of matching. Meanwhile, it is essential that the distribution of the interest points over the image should be spatially well distributed. To get the above-mentioned requirements, ANMS is used to get a desired number of interest points from an image.

Based on f_M, to select the desired number of interest points, points are suppressed. Those pixels are considered which are maximum in a neighbourhood having radius r. The suppression radius (r) is initially taken as zero, and it is increased gradually to obtain the desired number of keypoints (n_{ip}). The first interest point in the list is not suppressed at any radius, and it is global maxima. Interest points are added one by one to the list as the suppression radius decreases from the infinity. After all, if an interest point appears once in the list, it will remain in the list because if the obtained interest point is maximum in radius r, then it will also maximum in radius r', which is smaller than r. In practice, to make the non-maximal suppression more robust, neighbour should have sufficient larger strength. Hence, the minimum suppression radius (r_i) can be obtained by

$$r_i = \min_j |x_i - x_j|, s.t. f(x_i) < C_{robust} f(x_i), x_j \epsilon I \quad (6)$$

where x_i is the location of 2D interest point and I is the set of all interest points locations. $C_{robust} = 0.9$ is taken to ensure that a neighbour must have sufficient higher strength for suppression to take place.

3.3 *Feature Descriptor*

By extracting the keypoints, we have only obtained the information about their position and sometimes their coverage area (usually approximated by a circle or ellipse) in the image. Now the objective is to get the description of the local image structure for efficient and reliable matching of features in the whole image. Previously many feature vectors have been developed including Gaussian derivatives, local intensity patches, scale-invariant feature transform and affine-invariant descriptor for matching of features. According to Mikolajczyk and Schmid [14] survey, SIFT features generally give the best performance.

In SIFT, after applying ANMS algorithm, the set of interest points are obtained. A 16×16 window around these extracted points is applied. Within this window to select the keypoints descriptors, the subdivision of 4×4 window is considered. The magnitude and orientation of the gradient at extracted point are calculated and place them into eight buckets. This process is performed for each of the 16×16 windows which is further divided into 4×4 windows to obtain 128 features descriptors. Then, the magnitude is subtracted from the orientation of the interest points followed by normalization. A threshold of 0.2 is used for normalization of the vectors and renormalizes the vectors to the unit length to avoid any values which are significantly higher than the others [3].

3.4 Detection Algorithm

The proposed technique for image manipulation detection based on Harris interest points and adaptive non-maximal suppression(ANMS) is shown in Fig. 2.

Initially, the input image is taken and then Harris corner detection algorithm is used to detect the interest points which is described in Sect. 3. The number of interest points detected from Harris corner detection has high relevance. If the number of Harris points is small, it is difficult to detect the manipulated region and if the number of Harris points is greater, then it would be time-consuming to find the manipulated region. Hence, we have adopted ANMS to control the number of Harris points in an image. In this way, we can get a proper number of interest points for different image size and give the assurance for finding the manipulated region in manageable time.

Now for matching process of obtained descriptors, the outliers rejection with nearest neighbour is used. Because of smoothness of natural image generally the best match of a Harris point resides within its close spatial adjacency. Hence, it is important to restrict the searching of nearest neighbour of a Harris point. To restrict the searching, we use a certain distance like $10\sqrt{2}$ (as an example of 10×10 size region) between two matched Harris points. Still, many Harris points can be matched with each other but only those points are selected which have distinct similarities. The feature vector f' other than f', the distance between f and f' must satisfy $\| f' - f \|_2 < \varepsilon \| f' - f \|_2$, where $\varepsilon \epsilon (0, 1)$ and f is the Harris point location (x, y) and f' is the corresponding feature vector of Harris point at (x', y'). Those Harris

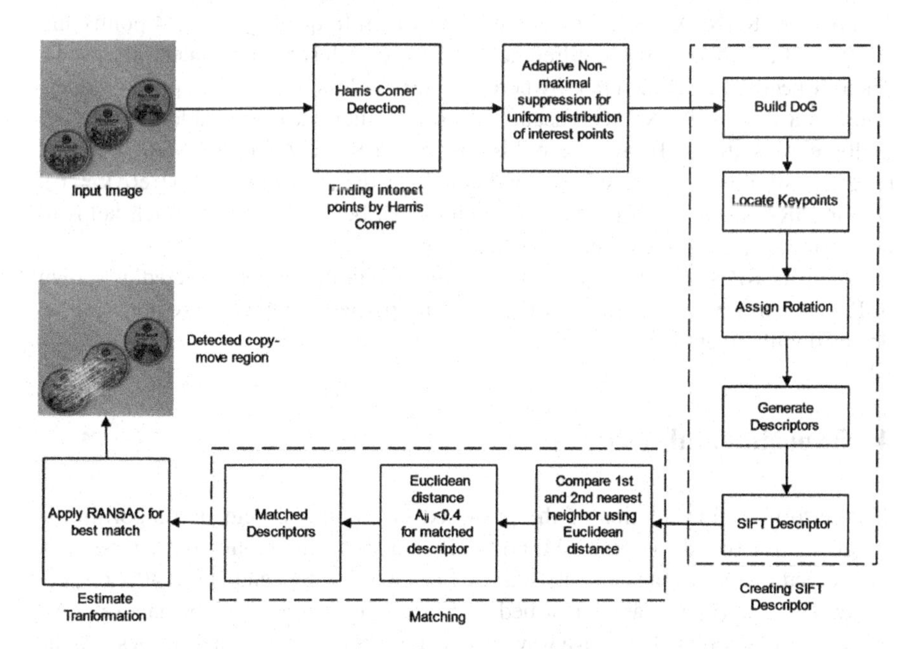

Fig. 2 Proposed technique

points, which are matched, are recorded pair-wise. In matching of obtained descriptors, outliers rejection method with nearest neighbour is used. The basic distance metric, Euclidean distance, is used to get first and second nearest neighbour for each obtained descriptor. The formula for Euclidean distance is given by

$$A_{ij} = \sqrt{\Sigma_{i,j=1}^{n}(d_i - d_j)^2} \tag{7}$$

where d_i and d_j are two different descriptor vectors. Now the ratio of the distance is taken, and it is assured that the obtained ratio should be less than the predefined threshold value 0.4 (which is optimal value for panorama stitching [4]), and then it is considered as the matched descriptor. Here the experiments are conducted on varying the value of threshold value. The match will be strong, and execution is subsequently faster if the threshold value is less than 0.4. Upon taking the value of threshold greater than 0.4, there will be more matches and which results greater execution time. In the proposed method, 0.4 is a good trade-off between speed and breadth of points. Apart from all these settings, the distance of descriptors from itself is set to infinity; otherwise, it will find itself as the nearest neighbour.

Additionally, filtering is used to improve the speed of the matching. One more step is taken to remove the false matching. There are few interest points obtained due to same texture. Those interest points are near to each other by few pixels, and hence, those interest points are removed which are 3 pixels or less apart. After matching process, many equivalent points are obtained which have similar features. To find the perfect match of the descriptors and locate the copy–move region, RANSAC [19] is used. RANSAC is an iterative method which frequently selects 4 points and computes the sum of squared differences (SSD) of each set of equivalent points [4]. Then, it keeps the set of inliers, where the SSD is less than some epsilon values. There is a flaw in RANSAC, because of single inlier set it is unable to scale the multiple translations. To overcome this flaw, multi-RANSAC (mRANSAC) is used here. It is similar to RANSAC, except that it saves the top k inlier sets (Here k = 6). Due to mRANSAC, multiple transformations can be captured and which helps to detect multiple copy–move regions in an image.

By using RANSAC, the multiple copy–move regions can be detected as shown in Fig. 3a. However, as seen from Fig. 3b, only partially duplicated regions can be detected without applying RANSAC.

4 Experimental Analysis

The performance evaluation of the proposed manipulation detection algorithm is conducted on two data sets CoMoFod [21] and image manipulation data set [6]. The forged images of the two data sets are of size 512 × 512 and 600 × 400, respectively. The experiments are performed on RGB images present in the data sets. All images are manipulated in such a way that it cannot be detected by naked eyes. All the

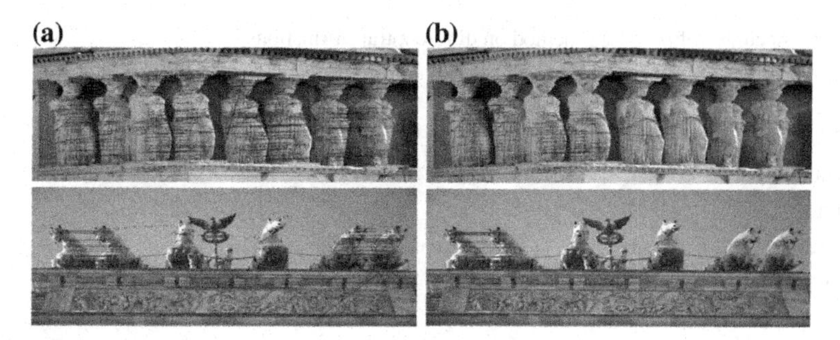

Fig. 3 Detection of copy–move attack **a** With RANSAC and **b** Without RANSAC

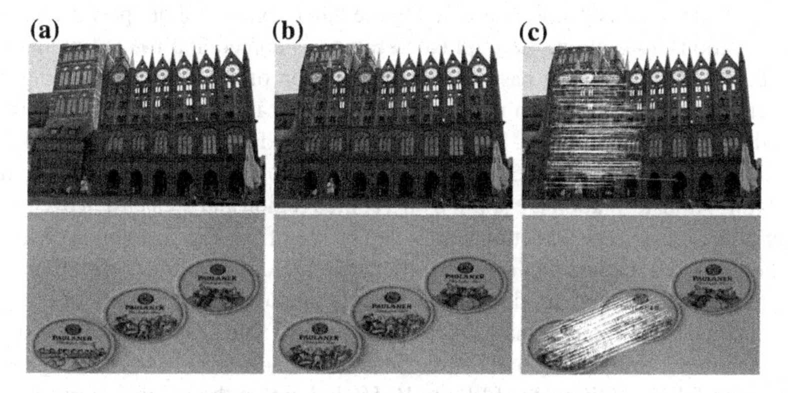

Fig. 4 Example of copy–move attack **a** Original image, **b** Manipulated image, **c** Detection result

Table 1 Accuracy of copy–move attack

Data set	Number of Images	Number of detected images	Accuracy%
CoMoFod [21]	200	175	87.5
IMD [6]	45	39	86.6

experiments are performed on MATLAB R2013a. The experimental results are discussed in the below section in detail.

Initially, the simple copy–move is presented here in which the duplicated region is copied and pasted over other region in the same image without using any post-processing operations. In CoMoFod [21] and image manipulation data set [6], there are 200 and 45 forged images in both of the data set of different size, respectively. The example of detection results can be seen from Fig. 4. The accuracy of detected copy–move attack is shown in Table 1. As it can be seen that the proposed method works good on both of the data sets and results are satisfactory. Some of the popular transformations are also applied here to the copied region before changing its location. Transformation like translation, rotation, scaling, distortion and

Table 2 Accuracy of proposed method on different transformations

Transformations	No. of correctly detected images	DAR (%)	FPR (%)
Translation	38	95	5
Rotation	32	80	20
Scaling	31	77.5	22.5
Distortion	40	100	0
Combination	34	85	15

combination are applied on the CoMoFod [21] data set in the experiment. In translation, a copied region is simply translated to the other region without performing any transformation. In rotation, the duplicated region is copied and rotated with some angle $\theta \epsilon (0°, 360°)$ and then pasted over other region of the image. In scaling, the duplicated region is scaled up or down with some scaling factor and then pasted over other region of the image. In distortion, the copied region is distorted and then pasted over other region of the image. For combination, two or more (like rotation and scaling) transformation are applied on a copied region and then pasted over other region of the image. The detection results can be seen from Fig. 5, and it is observed that the proposed method can successfully detect the different transformations like translation, rotation, scaling, distortion and combination. The accuracy of the proposed method on different transformation is listed in Table 2, and it is observed that the detection accuracy rate (DAR) of the detection results is quite satisfactory. Also, the average false positive rate (FPR) is 10.5% indicating the efficiency of the proposed algorithm.

To compare the effectiveness of the proposed method with the existing method, the proposed method is compared with L. Chen et al. [5] and Silva et al. [20] methods. L. Chen et al. [5] have proposed a method for duplicate region detection based on Harris corner points and step sector statistics. This method is robust enough to geometrical transformations (like JPEG compression and Gaussian noise). The step sector statistics method is proposed to give a unique representation of the each circle region around the interest points. Then matching of those interest points is performed to get the duplicated region. Silva et al. [20] proposed a method which is based on multi-scale analysis and voting process of a digital image. SURF is used to detect the keypoints, and then their descriptors are localized in the image. Next, these keypoints are matched and clustered. Multi-scale analysis is employed to detect the manipulated regions. This method is robust to rotation, scaling and partially robust to compression.

To control the number of interest points detected through Harris corner method, the thresholding is done which changes with image size. Because of this, the density of interest points cannot be controlled. However, in the proposed method the adaptive non-maximal suppression (ANMS) is used for controlling the density of interest points for any image size. The comparison results of proposed method with

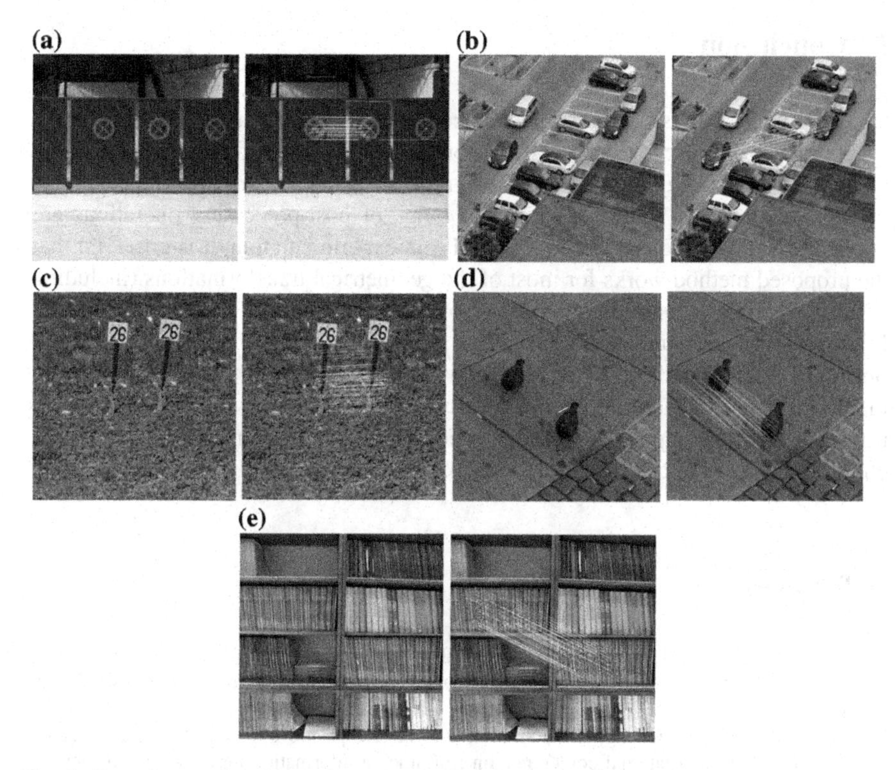

Fig. 5 Detection results of Transformations **a** Translation, **b** Rotation, **c** Scaling, **d** Distortion, **e** Combination

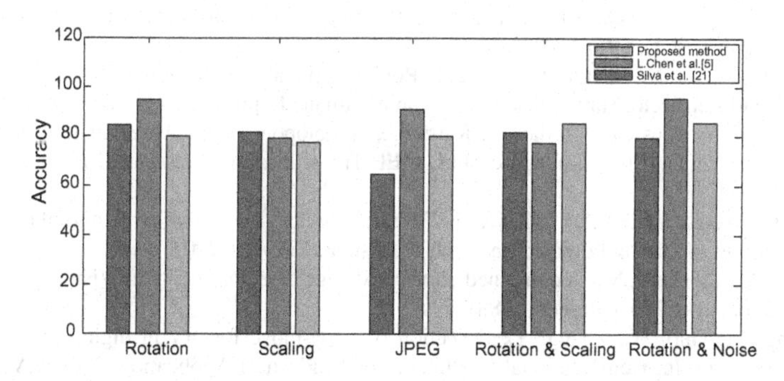

Fig. 6 Detection accuracy of L. Chen et al. [5], Silva et al. [20] and the proposed method

L.Chen [5] and Silva et al. [20] method are shown in Fig. 6, and it is observed that the proposed method gives quite satisfactory results in all post-processing operations as comparison to existing state-of-the-art methods.

5 Conclusion

Copy–move forgery is common and carried out easily and accurately. In this paper, a technique is proposed based on Harris corner and ANMS for forgery detection. By using ANMS, the keypoint density can be reduced and the distribution of keypoints throughout image is better. Here a wide range of post-processing operations are applied to validate the proposed method. From experimentation, it is observed that the proposed method works for most of the geometrical transformations (including rotation, scaling, translation, distortion and combination) with satisfactory results. It indicates the robustness and consistency of proposed method with less time complexity of $O(n^2)$ than the L. Chen et al. [5] methods. In future, we will try to improve the performance of the proposed method which can also deal with image degradations (like JPEG compression, Gaussian noise, brightness change, image blurring, colour reduction and contrast adjustments).

References

1. Amerini, I., Ballan, L., Caldelli, R., Del Bimbo, A., Serra, G.: A sift-based forensic method for copy-move attack detection and transformation recovery. IEEE Trans. Inf. Forensics Secur. **6**(3), 1099–1110 (2011)
2. Bo, X., Junwen, W., Guangjie, L., Yuewei, D.: Image copy-move forgery detection based on surf. In: 2010 international conference on Multimedia information networking and security (MINES), pp. 889–892. IEEE (2010)
3. Brown, M., Lowe, D.G.: Invariant features from interest point groups. In: BMVC, vol. 4 (2002)
4. Brown, M., Lowe, D.G.: Automatic panoramic image stitching using invariant features. Int. J. Comput. Vis. **74**(1), 59–73 (2007)
5. Chen, L., Lu, W., Ni, J., Sun, W., Huang, J.: Region duplication detection based on harris corner points and step sector statistics. J. Vis. Commun. Image Represent. **24**(3), 244–254 (2013)
6. Christlein, V., Riess, C., Jordan, J., Riess, C., Angelopoulou, E.: An evaluation of popular copy-move forgery detection approaches. IEEE Trans. Inf. Forensics Secur. **7**(6), 1841–1854 (2012)
7. Fridrich, A.J., Soukal, B.D., Lukáš, A.J.: Detection of copy-move forgery in digital images. In: Proceedings of Digital Forensic Research Workshop. Citeseer (2003)
8. Harris, C., Stephens, M.: A combined corner and edge detector. In: Alvey Vision Conference. vol. 15, pp. 10–5244. Citeseer (1988)
9. Huang, H., Guo, W., Zhang, Y.: Detection of copy-move forgery in digital images using sift algorithm. In: Computational Intelligence and Industrial Application, 2008. PACIIA'08. Pacific-Asia Workshop on. vol. 2, pp. 272–276. IEEE (2008)
10. Kang, X., Wei, S.: Identifying tampered regions using singular value decomposition in digital image forensics. In: 2008 International Conference on Computer Science and Software Engineering, vol. 3, pp. 926–930. IEEE (2008)
11. Li, G., Wu, Q., Tu, D., Sun, S.: A sorted neighborhood approach for detecting duplicated regions in image forgeries based on DWT and SVD. In: 2007 IEEE International Conference on Multimedia and Expo, pp. 1750–1753. IEEE (2007)
12. Luo, W., Huang, J., Qiu, G.: Robust detection of region-duplication forgery in digital image. In: 18th International Conference on Pattern Recognition, 2006. ICPR 2006. vol. 4, pp. 746–749. IEEE (2006)

13. Mahdian, B., Saic, S.: Detection of copy-move forgery using a method based on blur moment invariants. Forensic Sci. Int. **171**(2), 180–189 (2007)
14. Mikolajczyk, K., Schmid, C.: A performance evaluation of local descriptors. IEEE Trans. Pattern Anal. Mach. Intell. **27**(10), 1615–1630 (2005)
15. Moravec, H.P.: Obstacle avoidance and navigation in the real world by a seeing robot rover. Technical report, DTIC Document (1980)
16. Pan, X., Lyu, S.: Detecting image region duplication using sift features. In: 2010 IEEE International Conference on Acoustics Speech and Signal Processing (ICASSP), pp. 1706–1709. IEEE (2010)
17. Pan, X., Lyu, S.: Region duplication detection using image feature matching. IEEE Trans. Inf. Forensics Secur. **5**(4), 857–867 (2010)
18. Popescu, A., Farid, H.: Exposing digital forgeries by detecting duplicated image region [technical report]. 2004–515. Hanover, Department of Computer Science, Dartmouth College. USA p. 32 (2004)
19. Shivakumar, B., Baboo, L.D.S.S.: Detection of region duplication forgery in digital images using surf. IJCSI Int. J. Comput. Sci. Issues **8**(4) (2011)
20. Silva, E., Carvalho, T., Ferreira, A., Rocha, A.: Going deeper into copy-move forgery detection: exploring image telltales via multi-scale analysis and voting processes. J. Vis. Commun. Image Represent. **29**, 16–32 (2015)
21. Tralic, D., Zupancic, I., Grgic, S., Grgic, M.: CoMoFod new database for copy-move forgery detection. In: ELMAR, 2013 55th International Symposium, pp. 49–54. IEEE (2013)
22. Zhang, J., Feng, Z., Su, Y.: A new approach for detecting copy-move forgery in digital images. In: 11th IEEE Singapore International Conference on Communication Systems, 2008. ICCS 2008, pp. 362–366. IEEE (2008)

Spatial Co-location Pattern Mining Using Delaunay Triangulation

G. Kiran Kumar, Ilaiah Kavati, Koppula Srinivas Rao and Ramalingaswamy Cheruku

Abstract Spatial data mining is the process of finding interesting patterns that may implicitly exist in spatial database. The process of finding the subsets of features that are frequently found together in a same location is called co-location pattern discovery. Earlier methods to find co-location patterns focuses on converting neighbourhood relations to item sets. Once item sets are obtained then can apply any method for finding patterns. The criteria to know the strength of co-location patterns is participation ratio and participation index. In this paper, Delaunay triangulation approach is proposed for mining co-location patterns. Delaunay triangulation represents the closest neighbourhood structure of the features exactly which is a major concern in finding the co-location patterns. The results show that this approach achieves good performance when compared to earlier methodologies.

Keywords Delaunay triangulation · Co-location patterns · Participation ratio Participation index · Max PI · Medoid PI

G. Kiran Kumar (✉) · K. Srinivas Rao
Department of CSE, MLR Institute of Technology, Hyderabad 500043, India
e-mail: ganipalli.kiran@gmail.com

I. Kavati
Department of CSE, Anurag Group of Institutions, Hyderabad 500038, India
e-mail: kavati089@gmail.com

K. Srinivas Rao
e-mail: ksreenu2k@gmail.com

R. Cheruku
Department of Computer Science and Engineering, National Institute of Technology Goa, Ponda 403401, Goa, India
e-mail: rmlswamygoud@gmail.com

D. Reddy Edla et al. (eds.), *Advances in Machine Learning and Data Science*, Advances in Intelligent Systems and Computing 705,
https://doi.org/10.1007/978-981-10-8569-7_10

1 Introduction

Spatial co-location pattern mining is similar to association rule mining. Spatial data mining is the process of extracting implicit information, conceivably helpful patterns from spatial databases [1, 6, 8]. Co-location pattern mining is the method of discovering the feature subsets which are appeared together frequently in the same area. The co-location patterns have large impact in many applications including but not limited to location-based services, GIS, healthcare, business planning, transportation, traffic monitoring. For example, ATM machines are generally installed nearby hospitals, malls, and banks. In this context, we can consider ATM, mall, hospital as one of the co-location patterns. This implies ATMs can be installed wherever malls and hospitals present. A co-location rule is of the form: $L1 \rightarrow L2(p, cp)$ where L1 and L2 are co-locations, p and cp refers to prevalence measure and conditional probability [9].

Contribution of different authors on co-location pattern mining is as follows. Morimoto et al. propose a space partitioning and non-overlap grouping method for finding neighbouring features for a frequent neighbouring feature set [7]. However, there are number of outliers across the spatial space and false results. In another approach, a join-based algorithm was proposed for co-location pattern mining. With the ever-growing co-location patterns and their instance, the join-based approach is considered very computationally expensive. Reducing the number of expensive spatial join operations in finding co-location instances is very important which is proposed in a partial join approach. Join less algorithms have been proposed, but methods including the join less algorithm do not discover both single self-co-location patterns and complex self-co-location patterns [11].

Further, a novel constraint neighbourhood-based approach was developed by Tran Van Cahn et al. to find co-location patterns [10]. Their main concern is to discover different co-location patterns such as star, clique, including single and complex self-co-locations.

Rest of the paper is organized as follows. In Sect. 2, we presents the proposed approach for co-location pattern mining. Section 3 presents the experimental evaluation. In Sect. 4, the conclusion and future work are discussed.

2 Methodology

In our method, we use Delaunay triangulation approach for mining co-location patterns. The inspiration of utilizing Delaunay triangulation as part of this work is that it has some unique features, including:

- Delaunay triangulation divides the given region into smaller parts and exactly connects the neighbouring spatial features [4].
- Further, the computing complexity of the Delaunay triangulation is less as it generates only O(n) triangles.

The fundamental thought of utilizing Delaunay diagram is to stay away from the need of characterizing a separation edges for determining neighbourhoods as they do not need to repeat the way towards discovering neighbourhoods for different client characterized parameters and discovers neighbourhood powerfully.

2.1 Case Study

Let us consider the following spatial data with spatial features as F = {P, Q, R, S, T} as shown in Fig. 1. Different spatial features are represented with different symbols. The notations used to represent different features are given in Table 1.

Feature P contains four instances with the ids {p1, p4, p10, p14}, Object Q contains five instances with the ids {q2, q3, q15, q16, q19}, Object R contains 4 instances with ids {r7, r11, r13, r18}, Object S contains three instances with ids {s6, s9, s17}, and Object T contains four instances with ids {t5, t8, t12, t20}. In Fig. 2, the neighbourhood relation R is defined using Delaunay triangulation [2, 3] instead of using Euclidean distance. Neighbouring instances are connected

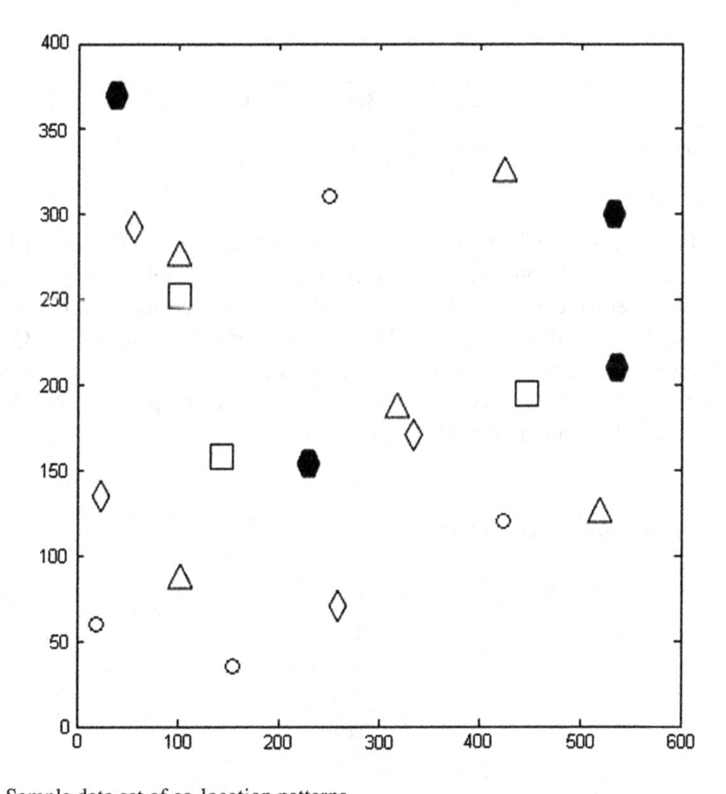

Fig. 1 Sample data set of co-location patterns

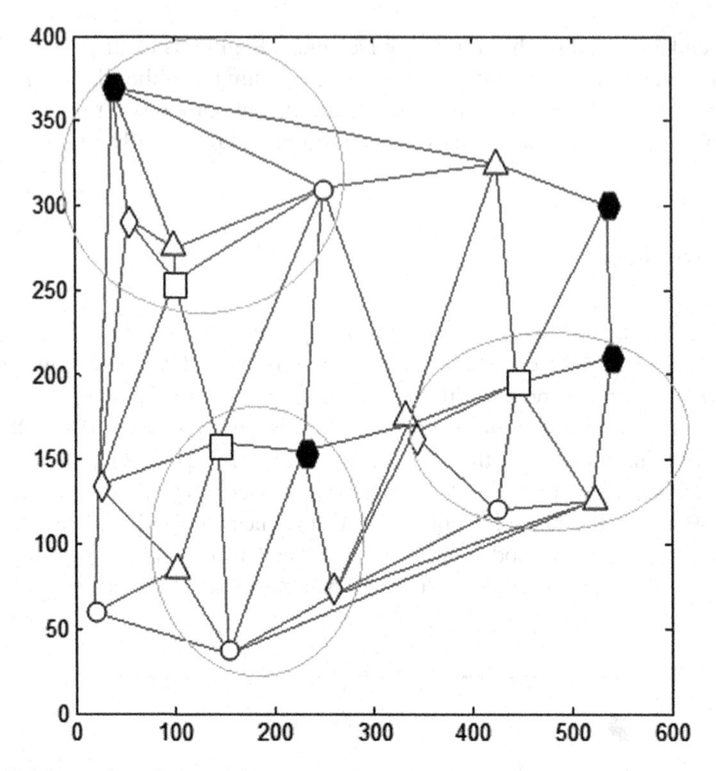

Fig. 2 Delaunay triangulation of the patterns

by edges. For instance, {p1, q3, r11, s6, t5}, {p4, q15, r18, s17, t8} and {p10, q2, r7, s9, t12} are all neighbour-sets because each set forms a clique. Here, we use the instance-id to refer to feature in Fig. 2. {P, Q, R, S, T} is a co-location pattern. The neighbourhood-set {p10, q2, r7, s9, t12} is a row instance of pattern {P, Q, R, S, T}, but the neighbourhood-set {p10, q2, r7, s9, t12, r13} is not a row instance of co-location {P, Q, R, S, T} because it has a proper subset {p10, q2, r7, s9, t12} which contains all the features in {P, Q, R, S, T}.

Table 1 Notations used for various features

Feature	Symbol
P	Circle
Q	Triangle
R	Hexagon
S	Square
T	Diamond

- **Conditional Probability**: If R: X → Y is a co-location rule, then the conditional probability cp(R) is defined as

$$\frac{|\{U \in rowset(X) \exists\, U' \; such\; that(U \subseteq U') \wedge U' \in rowset(X \cup Y)\}|}{|rowset(X)|} \quad (1)$$

Rowset(P, Q, R, S, T) contains following three patterns {pl, q3, r11, s6, t5}, {p4, q15, r18, s17, t8} and {p10, q2, r7, s9, t12} as shown in Fig. 2, and rowset (P, Q, R) contains following three patterns {p4, q15, r18}, {p4, q19, r11} and {p4, q16, r18}. Since lrowset(P, Q, R)l = 3 and only one row of {P, Q, R} satisfy the subset condition, {P, Q, R, S, T}. The conditional probability cp({P, Q, R}, {S, T}) = 1/3= 33.3%.
Let P, Q and R are objects which may cause traffic jam and S, T are objects which represent the violation of traffic rules. When all the these five occur simultaneously, then it may lead to a road accident. cp({P, Q, R} {S, T}) represent the conditional probability for happening of accidents on road, if there is a traffic jam. According above illustration, there are three cases where a traffic jam can happen. They are {p4, q15, r18}, {p4, q19, r11} and {p4, q16, r18}. Out of these three cases, accidents may happen only when both the traffic jam and the violation of road rules occurs, i.e. {p4, q15, r18}. Hence, according to our data, we can say that for every three traffic jams, there is a chance of occurring one accident, i.e. 33.3%.
- **Participation ratio (PR)**: "Given a spatial database S, to measure how a spatial feature f is co-located with other features in co-location pattern C, a participation ratio pr(C, f) can be defined as"

$$\frac{|\{r|(r \in S) \wedge (r.f = f) \wedge (r\; is\; in\; a\; row\; instance\; of\; C)\}|}{|\{r|(r \in S) \wedge (r.f = f)\}|} \quad (2)$$

By using the above formula, the participation ration pr({P, Q, R, S, T}, P) = 3/4, because rowset {P, Q, R, S, T} has three patterns. P has four instances, i.e. {p1, p4, p10, p14} Among four instances of P, three of them, namely {p1, p10, p4} has Q, R, S and T in a neighbour-set. Similarly, we can have pr({P, Q, R, S, T}, Q) = 3/5, pr({P, Q, R, S, T}, R) = 3/4, pr({P, Q, R, S, T}, S) = 3/3 and pr({P, Q, R, S, T}, T) = 3/4.
- **Participation index**: Participation index (PI) is the minimal of participation ratios of all the features in the event. The significance of PI is that when the least participated feature occurs, there may high possibility of remaining features also participate in the event. Here, for the event {P, Q, R, S, T}, the participation ratios of {P, Q, R, S, T} are {3/4, 3/5, 3/4, 3/3, 3/4}, respectively. Among these, feature Q has minimal participation ratio, i.e. 3/5. Hence, the participation index PI({P, Q, R, S, T}) is 3/5 which correspond to feature Q.
- **Max Participation index**: Max participation index (MaxPI) is the maximum of PRs of all the features in the event. MaxPI is helpful in knowing about which spatial features are impacting most on the event. The participation proportions of

P, Q, R, S and T are 3/4, 3/5, 3/4, 3/3, 3/4, respectively, for the event {P, Q, R, S, T}. Among these, S has the highest impact of 100%.

- **Medoid Participation index**: Generally, MaxPI is used to choose the most encouraging component or the feature that should be avoided. We will analyse the occurrence of an accident, i.e. event with an example. Let us say P represents a car, S represents a traffic police, and Q represents a traffic signal. In this case, we can conclude that the occurrence of feature S should be prevented. But it is not possible to prevent a traffic police to enter onto the road. For this situation, we cannot utilize the variable MaxPI to choose the element that must be taken care. Hence, for this situation, MaxPI has not working and we use another method known as "Medoid Approach" [5], which considers the medoid of the PRs of the features in the list {P, Q, R, S, T}. The participation ratios of {P, Q, R, S, T} are 3/4, 3/5, 3/4, 3/3, 3/4, respectively. Here, PI is 3/5 which represents the feature Q, MaxPI is 3/3 which represents the feature S, and finally, medoidPI is 3/4 which represents the features P, R and T.

3 Experimental Results

In this section, we report the results of the proposed approach. We examined this approach on a synthetic data set which is given in Fig. 1. We customized the co-location pattern mining calculation using Delaunay triangulation approach. We used the following measures to analyse the performance of the proposed approach which are explained in Sect. 3: (a) participation ratio, (b) participation index, (c) max participation index, (d) medoid participation index. It can be seen from Table 2 that the Delaunay triangulation approach achieves a PR of {0.75, 0.6, 0.75, 1, 0.75} for the co-location patterns {P, Q, R, S, T}, respectively. The PI, MaxPI, MedPI of the pro-

Table 2 Results of co-location for size 5

Measure	Co-location	Rowset	PR	PI	Max PI	Med PI
Euclidean distance [5]	P, Q, R, S, T	{t12, s9, q2, p10, r13} {t12, s9, q2, p10, r13} {t12, s9, q16, p10, r13} {t5, s6, q3, p10, r11} {t5, s6, q3, p1, r11} {t12, s9, q2, p10, r7} {t12, s9, q16, p10, r7} {t5, s6, q3, p10, r7}	{0.5, 06.7, 0.6, 0.75, 0.75}	0.5	0.75	0.67
Delaunay	P, Q, R, S, T	{p1, q3, r11, s6, t5} {p4, q15, r18, s17, t8} {p10, q2, r7, s9, t12}	{0.75, 0.6, 0.75, 1, 0.75}	0.6	1	0.75

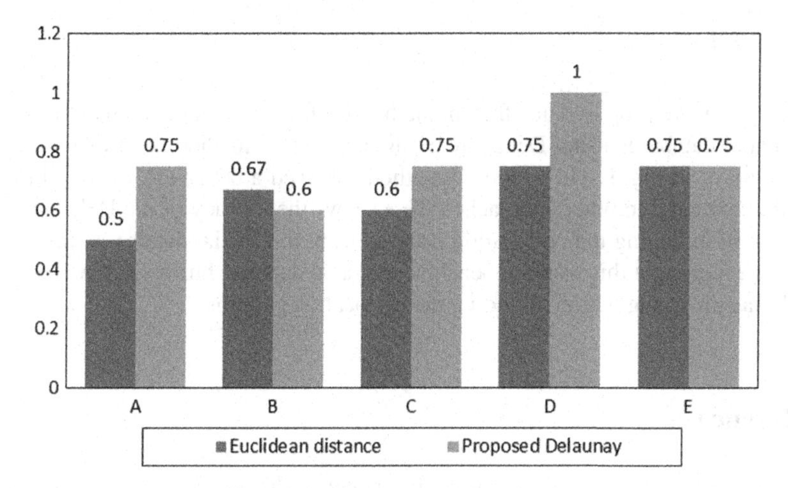

Fig. 3 Participation ratio (PR) comparison for features {P, Q, R, S, T}

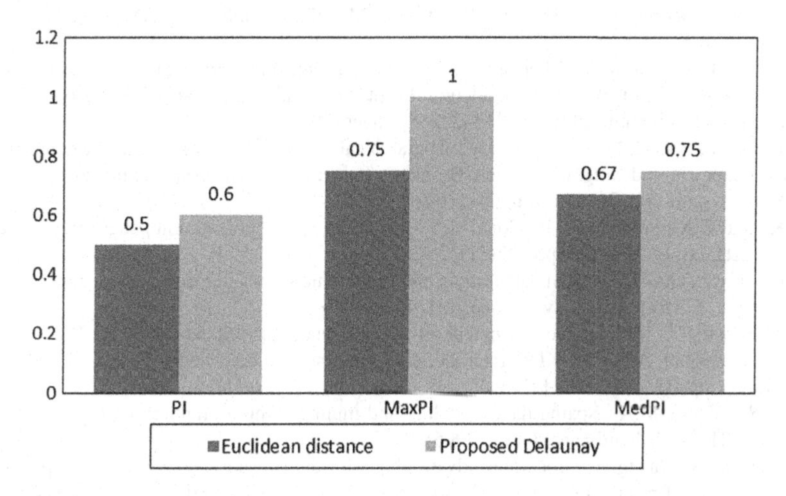

Fig. 4 PI, MaxPI and MedPI comparison

posed approach are 0.6, 1, 0.75, respectively, for the co-location patterns {P, Q, R, S, T}. Further, we also analysed this approach with our previous work [5] which is also shown in Table 2. It can be observed that the proposed approach performs better compared our previous work in terms of PR, PI, MaxPI and MedPI. The analysis charts from this work are also shown in Figs. 3 and 4. The participation ratios of different features {P, Q, R, S, T} with the proposed and the method in [5] are shown in Fig. 3. Further, PI, MaxPI and MedPI of these two approaches are given in Fig. 4.

4 Conclusion

In this paper, we proposed an efficient method for finding co-location patterns from the spatial databases using Delaunay triangulation. Our method is based on a case study shown in Fig. 1. The proposed method achieved a better PR, PI, MaxPI and MedPI compared to other approaches. This shows the efficacy of the Delaunay triangulation in finding the co-location patterns from the spatial databases. In future, we will experiment this work on benchmark spatial data set. Further, we will extend the Delaunay triangulation to finding the co-location patterns.

References

1. Gordon, A.D.: Hierarchical classification. In: Clustering and Classification, pp. 65–121. World Scientific (1996)
2. Kavati, I., Chenna, V., Prasad, M.V., Bhagvati, C.: Classification of extended Delaunay triangulation for fingerprint indexing. In: 2014 8th Asia Modelling Symposium (AMS), pp. 153–158. IEEE (2014)
3. Kavati, I., Prasad, M.V., Bhagvati, C.: Vein pattern indexing using texture and hierarchical decomposition of Delaunay triangulation. In: International Symposium on Security in Computing and Communication, pp. 213–222. Springer (2013)
4. Kavati, I., Prasad, M.V., Bhagvati, C.: Hierarchical decomposition of extended triangulation for fingerprint indexing. In: Efficient Biometric Indexing and Retrieval Techniques for Large-Scale Systems, pp. 21–40. Springer (2017)
5. Kumar, G.K., Premchand, P., Gopal, T.V.: Mining of spatial co-location pattern from spatial datasets. Int. J. Comput. Appl. **42**(21), 25–30 (2012)
6. Mennis, J., Guo, D.: Spatial data mining and geographic knowledge discoveryan introduction. Comput. Environ. Urban Syst. **33**(6), 403–408 (2009)
7. Morimoto, Y.: Mining frequent neighboring class sets in spatial databases. In: Proceedings of the Seventh ACM SIGKDD International Conference on Knowledge Discovery and Data Mining, pp. 353–358. ACM (2001)
8. Pech Palacio, M.A.: Spatial data modeling and mining using a graph-based representation. Ph.D. Thesis, Villeurbanne, INSA (2005)
9. Shekhar, S., Huang, Y.: Processing advanced queries-discovering spatial co-location patterns: a summary of results. Lect. Notes Comput. Sci. **2121**, 236–256 (2001)
10. Van Canh, T., Gertz, M.: A constraint neighborhood based approach for co-location pattern mining. In: 2012 Fourth International Conference on Knowledge and Systems Engineering (KSE), pp. 128–135. IEEE (2012)
11. Yoo, J.S., Boulware, D., Kimmey, D.: A parallel spatial co-location mining algorithm based on mapreduce. In: 2014 IEEE International Congress on Big Data (BigData Congress), pp. 25–31. IEEE (2014)

Review on RBFNN Design Approaches: A Case Study on Diabetes Data

Ramalingaswamy Cheruku, Diwakar Tripathi, Y. Narasimha Reddy and Sathya Prakash Racharla

Abstract Radial Basis Function Neural Networks (RBFNNs) are more powerful machine learning technique as it requires non-iterative training. However, the hidden layer of RBFNN grows on par with the growing dataset size. This results in increase in network complexity, training time, and testing times. It is desirable to design appropriate RBFNN which balance between simplicity and accuracy. In the literature, many approaches are proposed for reducing the neurons in the RBFNN hidden layer. In this paper, a comprehensive survey is performed on hidden layer reduction techniques with respect to Pima Indians Diabetes (PID) dataset.

Keywords Diabetes mellitus · RBFNN · Hidden layer size · RBFNN Design Hidden layer reduction

1 Introduction

Diabetes is a metabolic and hereditary disease that is caused due to deficiency of insulin hormone in human body. Insulin plays key role in conversion of food into

R. Cheruku (✉) · D. Tripathi
Department of Computer Science and Engineering, National Institute of Technology Goa, Ponda 403401, Goa, India
e-mail: rmlswamygoud@gmail.com

D. Tripathi
e-mail: diwakartripathi@nitgoa.ac.in

Y. Narasimha Reddy
Department of CSE, Brindavan Institute of Technology and Science, Kurnool 518218, Andhra Pradesh, India
e-mail: narasimhareddy.mtech@gmail.com

S. Prakash Racharla
Department of CSE, CVR College of Engineering, Hyderabad 501510, Telangana, India
e-mail: prakashscits@gmail.com

D. Reddy Edla et al. (eds.), *Advances in Machine Learning and Data Science*, Advances in Intelligent Systems and Computing 705,
https://doi.org/10.1007/978-981-10-8569-7_11

energy. Lack of sufficient insulin causes presence of excess sugar levels in the blood. As a result, the glucose levels in diabetic patients are more than normal ones. Many people referred diabetes as diabetes mellitus (DM). It has symptoms like frequent urination, increased hunger, increase thirst, and high blood sugar. Diabetes is the fastest rising long-term illness condition that impacts lots of people globally. The excess blood sugar within the blood vessels can harm the blood vessels; this kind of situation leads to various complications like cardiovascular damage, kidney damage, nerve damage, eye damage, and stroke [1, 2]. World Health Organization (WHO) statistics shows diabetes contributed major share in Non-Communal Disease (NCD) deaths across worldwide population [3].

Multi-Layer Feed Forward Neural Networks (MLFFNNs) and Multi-Layer Perceptron Neural Networks (MLPNNs) are most popular techniques for classification and use iterative process for training. Contrary to MLFFNNs and MLPNNs, Radial Basis Function Neural Networks (RBFNNs) are trained in single iteration and learn applications quickly. Thus, RBFNNs are drawing researchers attention for classification tasks. Moreover, performance of these neural networks is on par with MLFFNNs and MLPNNs.

2 Preliminaries

2.1 Dataset Description

We have used Pima Indians Diabetes (PID) dataset obtained from UCI machine learning repository [4] whose detail specifications are shown in Table 1. There are eight conditional attributes and one decision attribute. PID dataset consists of a total of 768 diabetes patients data in which 500 records are related to diabetes negative (class label 0) and 268 records are related to diabetes positive (class label 1) [1].

2.2 Radial Basis Function Neural Network

The RBFNN [5, 6] is an alternative model to the MLPNNs and MLFFNNs for the classification task. It is considered to be a four-layer network with input, hidden, output, and decision layers, respectively. It is explained in Fig. 1 for PID dataset, i.e., two-class problem.

- **Input layer**: The input layer is made up of eight (dimensionality) neurons. The input layer is connected fully feed forward manner to hidden layer. Moreover, there are no weights associated with links between input and hidden layers.
- **Hidden layer**: The hidden layer is made up of H neurons and complexly connected with two output layer neurons. At each hidden layer neuron, it is a nonlinear transformation because of Gaussian activation function. The output value of

each hidden layer neuron is computed using Eq. (1).

$$\varphi_i(X) = \frac{1}{\sqrt[D]{2\Pi}(\sigma_i)^D} e^{-\frac{((X-\mu_i).(X-\mu_i)^T)}{2(\sigma_i)^2}}, i = 1, 2, \ldots, H. \quad (1)$$

- **Output layer**: The output layer is made up of two (number of distinct classes) neurons. Response of the output layer neuron is computed using Eq. (2).

$$0_j(X) = \sum_{i=1}^{H} w_{ji}\phi_i(X), i = 1, 2, \ldots, H, j = 1, 2. \quad (2)$$

Table 1 Pima Indians Diabetes dataset attributes description

Feature s. no	Feature description	Feature s. no	Feature description
F_1	Number of times pregnant	F_5	Serum insulin
F_2	Plasma glucose concentration	F_6	Body mass index
F_3	Diastolic blood pressure	F_7	Diabetes pedigree function
F_4	Triceps skin-fold thickness	F_8	Age
Class label	0 if diabetes is negative		
	1 if diabetes is positive		

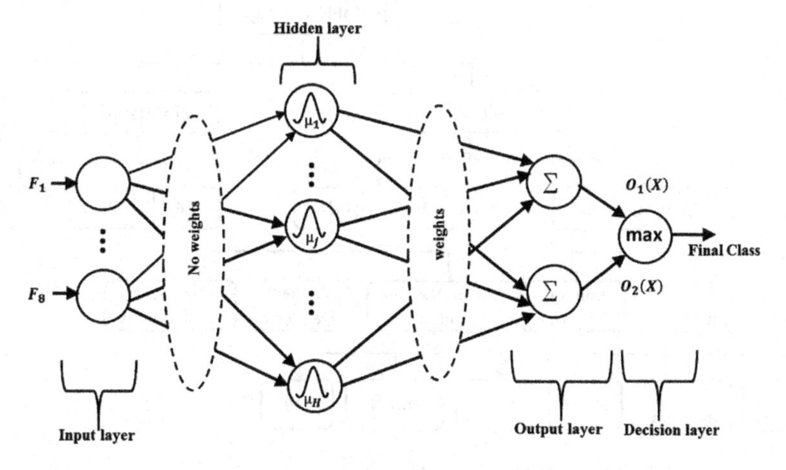

Fig. 1 RBFNN model for diabetes classification task

- **Decision layer**: The size of this layer is one. This layer determines the class label of given input vector (X) present at input layer.

$$class(X) = arg \max_{j} O_j(X), j = 1, 2. \quad (3)$$

The weight vector $[W]^T_{2\times H}$ between output layer and hidden layers is given by

$$W = \phi^{+}_{L\times H} * T \quad (4)$$

where ϕ^+ is the pseudoinverse of the ϕ matrix

$$\phi = \begin{bmatrix} \phi_1(I_1) & \dots & \phi_P(I_1) & \dots & \phi_H(I_1) \\ \phi_1(I_2) & \dots & \phi_P(I_2) & \dots & \phi_H(I_2) \\ & & \cdot & & \\ \phi_1(I_L) & \dots & \phi_P(I_L) & \dots & \phi_H(I_L) \end{bmatrix}_{L\times H} \quad (5)$$

where L and H are the number of training patterns and hidden layer neurons, respectively.

3 RBFNN Design Approaches

Taxonomy of RBFNN design approaches in the literature is shown in Fig. 2. The techniques are adopted for reducing the number of RBFNN hidden layer neurons in the literature shown chronological order in Fig. 3.

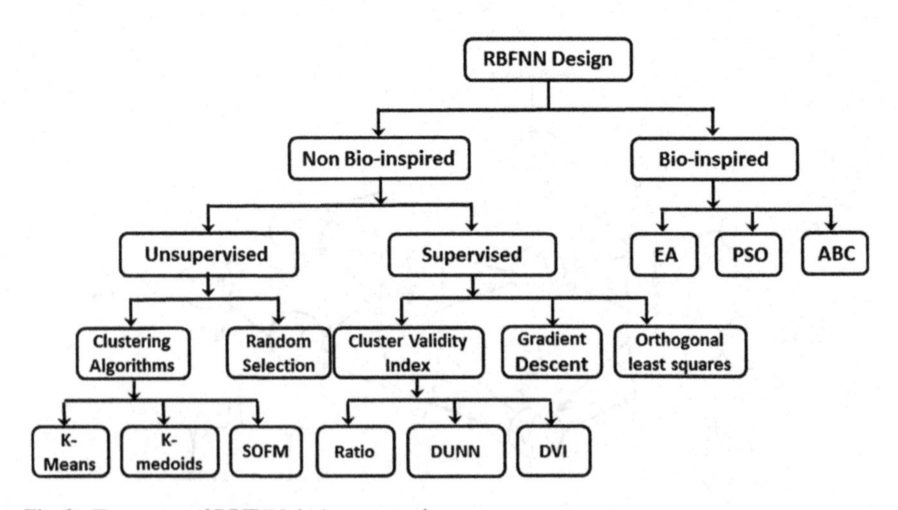

Fig. 2 Taxonomy of RBFNN design approaches

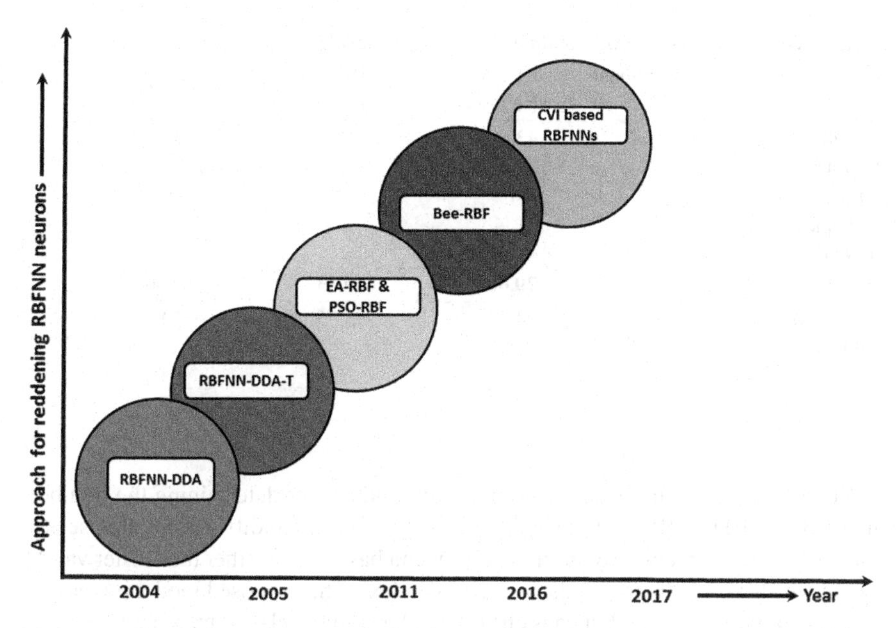

Fig. 3 Chronological order of RBFNN approaches in the literature

Table 2 Performance comparison of RBFNN variants on PID dataset

Method	Accuracy	Number of centers	References
Centers selected by randomly	73.15 ± 1.02	35	[7]
Centers selected by *k*-means	71.13 ±1.06	35	[7]

3.1 Unsupervised Approaches

In unsupervised approach, the clustering techniques are applied to whole dataset. Authors in [7] have used the *k*-means and random selection procedure to obtain the number, location, and dispersions of basis functions to be used in RBF networks. These experimental results on PID dataset are given in Table 2.

3.2 Supervised Approach

In supervised approach, the clustering techniques are applied over each class data, whereas in un supervised (direct) approach clustering algorithms are applied over the whole dataset. This supervised approach reduces the computational cost drastically (less number of patterns in each class as compared to whole dataset).

Table 3 Performance comparison of different validity indices [8]

	Conventional RBFN	RBFN + RatioIndex	RBFN + DunnIndex	RBFN + DVIndex
# Hidden layer neurons	768	43	48	24
# Links (Complexity of network)	7680	430	480	240
Accuracy (%)	68.53	**70.00**	69.33	69.56
% Reduction in network complexity (%)	0	94.40	93.75	**96.88**

Cheruku et al. [8] have used cluster validity indices for determining the number of neurons to be in RBFNN hidden layer. Usually, cluster validity indices are used to find the best cluster regions within the given data based on whether the cluster validity index is high or low (depends on validity index). The proposed model integrates cluster validity index with k-means clustering algorithm, and this integrated k-means clustering algorithm is applied over data in class-by-class fashion; i.e., integrated k-means is applied inside each class data. Whereas in direct approach, k-means clustering algorithm [9] is applied over the whole dataset. The experimental results on PID dataset using Ratio, Dunn, and Dynamic Validity (DV) indices [10] are shown in Table 3.

They have obtained the best cluster locations along with the number using integrated k-means algorithm. Once the cluster center locations are computed, each cluster standard deviation is computed. The computed cluster center and cluster standard deviation are given as inputs to corresponding basis function in the hidden layer.

3.3 *Artificial Bee Colony-Based RBFNN*

Ferreira et al. [7] have proposed the Bee-RBF which is inspired from ABC algorithm. In order to select the center location and dispersion of radial basis functions and number of basis functions in the RBFNN hidden layer, Ferreira et al. used cOptBees, plus a heuristic. The cOptBees are designed to solve data clustering problems. The presented approach, named Bee-RBF, is used to solve classification problems. It has obtained the 77.13% accuracy with nearly 35 cluster centers. The results are furnished in Table 4.

Table 4 Performance of Bee-RBF on PID dataset

Method	Accuracy	Number of centers	Reference
Bee-RBF	77.13	35.77	[7]

3.4 Particle Swarm Optimization Based RBFNN

Qasem et al. [11] have introduced a Time Variant Multi-Objective Particle Swarm Optimization (TVMOPSO) of radial basis function (RBF) network for diagnosing the medical diseases. These fine-tuned parameters are listed in Table 5.

RBFN-TVMOPSO uses multi-objective optimization approach for RBFN training. The algorithm will determine the set of connection (centers, weights) and its corresponding accuracy by minimizing its objective function. In each generation, every particle is evaluated based on the two objective functions, namely accuracy (based on training set mean squared error) and complexity (represents the smoothness of the model). The experimental results of RBFN-TVMOPSO on PID dataset are given in Table 6

Table 5 Parameters settings for RBFN-TVMOPSO

Parameter	Value parameter	Value	
Number of iterations	1000	Archive size	100
Objective functions	2	Inertia weights w_{max} and w_{min}	0.7 and 0.4
Limit of variable	[−0.5, 0.5]	Initial acceleration coefficients c_1 and c_2	2.5 and 0.5
Population size	100	Probability of mutation	0.5

Table 6 Results of RBFN-TVMOPSO on PID dataset

Method	Accuracy	Number of Pareto-optimal solutions	References
RBFN-TVMOPSO	78.02	96	[11]

Table 7 Parameters settings for evolutionary algorithms

Parameters	MEPGAN	MEPDEN
Population size	100	100
Probability of crossover (P_c)	0.9	0.9
Maximum number of iterations	1000	1000
Probability of mutation (P_m)	$\frac{1}{N}$	0.5
Distribution indices for crossover (η_c)	20	–
Distribution indices for mutation (η_m)	20	–

Table 8 Performance of EA-based RBFNNs on PID dataset

Method	Accuracy	Number of centers	References
MEPGANf1f2	72.78	5.4	[12]
MEPGANf1f3	68.35	5.8	[12]
MEPDENf1f2	77.07	3.1	[12]
MEPDENf1f3	78.76	3.1	[12]

3.5 Evolutionary Algorithm (EA) Based RBFNN

Qasem et al. [11] have used two evolutionary-based algorithms, namely memetic elitist Pareto non-dominated sorting genetic algorithm-based RBFN (MEPGAN) and memetic elitist Pareto non-dominated sorting differential evolution-based RBFN (MEPDEN). In this study, both MEPGAN and MEPDEN have been proposed to determine the best centers, widths, and weights of RBFN. In order to assist RBFN design, a rank-density-based EA to carry out fitness evaluation and selection schemes. In order to obtain best RBFNN architecture, it is necessary to find tuned EA parameters. These parameter's values are listed in Table 7. The performance of these EA-based RBFNNs on PID dataset is given in Table 8.

3.6 Other Techniques

There are other techniques available in the literature such as RBFNN-DDA-T [17], NSGA-II [20], Bat-Finder [19]. PSO-RBFNN [18]. These results are furnished in Table 9.

Table 9 Results of RBFN-TVMOPSO, RBFN-NSGA-II, and RBFN-MOPSO on diabetes dataset

Method	Accuracy	Number of neurons	References
RBFN-MOPSO	70.58	50	[11]
RBFN-NSGA-II	69.59	100	[11]
HMOEN-L2	78.50	–	[13]
HMOEN-HN	75.36	–	[13]
MPANN	74.90	–	[14]
MOBNET	77.85	–	[15]
RBFNN-DDA	74.35	288.5	[16]
RBFNN-DDA-T	73.50	65.6	[17]
PSO-RBFNN	72.60	66	[18]
Bat-finder	73.91	47	[19]

4 Conclusion and Future Directives

In this paper, we made comprehensive survey on techniques used for reduction of RBFNN hidden layer neurons. We have discussed how various bio-inspired algorithms are used for solving this NP-hard problem. We observed that to design RBFNN which balance between simplicity and accuracy is a tedious task. This survey helps researchers in machine learning and medical fields.

As future scope still we can apply latest bio-inspired techniques such as Spider Monkey, Bat, Whale Optimization to find better cluster number along with their location and spreads.

References

1. Cheruku, R., Edla, D.R., Kuppili, V.: Sm-ruleminer: spider monkey based rule miner using novel fitness function for diabetes classification. Comput. Biol. Med. **81**, 79–92 (2017)
2. Cheruku, R., Edla, D.R., Kuppili, V., Dharavath, R.: Rst-batminer: a fuzzy rule miner integrating rough set feature selection and bat optimization for detection of diabetes disease. Appl. Soft Comput. (2017)
3. WHO: World Health Organization. http://www.who.int/diabetes/action_online/basics/en/. Accessed 30 Sept 2016
4. Lichman, M.: UCI machine learning repository (2013). http://archive.ics.uci.edu/ml
5. Broomhead, D.S., Lowe, D.: Radial Basis Functions, Multi-variable Functional Interpolation and Adaptive Networks, Tech. rep. Royal Signals and Radar Establishment, Malvern (United Kingdom) (1988)
6. Yegnanarayana, B.: Artificial neural networks. PHI Learning Pvt. Ltd. (2009)
7. Cruz, D.P.F., Maia, R.D., da Silva, L.A., de Castro, L.N.: Beerbf: a bee-inspired data clustering approach to design RBF neural network classifiers. Neurocomputing **172**, 427–437 (2016)

8. Cheruku, R., Edla, D.R., Kuppili, V.: Diabetes classification using radial basis function network by combining cluster validity index and bat optimization with novel fitness function. Int. J. Comput. Intell. Syst. **10**(1), 241–265 (2017). https://doi.org/10.2991/ijcis.2017.10.1.17
9. Hartigan, J.A., Wong, M.A.: Algorithm as 136: a k-means clustering algorithm. J. R. Stat. Soc. Ser. C (Appl. Stat.) **28**(1), 100–108 (1979)
10. Shen, J., Chang, S.I., Lee, E.S., Deng, Y., Brown, S.J.: Determination of cluster number in clustering microarray data. Appl. Math. Comput. **169**(2), 1172–1185 (2005)
11. Qasem, S.N., Shamsuddin, S.M.: Radial basis function network based on time variant multi-objective particle swarm optimization for medical diseases diagnosis. Appl. Soft Comput. **11**(1), 1427–1438 (2011)
12. Qasem, S.N., Shamsuddin, S.M., Zain, A.M.: Multi-objective hybrid evolutionary algorithms for radial basis function neural network design. Knowl. Based Syst. **27**, 475–497 (2012)
13. Goh, C.K., Teoh, E.J., Tan, K.C.: Hybrid multiobjective evolutionary design for artificial neural networks. IEEE Trans. Neural Netw. **19**(9), 1531–1548 (2008)
14. Abbass, H.A.: Speeding up backpropagation using multiobjective evolutionary algorithms. Neural Comput. **15**(11), 2705–2726 (2003)
15. Garcıa-Pedrajas, N., Hervás-Martınez, C., Munoz-Pérez, J.: Multi-objective cooperative coevolution of artificial neural networks (multi-objective cooperative networks). Neural Netw. **15**(10), 1259–1278 (2002)
16. Paetz, J.: Reducing the number of neurons in radial basis function networks with dynamic decay adjustment. Neurocomputing **62**, 79–91 (2004)
17. Oliveira, A.L., Melo, B.J., Meira, S.R.: Improving constructive training of RBF networks through selective pruning and model selection. Neurocomputing **64**, 537–541 (2005)
18. Cheruku, R., Edla, D.R., Kuppili, V., Dharavath, R.: Pso-rbfnn: a pso-based clustering approach for rbfnn design to classify disease data. In: International Conference on Artificial Neural Networks. Springer (2017). https://doi.org/10.1007/978-3-319-68612-729
19. Edla, D.R., Cheruku, R.: Diabetes-finder: a bat optimized classification system for type-2 diabetes. Procedia Comput. Sci. **115c**, 235–242 (2017)
20. Ded, K., Pratap, A., Agarwal, S., et al.: A fast and elitist multi-objective genetic algorithm: Nsga2. IEEE Trans. Evolut. Comput. **6**(2), 149–172 (2002)

Keyphrase and Relation Extraction from Scientific Publications

R. C. Anju, Sree Harsha Ramesh and P. C. Rafeeque

Abstract This paper proposes a detailed view of extracting keyphrases and its relations from scientifically published articles such as research papers using conditional random fields (CRF). Keyphrase is a word or set of words that describe the close relationship of content and context in particular documents (Sharan, International conference on advances in computing communications and informatics (ICACCI), 2014) [1]. Keyphrases may be the topics of the document which represent the key logic of the document. Automatic keyphrase extraction has a major role in automatic systems like independent summarization, query or topic generation, question-answering system, search engine, information retrieval, document classification, etc. The relationships of the keyphrases are also extracted. Two types of relations are considered—synonym and hyponyms. The result shows that our proposed system outperforms the existing systems.

Keywords Keyphrase extraction · Topic extraction · Information extraction (IE)
Summarization · Question answering (QA) · Document classification

1 Introduction

Keyphrase is a word or set of words that describe the close relationship of content and context in the document [1]. Keyphrases are sometimes simple nouns or noun phrases (NPs) that represent the key ideas of the document, i.e., topic. Document keyphrases have enabled fast and accurate searching for a given document from a large text collection and have exhibited their potential in improving many natural language processing (NLP) and information retrieval (IR) tasks, such as text summarization, text categorization, opinion mining, and document indexing.

R. C. Anju (✉) · P. C. Rafeeque
Government Engineering College, Palakkad, Kerala, India
e-mail: anju.malu5000@gmail.com

S. H. Ramesh
Surukam Analytics Pvt. Ltd, Chennai, Tamilnadu, India

D. Reddy Edla et al. (eds.), *Advances in Machine Learning and Data Science*,
Advances in Intelligent Systems and Computing 705,
https://doi.org/10.1007/978-981-10-8569-7_12

Scientific research is the systematic investigation of scientific theories and hypothesis such as gaining, maintaining, and understanding the body of existing work in specific areas related to such fundamental objects. In such cases, researchers and practitioners face some typical questions.

- Which paper has addressed a specific task?
- Which paper has studied a process or variants?
- Which papers have utilized such materials?
- Which paper has addressed this task using variants of this process?

The existing works in this area do not address the above-mentioned problems efficiently. Hence, utilities are required to identify the keyphrases and relations. This paper proposes the task of keyphrase extraction from scientific documents using conditional random field (CRF). It also addresses the task of extracting type of keyphrases such as PROCESS, TASK, and MATERIAL, and relation between keyphrases. Mainly two types of relations are considered.

Synonym_Of (abbreviation): The relationship between two keyphrases A and B is said to be A => Synonym_Of (B) if they both denote the some semantic field.

e.g., Machine Learning => Synonym_Of (ML).

Hyponym_Of: The relationship between two keyphrases A and B is said to be A => Hyponym_Of (B) if semantic field of A is included within that of B.

e.g., Red => Hyponym_Of (colors).

Our proposed methods are compared with existing systems like AlchemyAPI [1] and RAKE [2], and it shows better result than both of the systems.

2 Literature Review

The methods for keyphrase extraction can be classified into supervised and unsupervised approaches. Mihlceva and Tarau (2004) [3] proposed an unsupervised approach that considered single tokens are vertices of a graph and co-occurrence relation between tokens are edges. Candidate can be ranked using PageRank, and adjacent keywords are merged into keyphrases in post-processing step. Keyphrase extraction algorithm (KEA) is a supervised system that used all n-grams of a certain length, a Naive Bayes classifier, and tf-idf and position features (Frank et al. 1999) [4]. Turney (2000) [5] introduced extractor, a supervised system that selects a stem and stemmed n-grams as candidate and turns its parameter (mainly related to frequency, position, and length) with a generic algorithm. Pinaki, Kishorjit, Sivaji (2002) [6] introduced a supervised method for keyword extraction as a part of SemEval 2010. They used CRF tool for finding keywords from the scientific publications.

KP_Miner system (2010) [7] is used for extracting the keyphrases in English and Arabic document. When they are developing the system, the goal was to create a general-purpose keyphrase extraction system that can be configured, by user based on their understanding of the documents or the use of any sophisticated natural language processing or linguistics tools. KP_Miner system is an unsupervised system. Keyphrase extraction in the system is a three-step process: candidate keyphrase selection, candidate keyphrase weight calculation, and finally keyphrase refinement. This system gives the result that precision is 24.9%, recall is 25.5%, and F-measure is 25.02%.

WINGNUS (2010) [8] developed a method for SemEval 2010. In this method, they used test input format of PDF because logical structure recovery is much more robust. For this, they used Google Scholar base Crawler to find the PDFs given plain text. For logical structure extraction Sect Label (Luong et al.) [8] used. Precision, recall, and F-measure for the system were 24.9%, 25.5%, and 25.2%, respectively.

In 2010, Lahiri and Mihalcea developed a system for extracting keywords from emails [9]. They proposed a supervised keyword extraction system. The system contains mainly five steps: e-mail preprocessing, candidate extraction, preprocessing, ranking/classification, and post-processing. They got the accuracy of precision—24.8%, recall—25.4%, F-measure—25.1%.

In 2013, Kamal Sankar proposed a hybrid approach to extract keyphrase from medical document [10]. In this paper, they propose amalgamation of two methods: First one assigns weights to candidate keyphrases based on effective combination of features such as position, term frequency, inverse document features, and second one assign weights to candidate keyphrases using some knowledge about their similarities to the structure and characteristics of keyphrases available in the memory. This keyphrase extraction method consists of three primary components: document preprocessing, candidate keyphrase identification, and assigning scores to the candidates for ranking.

Rapid automatic keyword extraction (RAKE) [11] is based on the keywords frequently contain multiple words but rarely contain standard punctuation or stop words. The input parameters for RAKE comprise a list of stop words, a set of phrase delimiters, and a set of word delimiters. RAKE uses stop words and phrase delimiters to partition the document text into candidate keywords, which are sequences of content words as they occur in the text. Co-occurrences of words within these candidate keywords are meaningful and allow us to identify word co-occurrence without the application of an arbitrarily sized sliding window. Word associations are thus measured in a manner that automatically adapts to the style and content of the text, enabling adaptive and fine-grained measurement of word co-occurrences that will be used to score candidate keywords.

AlchemyAPI [12] is developed by IBM, and they used a deep learning method to extract the keywords or keyphrases from the document.

Textacy [13] is a Python library for performing higher-level natural language processing (NLP) tasks, built on the high-performance spacy [14] library. This extracts the abbreviations from the documents with use of basics tokenization, part-of-speech tagging, dependency parsing, etc.

3 Keyphrase and Relation Extraction

3.1 System Design

The proposed architecture consists of three phrases: keyphrase extraction, abbreviation extraction, and hyponym extraction. Figure 1 shows the overall system diagram for the proposed system. Keyphrase extraction describes how CRF [12] used to extract the keyphrases and the type of the keyphrases from scientific documents. Two-pass method is used to extract abbreviation and definition from the document. Finally, rule-based system is used for hyponym extraction.

The proposed work implements a system to extract all the keyphrases from the documents and also find out the type of keyphrases such as PROCESS, TASK, and MATERIAL.

PROCESS: Keyphrases related to some scientific models, algorithm, and process should be labeled as PROCESS.

TASK: Keyphrases related to application, end goal, problem, and task should be labeled as TASK.

MATERIAL: Keyphrases identify the resource used in the paper.

Then, system finds all the abbreviation and its expansion and hyponyms from the document.

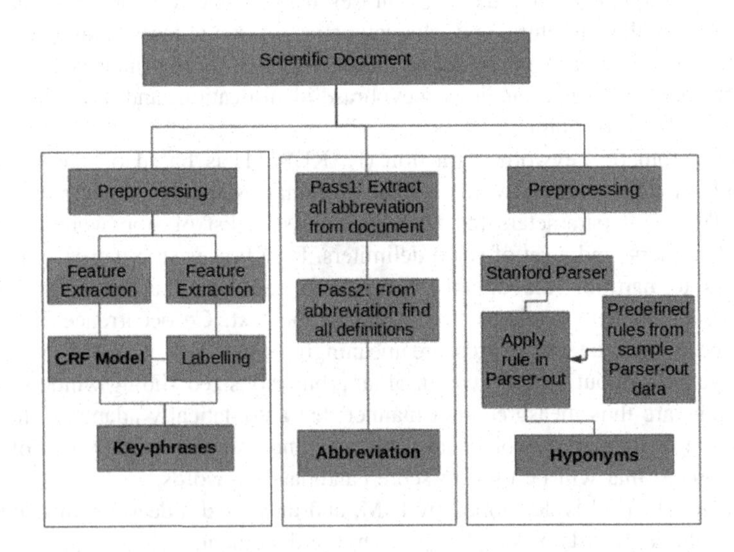

Fig. 1 Overall system architecture

3.2 Keyphrase Extraction

The proposed system extracts the keyphrases by using conditional random field (CRF). From the 500 collected documents, 350 are taken for the training purpose. After training, a CRF Model is created for generating keyphrases. Fifteen features are considered for training.

1. POS tags: Find out the part-of-speech (POS) tags for all the words in the document. Nouns (NN, NNP, NNS, etc.) can be most probable of a keyphrases. Then, that word feature is 1, and others are 0.
2. Named entities (NE): Find out all the named entities. From that, if a word will part of NE, then its feature value is 1, otherwise 0.
3. Upper case strings: Upper case words are 1, others are 0.
4. Term frequency greater than 2: If a word occurs more than 2 times, then it will 1, otherwise 0.
5. TF value normalized: For each word, find out the term frequency (TF) value. Normalized term frequency value is this feature.
6. IDF value: For each word, find the inverse document frequency (IDF) value.
7. Chunking: Finds all chunks from the document. If the word is part of that chunk, then it will be 1, otherwise 0.
8. Position of the word in the document.
9. Capitalized words: Capitalized word will be 1, otherwise 0.
10. Reference in a sentence: A sentence will contains the reference.
11. Section of text: Word will be in abstract, introduction, or body. If word in abstract feature value is 1. If word in introduction section then feature value is 2. Otherwise feature value is 3.
12. Article type: The document will occur in which article type. Eg., Corrosion Science, Environmental science.
13. Contains Greek letter and chemical symbol: If the word is Greek letter or chemical symbol, then feature value is 1, otherwise 0.
14. Contains hyphenated words: Feature value is 1 for hyphenated words.
15. Contains vocabulary list: Word contains in python vocabulary 0.0.5 [15] then feature value is 1.

3.3 Abbreviation and Expansion Extraction

Abbreviation and expansion extraction are a two-pass process. During the first pass, the program will extract all the abbreviations from the document and it saves in a temporary list. In second pass, take each abbreviation from the list and find its expansion.

3.4 Hyponym Extraction

A => *Hyponyms_Of (B)* means the semantic field of A is included within that of B, e.g., Red => *Hyponym_Of* (Color). General rules are generated by the parsed sample documents. These general rules applied to the new document and words that satisfy the rules are the hyponyms.

4 Evaluation and Discussion

4.1 Corpus Collection

Data sets for this project are built from Science Direct open access publications. It consists of 500 journal articles evenly distributed among the Domains Computer Science, Material Sciences, and Physics.

- 350 labeled documents from different scientific publications are collected for training purposes.
- 50 labeled document for testing case.
- 100 unlabeled document for testing case.

4.2 Evaluation and Results of Keyphrase Extraction

The evaluation of keyphrase extraction is done by comparing the proposed system with existing system such as AlchemyAPI [12] and rapid automatic keyword extraction (RAKE) [11]. Table 1 shows the recall, precision, F-measure result for three systems. After analyzing the table, it will be clear that my proposed system gives much good result than AlchemyAPI and RAKE system.

Table 1 Evaluation result for keyphrase extraction (top 15 candidates)

System	Recall	Precision	F-measure
AlchemyAPI	0.3002	0.3006	0.3003
Proposed system	0.3305	0.3462	0.3381
RAKE	0.3102	0.3130	0.3115

Table 2 Evaluation result for abbreviation extraction

System	Recall	Precision	F-measure
Textacy	0.6002	0.6080	0.6040
Proposed system	0.6505	0.6888	0.6991

4.3 Evaluation and Results of Abbreviation and Expansion Extraction

Table 2 shows the evaluation result for abbreviation extraction. The proposed system is compared with Textacy [13]. The results show that our system performs better than Textacy.

5 Conclusion and Future Works

Keyphrases have a very important role in most of our applications. CRF-based approach to keyphrase extraction has been included in the paper. Hyponyms and abbreviations are the two relations of the keyphrases. Rule-based approach is used for the extraction of hyponyms, and two-pass process is used in abbreviation and definition extraction. Proper cleaning of the input documents and identification of more appropriate features could have improved the score.

One major limitation of Stanford Parser is time consumption. So in future, use better parser for parsing. Statistical and hybrid approaches are other alternatives for rule-based method for hyponym extraction. Such method may give better result.

References

1. Sharan, A., Siddiqi, S.: A supervised approach to distinguish between keywords and stopwords using probability distribution functions. In: 2014 International Conference on Advances in Computing Communications and Informatics (ICACCI) (2014)
2. Python Vocabulary 0.0.5. http://pypi.python.org/pypi/vacabulary/0.0.5. Accessed 01 Feb 2017
3. Mihalcea, R., Tarau, P.: Text Rank: bringing order into texts. In: Conference on Empirical Methods in Natural Language Processing (2004)
4. Witten, I.H., Paynter, G.W., Frank, E., Gutwin, C., Craig, G.: Nevill-Manning KEA: Practical Automatic Keyphrase Extraction (1999)
5. Turney, P.D.: Learning algorithms for keyphrase extraction. Information Retrieval-INRT, pp. 34–99 (2000)
6. Bhaskar, P., Nongnieikapam, K., Bandyopadhyay, S.: Keyphrase extraction in scientific articles: a supervised approach. In: Proceeding of COLING 2012, pp. 17–24. Mumbai (2012)
7. El-Beltagy, S.R., Rafea, A., Miner, K.P.: Participated in SemEval-2 Proceeding the 5th International Workshop on Semantic Evaluation, pp. 190–193. ACL, Uppsala, Sweden (2010)

8. Nguyen, T.D., Kan, M.Y.: WINGNUS: keyphrase extraction utilizing document logical structure. In: Proceeding the 5th International Workshop on Sematic Evaluation, pp. 166–169. ACL, Uppsala, Sweden (2010)
9. Lahiri, S., Mihalcea, R., Lai, P.H.: Keyword extraction from emails. In: Proceedings of 5th International Workshop on Semantic Evaluation, 2016, pp. 1–24. ACL, Cambridge University Press, UK (2010)
10. Sarkar, K.: A hybrid approach to extract keyphrases from medical documents. Int. J. Comput. Appl. (0975-8887) **63**(18) (2013)
11. Rake: Rapid Automatic Keyword Extraction. https://hackage.haskell.org/package/rake. Accessed 23 Mar 2017
12. AlchemyAPI. https://www.ibm.com/watson/alchemy-api.htm. Accessed 20 Mar 2017
13. Textacy. https://pypi.python.org/pypi/textacy. Accessed 13 Mar 2017
14. Spacy. https://pypi.python.org/pypi/spacy. Accessed 25 Feb 2017
15. NLTK. http://www.nltk.org. Accessed 27 Jan 2017
16. Huang, C., Tian, Y., Zhou, Z., Ling, C.X., Huang, T.: Keyphrase extraction using semantic networks structure analysis. In: IEEE International Conference on Data Mining, pp. 275–284 (2006)
17. Marsi, E., Ozturk, P.: Extraction and generalization of variables from scientific publications. In: Proceedings of the 2015 Conference on Empirical Methods in Natural Language Processing, pp. 505–511. Lisbon, Portugal (2015)
18. Bordea, G., Buitelaar, P.: DERIUNLP: a context based approach to automatic keyphrase extraction. In: Proceedings of the 5th International Workshop on Semantic Evaluation ACL, pp. 146–149. Uppsala, Sweden (2010)
19. Palshikar, G.K.: Keyword extraction from a single document using centrality measures. In: 2nd International Conference PReMI 2007 LNCS 4815, pp. 503–510 (2007)
20. Eichler, K., Neumann, G.: DFKIKeyWE: ranking keyphrases extracted from scientific articles. In: Proceedings of the 5th International Workshop on Sematic Evaluation, pp. 150–153. ACL, Uppsala, Sweden (2010)
21. Haddoud, M., Mokhtari, A., Leiroq, T., Abdeddaim, S.: Accurate keyphrase extraction from scientific papers by mining linguistics information, CLBib (2015)
22. Litvak, M., Last, M., Aizenman, H., Gobits, I., Kande, A.: l DegExt: a language-independent graph-based keyphrase extractor. Adv. Intel. Soft Comput. **86**, 121–130 (2011)
23. Litvak, M., Last, M.: Graph-based keyword extraction for single-document summarization. In: Proceedings of the 2nd Workshop on Multi-source Multilingual Information Extraction and Summarization, pp. 17–24. Manchester, UK (2008)
24. Nallapati, R., Allan, J., Mahadevan, S.: Extraction of keywords from news stories. CIIR Technical Report, IR(345) (2013)
25. CRF++. http://taku910.github.io/crfpp. Accessed 28 Jan 2017
26. Stanford CoreNLP. https://stanfordnlp.github.io/CoreNLP. Accessed 27 Feb 2017

Mixing and Entrainment Characteristics of Jet Control with Crosswire

S. Manigandan, K. Vijayaraja, G. Durga Revanth and A. V. S. C. Anudeep

Abstract This paper aims to study the effect of passive control on elliptical jet at different levels of nozzle pressure ratio. This experiment is carried out for three different types of configurations at two, four, five, and six NPRs. The results are captured and compared to one another. The rectangular crosswire is used as a passive control and tested at Mach number of two. The crosswire running along the major axis of the elliptical jet exits. The pitot pressure decay and the pressure profiles are plotted for various nozzle expansions. The crosswire is placed at three different positions ¼, ½, and ¾ to alter the shock wave successfully and to promote the mixing of jet. The shock waves are captured using numerical simulations. Due to the introduction of passive control at the exit of issuing jet, the shock wave weakens effectively, which stimulates the mixing promotion of jet by providing a shorter core length. It is witnessed that the efficiency of the mixing is superior when the crosswire is placed at ½ positions than ¼ and ¾. In addition, we also had seen a notable change in axis switching of the jets.

Keywords Nozzle · Supersonic jet · Crosswire · Core length

1 Introduction

Among various non-circular and circular jets, elliptical jet witnessed huge exciting features. There are several notable works made on the elliptical jet with different types of passive control [1–10]. This paper also deals about the passive control called crosswire, which is placed at the exit of the nozzle. The crosswire is tested at three different positions to study the ability of jet development process. This paper explains the dark side of the passive control since they possesses high value in

S. Manigandan (✉) · G. Durga Revanth · A. V. S. C. Anudeep
Sathyabama University, Chennai, India
e-mail: manisek87@gmail.com

K. Vijayaraja
KCG College of Technology, Chennai, India

D. Reddy Edla et al. (eds.), *Advances in Machine Learning and Data Science*,
Advances in Intelligent Systems and Computing 705,
https://doi.org/10.1007/978-981-10-8569-7_13

aviation and automobile sectors. Passive control by tabs was studied by several pioneers due to its simple arithmetic. Several researchers concluded that introduction of tabs alters the shock wave and increases the mixing promotion without disturbing the explicit property of the jet [6–10]. Further, Rouly et al. [2] studied the effect of jet at low divergence. They conclude that divergence is also responsible for the mixing promotion. Quinn [3] investigated the effect of the square jet and found the vortices are shed due to skewing of shear layer. Samimy et al. [4] noticed the jet emits three unique layers which are responsible for shorter core length is potential core decay, pitot pressure, and far filed decay. Similar work is carried out by Kumar and Rathakrishnan [1]. They found elliptical jet gives profound effect than circular jet which is highly appreciable.

Arvindh et al. [5] investigated the elliptical jet at Mach number of two on CD elliptical nozzle at different levels of expansion. They proved that the elliptical jet performs better than other circular and non-circular jet. In addition, they also witnessed the decay of the jet is faster compared to other profiles. Recently, Manigandan and Vijayaraja [8] investigated the effect of tabs on CD nozzle. They found that tabs are performing better when they placed diametrically opposite. They also studied the acoustics details of the jet. Recently, a similar study was carried by Manigandan and Vijayaraja [9], they investigated the flow field and sound relationship of the jet passing elliptical throat. They found mixing promotion caused by the elliptical throat is higher than the circular throat due to the weakening of the shock waves. Further, they seen at NPR 6, the mixing efficiency is very rich and dynamic when compared to other nozzle pressure ratio. In addition, they also found the axis switching is the main factor to stimulate the efficiency of the jet at all expansion ratio. Srivastava and Kaushik [11] have demonstrated experiment on rectangular jet with crosswire. They proved crosswire has an ability to reduce the core length by 88% compared to uncontrolled jet.

Introduction of tabs at the nozzle lip is done by many scholars and researchers. Here, tabs are replaced with crosswire to achieve higher mixing promotion than tabs.

The crosswire is placed at one quarter, half quarter, and three quarter (¼, ½, and ¾) to enhance the generation of large and small vertical structures which is responsible for mixing promotion. We have taken the jet as elliptical, and the quantum of work is carried out by several researchers. All literatures concluded that non-circular jet gives better mixing than circular jet due to shed of vortices. This study concentrates on the effect of crosswire on issuing jet.

2 Experimental Details

In this present study, we investigated an elliptical nozzle jet with crosswire. The crosswire is a thin rectangular structure made of Kevlar composites. The elliptical passive control jet tested at two Mach number at two, four, five, and six NPRs, respectively. The blockage of the tabs is 3%. The experiments are carried out at

Aerodynamics Laboratory, Jeppiaar Engineering College, Chennai, India. The experiment was demonstrated at Mach number of two. The desired NPR is achieved by controlling the flow regulator. The pressure gauge helps us to adjust the regulator to the appropriate levels of NPRs (Figs. 1 and 2).

The distance between the settling chamber and the nozzle is approximately 2 m, which is sufficiently adequate for mixing [9]. This facility utilizes the honeycomb structure on the lip of the settling chamber to achieve the smooth flow without turbulence. The elliptical jet with rectangular crosswire is tested at two, four, five, and six NPRs, and the values are plotted. After postprocessing, the results are compared to find optimized position to place the crosswire. The nozzle where designed at aspect ratio of two. The length of nozzle is 150 mm. The distance between the inlet geometry and the throat section is 65 mm. The nozzle and crosswire were made of mild steel using particle energy discharge process. The crosswire is bonded along the major axis of the jet at three different positions.

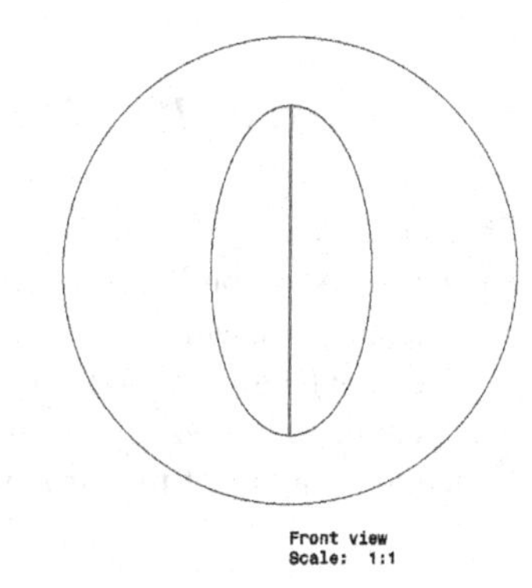

Fig. 1 Nozzle CAD drawing

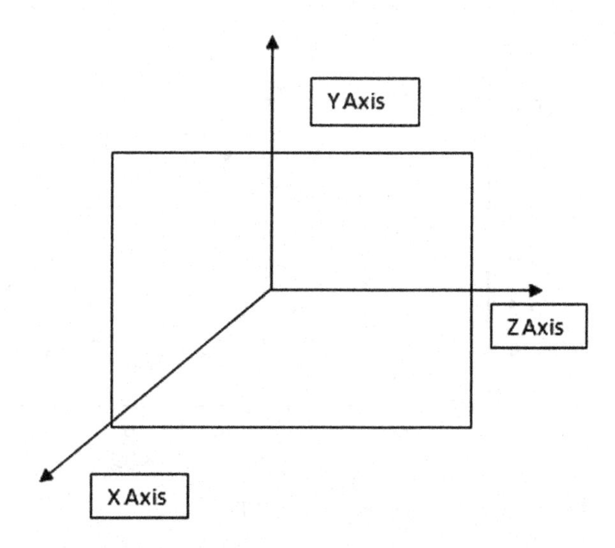

Fig. 2 Schematic diagram of jet axis

The flow is made to pass through at different levels of expansion, and the magnitude of Pt/Po is measured with respect to X/D. The values of the pressure are calibrated using three-dimensional transverse mechanisms. The effectiveness of the crosswire is analyzed using numerical simulation technique.

3 Results and Discussion

Estimating the supersonic jet core remains clumsy task for all pioneers. Precise calculation of the pitot pressure is very important to derive the accurate result. Hence, it is necessary to calibrate the pitot pressure accurately. Pitot pressure is plotted against X-axis alone to avoid such unwieldy conditions and errors. The X-axis distance made non-dimensional by dividing with the diameter of the nozzle exit.

3.1 Centerline Pitot Decay

The pitot pressure decay for three configurations is calibrated at under-expanded condition corresponding to NPR 2 which is plotted in Fig. 3. The shock strength is considerably less when crosswire is placed at ½ positions than compared to ½ and ¾. The core length of ½ position extends up to X/D = 10. Further, it is seen that the core length of the ½ and ¼ is X/D = 21 and 22 which is considerably higher. Hence, it is apparent that crosswire has the ability to generate higher efficiency in

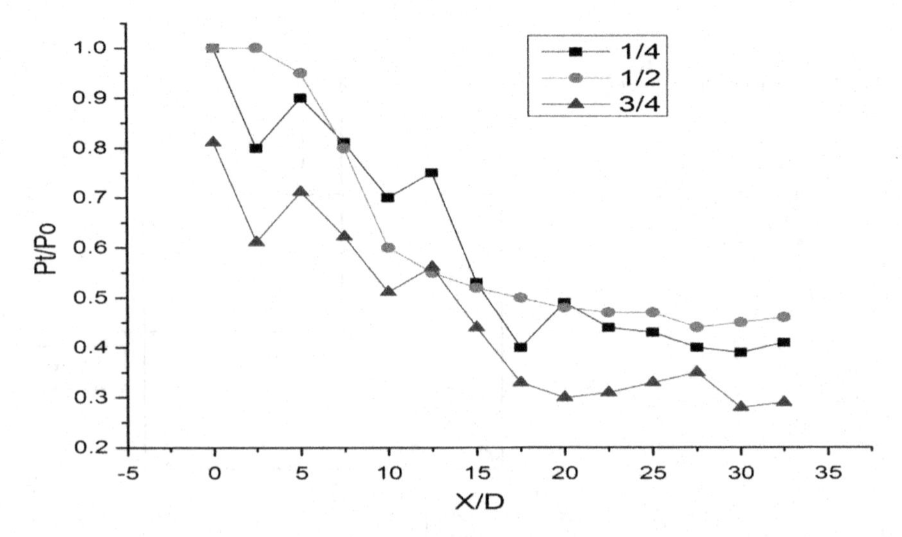

Fig. 3 Centerline pressure decay at NPR 2

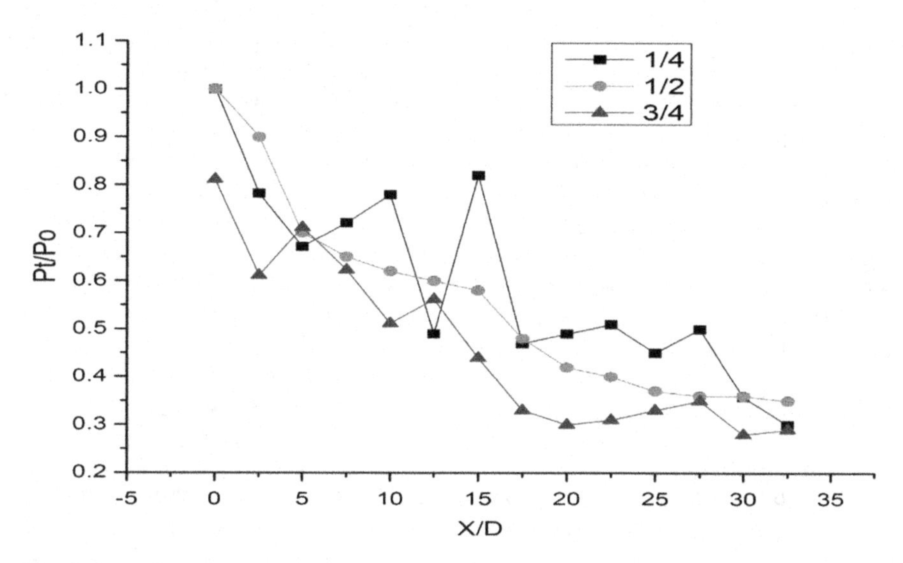

Fig. 4 Centerline pressure decay at NPR 4

mixing when they placed at center axis of the major axis of the jet due to the shock cell strength.

At NPR 4, the flow is marginally expanded. From Fig. 4, it is evident that the shock wave shows an appreciable change due to the presence of crosswire. However, the performance of crosswire which is positioned at ½ performs better by having the 50% reduced core length. The core length of ½, ¼, and ¾ are 5, 18, and 19, respectively. The configuration ¼ and ¾ witnesses small notable change in core length compared to NPR 2 which is appreciable.

At NPR 5, the flow is almost partially expanded. The pitot pressure decay is plotted in Fig. 5. It is seen that core length of the crosswire at ½ position is X/D = 4.5, whereas for other two configurations are X/D of 21 and 23. The crosswire at center of major axis performs superior than ¼ and ¾ positions. At NPR 6 Fig. 6, the flow is perfectly expanded for Mach number two. The core length of ½ position is X/D = 2, which is nearly 60% lower than NPR 5. This welcome change takes place due to the zero pressure gradients and weakening of shock waves. The core length for the other two positions remains identical to one another at X/D = 21.

From the above plots and discussions, it is crystal clear that the crosswire has an ability to reduce the core length and increase the mixing promotion effectively, Fig. 7. The crosswire at center position produces high mixing compared to ¼ and ¾. This occurs due to the shock cell growth and oscillations of the jet. The effective core length is achieved at NPR 6 for passive control jet.

The axis switching of the jet is seen in Figs. 8 and 9. The axis switching is captured numerically based on the few literatures. The flow property and the mathematic formulations are derived from the literature [11, 12]. The visualization

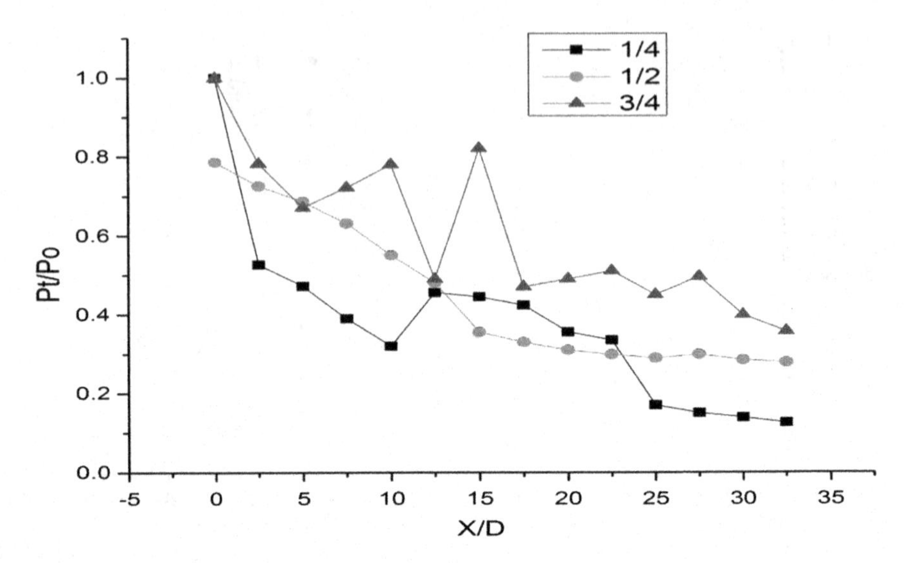

Fig. 5 Centerline pressure decay at NPR 5

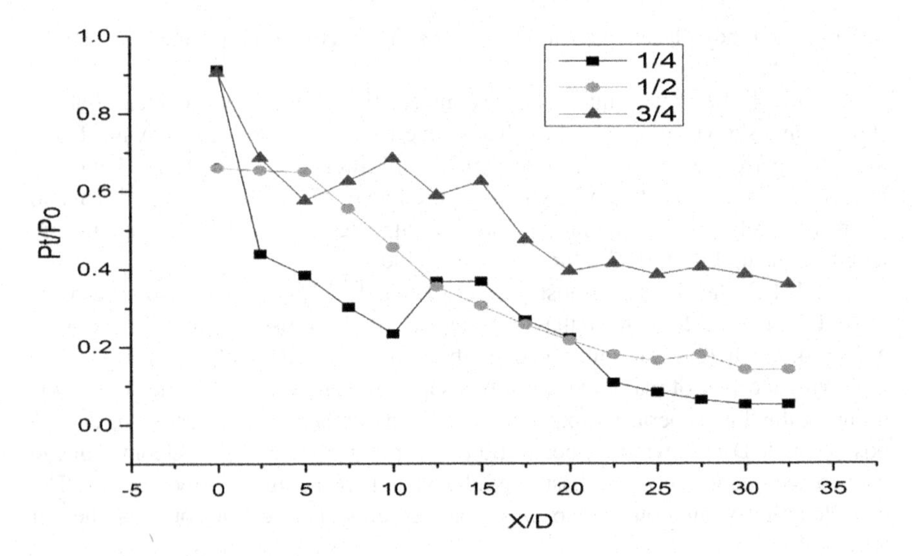

Fig. 6 Centerline pressure decay at NPR 6

image for the three types of configurations is shown in figure and is compared with each other to find the upstream and downstream flow property. Figures 8 and 9 show the axis switching of the jet at X/D of 5 and 10. Figures 8 and 9 show the flow is decayed from elliptic to circular due to the mechanism of the axis switching. It is evident that axis switching is the ultimate reason to achieve high mixing promotion using passive control.

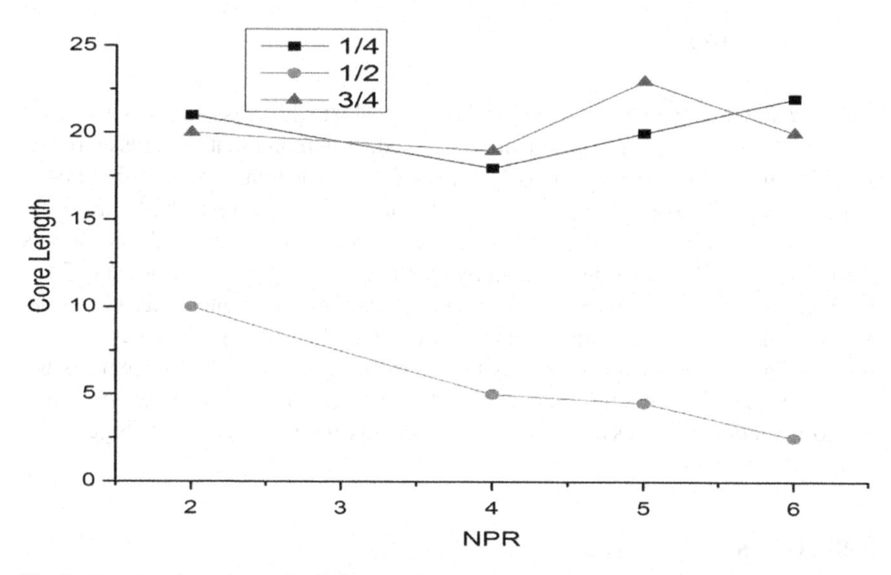

Fig. 7 Core length variation for different NPRs

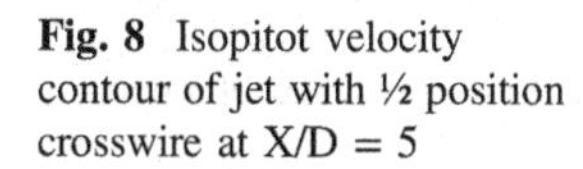
Fig. 8 Isopitot velocity contour of jet with ½ position crosswire at X/D = 5

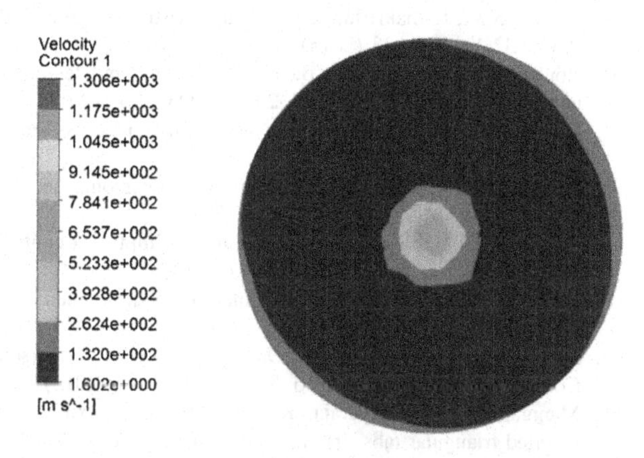

Fig. 9 Isopitot velocity contour of jet with ½ position crosswire at X/D = 10

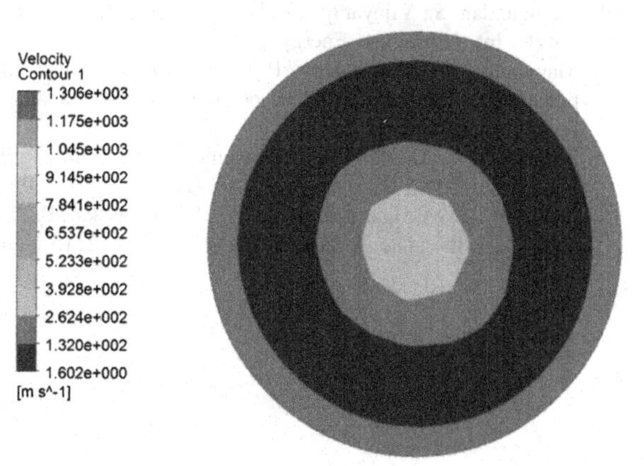

4 Conclusion

The flow and mixing characteristics of the elliptical jet with crosswire under various levels of NPR have been experimentally studied. It is found that the crosswire has an ability to modify the shockwave due to which core length of the configurations shows an appreciable change. It is also found that crosswire at half quarter ½ produces a higher result compared to ¼ and ¾. As NPR is increased, the flow is expanded; at NPR 6, the flow is perfectly expanded leading to reduction up to 75%. Further, it is seen that when the crosswire positioned at center axis of the jet provides maximum mixing promotion as well as higher mixing efficiency in huge margin. The numerical result reveals that weakening of shock cell strength is the ultimate reason for reduction in core length. The visualization image exposed that due to reduction of shockwave at the jet, core mixing augmentation is bigger.

References

1. Kumar, S.A., Rathakrishnan, E.: Characteristics of controlled Mach 2 elliptic jet. J. Propul. Power **32**(1), 121–133 (2015)
2. Rouly, E., Warkentin, A., Bauer, R.: Design and testing of low-divergence elliptical-jet nozzles. J. Mech. Sci. Technol. **29**(5), 1993–2003 (2015)
3. Quinn, W.R.: Streamwise evolution of a square jet cross section. AIAA J. **30**(12), 2852–2857 (1992)
4. Samimy, M., Reeder, M., Zaman, K.: Supersonic jet mixing enhancement by vortex generations. AIAA Paper, 91-2263 (1991)
5. Aravindh Kumar, S.M., Rathakrishnan, E.: Elliptic jet control with triangular tab. Proc. Inst. Mech. Eng. Part G: J. Aerosp. Eng. (2016)
6. Mitruka, J., Singh, P.K., Rathakrishnan, E.: Exit geometry effect on jet mixing. Appl. Mech. Mater. **598**, 151–155 (2014)
7. Hassan, E., Boles, J., Aono, H., Davis, D., Shyy, W.: Supersonic jet and crossflow interaction. Comput. Model. Prog. Aerosp. Sci. **28**(57), 1–24 (2013)
8. Manigandan, S., Vijayaraja, K.: Acoustic and mixing characteristic of CD nozzle with inverted triangular tabs. Int. J. Ambient Energy 1–9 (2017)
9. Manigandan, S., Vijayaraja, K.: Flow field and acoustic characteristics of elliptical throat CD nozzle. Int. J. Ambient Energy 1–9 (2017)
10. Manigandan, S., Gunasekar, P., Devipriya, J., Anderson, A., Nithya, S.: Energy-saving potential by changing window position and size in an isolated building. Int. J. Ambient Energy 1–5 (2017)
11. Srivastava, S., Kaushik, M.: Supersonic square jet mixing in presence of cross-wire at nozzle exit. Am. J. Fluid Dyn. **5**(3A), 19–23 (2015)
12. Devipriya, J., Manigandan, S., Nithya, S., Gunasekar, P.: Computational investigation of flow over rough flat plate. J. Chem. Pharm. Sci. **974**, 2115 (2017)

GCV-Based Regularized Extreme Learning Machine for Facial Expression Recognition

Shraddha Naik and Ravi Prasad K. Jagannath

Abstract Extreme learning machine (ELM) with a single-layer feed-forward network (SLFN) has acquired overwhelming attention. The structure of ELM has to be optimized through the incorporation of regularization to gain convenient results, and the Tikhonov regularization is frequently used. Regularization benefits in improving the generalized performance than traditional ELM. The estimation of regularization parameter mainly follows heuristic approaches or some empirical analysis through prior experience. When such a choice is not possible, the generalized cross-validation (GCV) method is one of the most popular choices for obtaining optimal regularization parameter. In this work, a new method of facial expression recognition is introduced where histogram of oriented gradients (HOG) feature extraction and GCV-based regularized ELM are applied. Experimental results on facial expression database JAFFE demonstrate promising performance which outperforms the other two classifiers, namely support vector machine (SVM) and k-nearest neighbor (KNN).

Keywords Facial expression recognition (FER) · Extreme learning machine (ELM) · Regularization · Generalized cross-validation (GCV) · Classification

1 Introduction

The inner mind is represented by variations in the gesture, gait, voice and slightest variation in eye and face. Important facial variations are due to face expressions which are categorized into six basic universal expressions which reflect intentional and emotional state of a person. The six basic expressions are happiness, sadness, surprise, fear, anger, and disgust according to Ekman [5]. Extensive research has

S. Naik (✉) · R. P. K. Jagannath
National Institute of Technology Goa, Ponda, Goa, India
e-mail: shraddha@nitgoa.ac.in

R. P. K. Jagannath
e-mail: k.j.raviprasad@nitgoa.ac.in

D. Reddy Edla et al. (eds.), *Advances in Machine Learning and Data Science*, Advances in Intelligent Systems and Computing 705,
https://doi.org/10.1007/978-981-10-8569-7_14

been carried out so far in the field of facial expression recognition (FER) in numerous domains [3, 6, 18].

In FER, face analysis is broadly carried out in two ways. One way is to describe facial expression measurements using action unit (AU). An AU is an observable component of facial muscle movement. Ekman and Friesen proposed Facial Action Coding System (FACS) in 1978 [4]. All expressions are broken down using different combinations of the AUs [17]. Using another way, the analysis of face was carried out directly based on the facial effect produced or emotions. In the literature, plenty of algorithms have been discussed to solve this problem. In this work, more focus is on the expression classification phase. The "extreme learning machine (ELM)" classifier is used for classification. ELM has a single hidden layer in its structure. Therefore, the whole system boils down to solving linear systems of equations which makes it easy to obtain global optimum value in cost function [10].

Regularized ELM has higher efficiency and a better generalization property. Different types of regularization techniques are possible. ℓ_2-regularization also known as Tikhonov regularization or ridge regression is one of the commonly used regularization techniques [14]. The best part of ℓ_2-regularization is that prediction performance increases compared to the traditional ELM. The choice of an appropriate regularization parameter is one of the most daunting tasks. Usually, this choice is accomplished by empirical or heuristic methods.

In this work, FER experiment was performed on JAFFE dataset. Face detection was carried out using Viola–Jones face detection algorithm [19]. "Histogram of oriented gradients (HOG)" technique was applied to obtain feature descriptors. Obtained features were classified using regularized ELM, where regularization parameter (λ) was calculated using "generalized cross-validation (GCV)" method.

The rest of this paper has been organized as follows: In Sect. 2, we elaborate discussion on the FER, formation of regularized ELM. We also discuss GCV in brief. Proposed methodology is explained briefly in the Sect. 3. Section 4 sheds light on experimental results and discussion. Section 5 concludes the discussion.

2 Methods

2.1 *Facial Expression Recognition*

"Facial expression recognition (FER)" system begins with the acquisition of face in image or video frames, followed by feature extraction and expression classification.

Face acquisition phase also allows to detect and locate faces in the complex background images. Many techniques are available for automatic face detection [19, 20]. Viola–Jones face detection algorithm is one of the robust and rapid visual detectors. Haar-like features are used for face detection. Features handled by detectors are computed very quickly using integral image representation. Algorithm trains a cascade of classifiers with AdaBoost, where AdaBoosting is used to eliminate the redundant features [19].

Next phase of FER is facial feature extraction. It is commonly categorized into geometry-based method and appearance-based method. In the geometry-based method, facial configuration is captured and a set of fiducial points are used to categorize the face into different categories. In the appearance-based method, feature descriptors encode important visual information from the facial images [15].

In HOG feature descriptor, distribution of intensity gradient or edge directions is used to describe local object appearance and shape within an image. Practically, it is accomplished by dividing an image into small cells and gradient is calculated over every pixel of the cell. Combining all the histogram entries together forms the facial feature descriptor [2]. Expression classification is the last phase of FER. Several classifiers are available to classify feature descriptor of images like support vector machine (SVM), artificial neural network (ANN), linear discriminant analysis (LDA) [1, 6].

2.2 *Extreme Learning Machine*

"Extreme learning machine (ELM)" was devised by Huang and Chen [10]. Input weight matrix and biases are being randomly initialized. The time needed for the recursive parameter optimization is excluded. Therefore, computational time of ELM is less. Transcendent characteristics of ELM provide efficient results. The classic ELM model and its few variants are discussed briefly in [10]. ELM training basically consists of two stages. In stage one, random feature map is generated and the second stage solves linear equations to obtain hidden weight parameters. General architecture of ELM is shown in Fig. 1; it is visible that ELM is a feed-forward network with a single hidden layer but input neurons may not act like other neurons [9, 11, 12]. Predicted target matrix $\hat{T}$ is calculated as

$$\hat{T} = f_L(x) = \sum_{j=1}^{L} \beta_i H_i(x) = H\beta, \tag{1}$$

where d is the total number of features, L is the total number of hidden units, c is the total number of classes, $\beta = [\beta_1, \beta_2, \ldots, \beta_L]^T$ is the output weight vector/matrix between the hidden layer and output layer, T is the given training data target matrix, and $H(x) = [H_1(x), H_2(x), \ldots, H_L(x)]$ is ELM nonlinear feature mapping.

Sigmoid, hyperbolic tangent function, Gaussian function, multi-quadratic function, radial basis function, hard limit function, etc., are the commonly used activation functions [9, 11]. For a particular ELM structure, the activation function of hidden units may not be unique. In our experiments, the activation function used is "hard limit function." The output weight matrix β can be obtained by minimizing the least square cost function which is given as

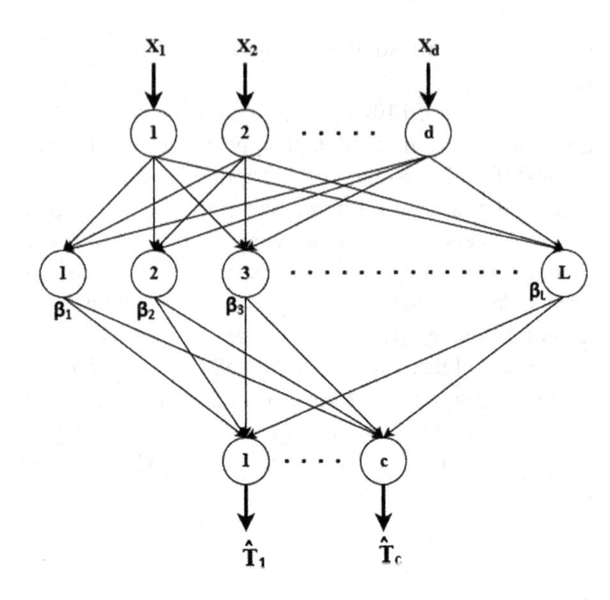

Fig. 1 General architecture of ELM

$$\min_{\beta \in R^{L\times c}} ||H\beta - T||^2. \tag{2}$$

The optimal solution can be obtained by $\beta^* = H^{\dagger}T$, where $H^{\dagger}$ is Moore–Penrose generalized inverse of matrix H and β^* is an optimal β solution.

2.3 *Regularized ELM*

In case of overdetermined or underdetermined system of linear equations, optimal solutions may not be obtained using the above method. In such cases, regularized ELM enhances the stability and generalization performance of ELM [10, 12]. Selection of regularization parameter plays a vital role. Basically, regularization parameter is calculated empirically or using some heuristic methods based on the prior experience. There are many variants of Tikhonov-regularized ELM which make accuracy computation more accurate. In this paper, the analysis is done using ℓ_2-regularization which is as follows:

$$\min_{\beta \in R^{L\times c}} ||H\beta - T||^2 + \lambda||\beta||^2. \tag{3}$$

Now, after differentiating (3) we get an optimal β^* as displayed

$$\beta^* = (\lambda I + H^T H)^{-1} H^T T. \tag{4}$$

The preeminent task is to predict regularization strength that is also known as regularization parameter λ in the above (3). Basically, the value of λ is chosen heuristically. Heuristic method follows trial and error technique, and therefore, the possibility of finding minima is low. There are many other methods to predict λ. It takes some range of values for λ, and after substituting these values in (3), the better value λ will be estimated which gives better accuracy. This is called coordinate descent method [7, 12].

2.4 Generalized Cross-Validation

"Generalized cross-validation (GCV)" estimates λ of ℓ_2-regularization [8]. The fundamental thing used in GCV is a concept of singular value decomposition (SVD). For any given problem, the value of $\lambda > 0$ for which expected error is less than Gauss Markov estimate. Regularization parameter λ minimizes loss function which is non-trivial quadratic depends on variance and β. However, in GCV there is no need of estimating the variance of errors. Here, $\hat{\lambda}$ which is the optimal value for λ is predicted by using

$$G(\lambda) = \frac{\frac{1}{n}||(I - A(\lambda))T||^2}{\left[\frac{1}{n}\operatorname{tr}(I - A(\lambda))\right]^2}, \quad (5)$$

where $A(\lambda) = H(H^T H + n\lambda I)^{-1} H^T$.

Using the above formula, we get value of $\hat{\lambda}$, and substituting this $\hat{\lambda}$ value into the ℓ_2-regularization, β is obtained. GCV method is computationally expensive because $A(\lambda)$ calculation requires finding of an inverse, multiple times.

3 Proposed Methodology

The basic FER system is divided into three subsystems as seen in Sect. 2. In the proposed work, neutral images from the dataset are subtracted from other expression images. Face was detected using Jones and Viola Face detection algorithm. Once face was extracted from the given images, HOG features were extracted from the subtracted images.

Furthermore, extracted features are classified using GCV-based regularized ELM. The problem of facial expression analysis is an ill-posed problem. The number of hidden layers used is more than the number of features which is nothing but an input vector. The FER system has become linear, and ℓ_2-regularization is used to improve generalization. In ℓ_2-regularization, the choice of regularization parameter plays an important role. Usually, the value of regularization parameters λ is chosen randomly or with some heuristic method. In the proposed method, GCV-based regularization of ELM is used to improvise the performance as shown in Sect. 4.

4 Experimental Results and Discussion

In this work, the effectiveness of GCV-based regularized ELM classifier on FER is demonstrated. In the experiments, Japanese Female Facial Expression (JAFFE) [13] database was used which contains a total of 213 images with 256×256 pixel. Ten Japanese female models posed for six basic expressions (anger, fear, disgust, happy, sad, and surprise), and one was neutral. From training dataset images, neutral images are subtracted. HOG features are extracted from obtained images. Regularized ELM is used to classify. The above process has been explained pictorially in Fig. 2.

GCV-based regularized ELM improves the generalization property. And due to this, recognition accuracy gets enhanced. The selection of hidden number of units (L) is done by choosing different values and based on that accuracies are calculated. The plot of ELM accuracy with the varying hidden number of units has been shown in Fig. 3. Finally, results are compared with non-regularized ELM. The confusion matrices of traditional ELM, GCV-based regularized ELM, SVM and KNN are demonstrated in Table 1. This demonstrates that GCV-based regularized ELM gives better results than other compared classifiers.

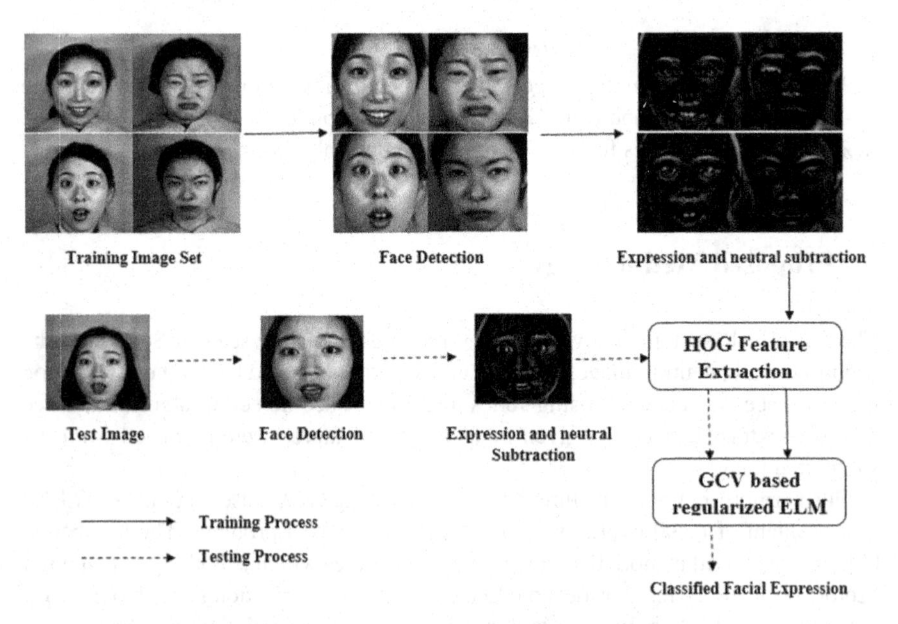

Fig. 2 Pictorial diagram of experiments using GCV-based regularized ELM

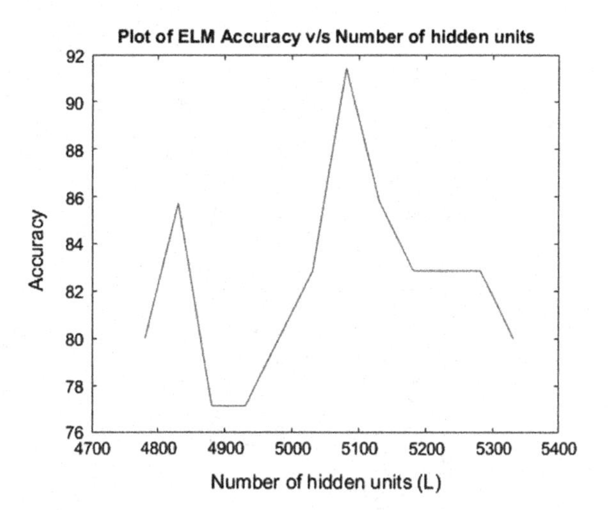

Fig. 3 The plot depicts ELM testing accuracy (%) versus number of hidden units (L) for underdetermined FER

Experiments are executed using different classifiers like classic ELM, GCV-based regularized ELM, SVM, and KNN. Various performance measures are available to measure the performance of the classification task [16]. The important measures used here are accuracy, precision, recall, and G-measure to demonstrate how the performance of different classifiers varies, and this is given in Table 2.

The solution for FER system can be devised in multiple ways. The main motive of this work is to demonstrate that regularized ELM can be applied to ill-posed problems like FER. ELM can be a good choice as it is easy to design, has a property of online learning, has strong tolerance to input noise, and has good generalization capability. Performance results can be obtained if we obtain correct value of λ. The correct value of λ can be obtained using GCV method. The proposed architecture for FER is one of the simplest and fastest methods. Better techniques can be devised to obtain the regularization parameter which is the strength of regularization. Existing work can be extended on different types of ill-posed problems. This regularization methodology can also be tested on other types of classifiers. Further improvement can also be obtained by pruning the existing architecture. The accuracy of FER system can be further improved by using different better feature extraction techniques, and also by using different feature selection methods. Our main focus in this work is applying automatic computation of regularization parameter for ELM.

5 Conclusion

"Extreme learning machine (ELM)" is a prominent feed-forward neural network with three layers. Regularization improves the generalization capability and strong tolerance to input noise which is suitable for ill-posed problems like "facial expression recognition (FER)." The selection of regularization parameter plays a vital role, and

Table 1 Confusion matrix for facial expression recognition using HOG for traditional ELM, GCV-based regularized ELM, SVM, and KNN

	Anger	Fear	Disgust	Happy	Sad	Surprise	Anger	Fear	Disgust	Happy	Sad	Surprise
	Traditional ELM						GCV-based regularized ELM					
Anger	**100**	0	0	0	0	0	**100**	0	0	0	0	0
Fear	4	**96**	0	0	0	0	4	**96**	0	0	0	0
Disgust	0	36.67	**63.33**	0	0	0	3.33	6.67	**90**	0	0	0
Happy	0	0	0	**100**	0	0	0	0	0	**100**	0	0
Sad	0	0	33.33	16.67	**50**	0	0	0	26.67	16.67	**56.67**	0
Surprise	0	0	0	0	0	**100**	0	0	0	0	0	**100**
Accuracy	84.89						90.44					
	SVM						KNN					
Anger	**81.48**	7.40	3.70	0	3.70	0	**77.27**	13.63	0	0	4.54	4.54
Fear	0	**70.83**	4.16	0	8.33	16.66	17.64	**52.94**	17.64	0	11.76	0
Disgust	0	5	**55**	0	15	25	0	14.81	**81.48**	0	3.76	0
Happy	0	0	0	**86.66**	0	13.33	15.38	0	0	**80.76**	0	3.86
Sad	0	8	0	4	**52**	36	6.66	3.33	10	6.66	**63.33**	10
Surprise	0	0	0	0	0	**100**	0	0	0	0	0	**100**
Accuracy	74.79						75.96					

Table 2 The performance of different classifiers are listed using various measures

Classifiers	Accuracy	Precision	Recall	G-measure
Traditional ELM	84.89	47.57	100	68.97
GCV-based regularized ELM	**90.44**	**60.68**	**100**	**77.90**
SVM	74.79	80.70	45.80	60.81
KNN	75.96	50	72.17	60.07

it is achieved by "generalized cross-validation (GCV)" method. GCV-based regularization parameter is one of the ways which is applied to avoid heuristics. Using this methodology in FER, recognition rate can be improved. This paper is aimed at proper classification of extracted HOG features in FER which is an ill-posed problem.

References

1. Bartlett, M.S., Littlewort, G., Frank, M., Lainscsek, C., Fasel, I., Movellan, J.: Recognizing facial expression: machine learning and application to spontaneous behavior. In: IEEE Computer Society Conference on Computer Vision and Pattern Recognition, 2005, CVPR 2005, vol. 2, pp. 568–573. IEEE (2005)
2. Dalal, N., Triggs, B.: Histograms of oriented gradients for human detection. In: IEEE Computer Society Conference on Computer Vision and Pattern Recognition, 2005, CVPR 2005, vol. 1, pp. 886–893. IEEE (2005)
3. Deshmukh, S., Patwardhan, M., Mahajan, A.: Survey on real-time facial expression recognition techniques. IET Biom. **5**(3), 155–163 (2016)
4. Ekman, P., Friesen, W.V.: Facial action coding system (1977)
5. Ekman, P., Oster, H.: Facial expressions of emotion. Annu. Rev. Psychol. **30**(1), 527–554 (1979)
6. Fasel, B., Luettin, J.: Automatic facial expression analysis: a survey. Pattern Recognit. **36**(1), 259–275 (2003)
7. Friedman, J., Hastie, T., Tibshirani, R.: Regularization paths for generalized linear models via coordinate descent. J. Statist. Softw. **33**(1), 1 (2010)
8. Golub, G.H., Heath, M., Wahba, G.: Generalized cross-validation as a method for choosing a good ridge parameter. Technometrics **21**(2), 215–223 (1979)
9. Huang, G.B., Chen, L.: Convex incremental extreme learning machine. Neurocomputing **70**(16), 3056–3062 (2007)
10. Huang, G.B., Chen, L.: Enhanced random search based incremental extreme learning machine. Neurocomputing **71**(16), 3460–3468 (2008)
11. Huang, G., Huang, G.B., Song, S., You, K.: Trends in extreme learning machines: a review. Neural Netw. **61**, 32–48 (2015)
12. Huang, G.B., Zhou, H., Ding, X., Zhang, R.: Extreme learning machine for regression and multiclass classification. IEEE Trans. Syst. Man Cybern. Part B (Cybern.) **42**(2), 513–529 (2012)
13. Lyons, M., Akamatsu, S., Kamachi, M., Gyoba, J.: Coding facial expressions with gabor wavelets. In: Third IEEE International Conference on Automatic Face and Gesture Recognition, 1998. Proceedings, pp. 200–205. IEEE (1998)

14. MartíNez-MartíNez, J.M., Escandell-Montero, P., Soria-Olivas, E., MartíN-Guerrero, J.D., Magdalena-Benedito, R., GóMez-Sanchis, J.: Regularized extreme learning machine for regression problems. Neurocomputing **74**(17), 3716–3721 (2011)
15. Shan, C., Gong, S., McOwan, P.W.: Facial expression recognition based on local binary patterns: a comprehensive study. Image Vis. Comput. **27**(6), 803–816 (2009)
16. Sokolova, M., Lapalme, G.: A systematic analysis of performance measures for classification tasks. Inf. Process. Manage. **45**(4), 427–437 (2009)
17. Tian, Y.I., Kanade, T., Cohn, J.F.: Recognizing action units for facial expression analysis. IEEE Trans. Pattern Anal. Mach. Intell. **23**(2), 97–115 (2001)
18. Tian, Y., Kanade, T., Cohn, J.F.: Facial expression recognition. In: Handbook of Face Recognition, pp. 487–519. Springer (2011)
19. Viola, P., Jones, M.J.: Robust real-time face detection. Int. J. Comput. Vision **57**(2), 137–154 (2004)
20. Yang, M.H., Kriegman, D.J., Ahuja, N.: Detecting faces in images: a survey. IEEE Trans. Pattern Anal. Mach. Intell. **24**(1), 34–58 (2002)

Prediction of Social Dimensions in a Heterogeneous Social Network

Aiswarya and Radhika M. Pai

Abstract Advancements in communication and computing technologies allow people located geographically apart to meet on a common platform to share information with each other. Social networking sites play an important role in this aspect. A lot of information can be inferred from such networks if the data is analyzed appropriately by applying a relevant data mining method. The proposed work concentrates on leveraging the connection information of the nodes in a social network for the prediction of social dimensions of new nodes joining the social network. In this work, an edge clustering algorithm and a multilabel classification algorithm are proposed to predict the social dimensions of the nodes joining an existing social network. The results of the proposed algorithms are found out to be satisfactory.

1 Introduction

The communication and computing technologies have become so advanced today that it enables people geographically apart to meet on a common platform and share information with each other. Social media networking [1] sites play an important role in this aspect.

The people using the social networking sites form a social network if they are connected to each other in some manner. So the social network can be viewed as a graph, with the people in the network represented as nodes of the graph and the connection between the people as edges of the graph. So, leveraging these sites will provide a great amount of information about the people around the world, their behavior, news happening around the world, and so on. It also allows sharing of information like text, photographs, videos with other people in the social network.

Aiswarya · R. M. Pai (✉)
Department of Information and Communication Technology,
Manipal Institute of Technology, Manipal Academy of Higher Education,
Manipal 576104, India
e-mail: radhika.pai@manipal.edu

Aiswarya
e-mail: aiswarya.bhat@manipal.edu

D. Reddy Edla et al. (eds.), *Advances in Machine Learning and Data Science*,
Advances in Intelligent Systems and Computing 705,
https://doi.org/10.1007/978-981-10-8569-7_15

Most of the time the connection between nodes in the social media networks are not homogeneous. Different types of connections are possible between different people in a social media network which makes it heterogeneous in nature. If the social network is considered as a graph, it can be said that one node may be connected to another node because that node maybe a friend, a family member, or maybe a colleague. The connections and behavior related to each connection will be different. So these connections cannot be considered as one and the same or homogeneous.

There is a lot of scope for using the social network data in getting knowledge about the interests of the people or influence of one node over other nodes in the social network. These kind of data are very useful for the advertising companies to market their products to the interested people in the social network.

Most of the time, complete data about the nodes are not available in the social network, but the connection information between the nodes may be available. So data has to be inferred from the connection information only. Sometimes there may be a case where the relationship information between the nodes may not be available in social media; hence, the heterogeneous relationships need to be considered as homogeneous relationships, which lead to inaccurate inference of information.

In the proposed work, the concentration is on prediction of social dimensions of a new node which connects to the network, given only its connection information. Social dimensions represent the interests of a node which connects to the network and in the proposed work, and a blog network is considered. Since the blog network is heterogeneous in nature, two algorithms are proposed for prediction of social dimensions taking the network heterogeneity into account. One algorithm is an edge clustering algorithm, and the other algorithm is a multilabel classification algorithm. The proposed algorithms are tested on a blog network dataset, and the results are found out to be satisfactory.

Rest of the paper is organized as follows. Section 2 discusses the related work, Sect. 3.1 describes the methodology, Sect. 4 presents the implementation and results and finally conclusion and future work are presented in Sect. 5.

2 Related Work

The connections in the social network are inherently heterogeneous representing various types of relationships between the users. In the method known as collective inference [2], the heterogeneous connections between the nodes of the network were considered as homogeneous, which reduces the quality of the clusters formed.

Lei Tang and Huan Liu in their work [3] have presented a relational learning framework based on latent social dimensions. They have used a soft clustering method known as spectral clustering, identifying dense social dimensions, which causes computational problems and huge memory requirements.

Lei Tang and Huan Liu in another work [4] have used modularity maximization, which is again a soft clustering method with dense representation of social dimensions and is computationally intensive.

Lei Tang et. al in their work [5] predict the collective behavior in the social network. Here an effective edge-centric approach, i.e., a variant of k-means algorithm is proposed to extract the social dimensions, which are sparse compared to the dense representations in spectral clustering or modularity maximization clustering methods. So, the method was not computationally intensive and could handle large networks. But, here the social dimensions to be extracted had to be given as input, which was not very effective.

Lei Tang and Huan Liu in their work [6] examine how online behaviors of users in a social network can be predicted, given the behavior of few actors in the network. Both node view and edge view methods are used to extract the social dimensions. But here, the number of social dimensions to be extracted had to be given as input by the user. So, it would be practically useful to develop an approach which can automatically determine the optimal number of social dimensions.

3 Methodology

Consider an example of network given in Fig. 1. Node 1 connects to Node 2 because they go to the same music class and to Node 3 because they work at the same library.

Given the label information that Node 1 is interested in both Music and Reading, the labels of Node 2 and Node 3 have to be inferred. If these two connections are treated homogeneously, then it is inferred that Node 2 and Node 3 are interested in both Music and Reading. But the reality may not be so. This shows the drawback of the collective inference [2] method where the connections are treated homogeneously. But if we know how Node 1 connects to Node 2 and Node 3, it is more reasonable to conclude that Node 2 is interested in Music and Node 3 is interested in Reading. Figure 2 shows an example toy network with nine nodes. If the connection information is not available, then somehow the connections have to be differentiated into different affiliations as shown in Fig. 3, and it has to be found out which affiliation is more correlated with the targeted class label, so that class membership of each actor can be inferred more precisely. The Node 1 belongs to two affiliations as depicted in Fig. 3.

Social dimensions of the nodes of the toy network are shown in Table 1. Social dimensions are the affiliations of the node.

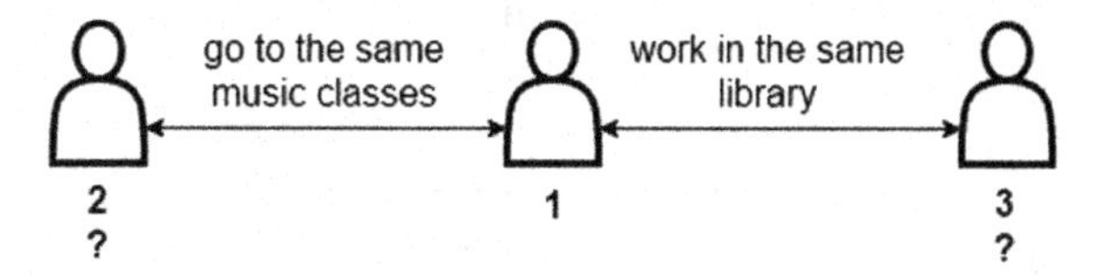

Fig. 1 Network example

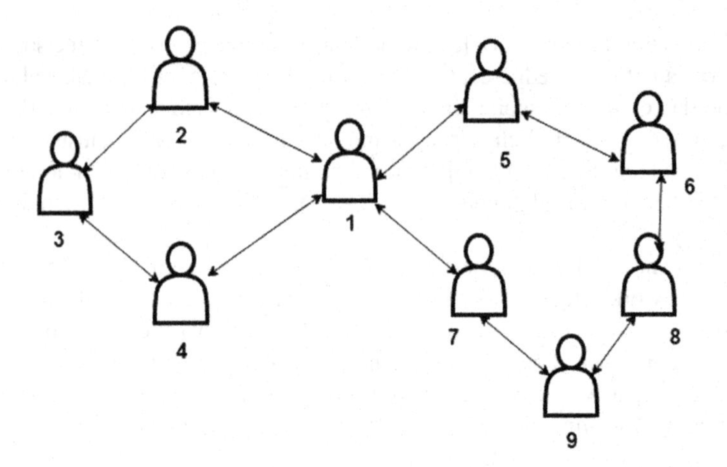

Fig. 2 Local network of Node 1

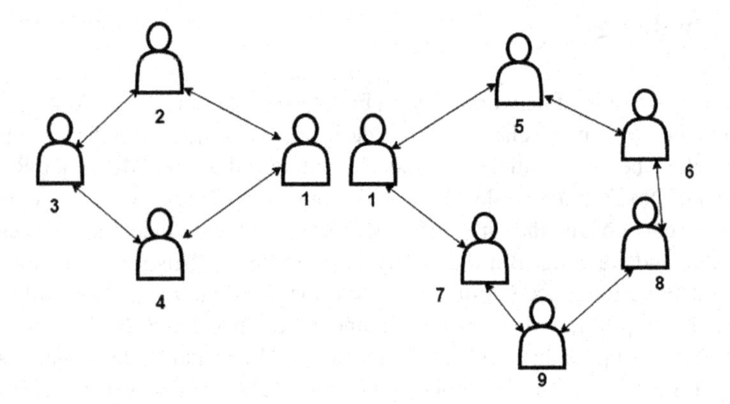

Fig. 3 Different affiliations

Table 1 Social dimensions of the nodes in the toy network

Node	Affiliation: Music	Affiliation: Reading
1	1	1
2	1	0
3	1	0
4	1	0
5	0	1
6	0	1
7	0	1
8	0	1
9	0	1

Table 2 Tag to category mapping

Tag	Category
Jazz	Music
mp3	Music
Tennis	Sports
Apparel	Fashion
Vacation	Travel

3.1 Edge Clustering

Instead of clustering the nodes in the network to form overlapping clusters, it would be much easier to partition the edges of the network into disjoint clusters. This would automatically give the overlapping node clusters. In this work, an edge clustering algorithm given in Algorithm 1 is proposed which partitions the edges into different clusters with different categories as labels. The algorithm makes use of a table, which is manually created and maps the tags associated with a node to a particular category. Tags represent words which belong to a certain category. The network considered had 82 categories. Example with four categories is shown in Table 2.

3.2 Multilabel Classification

In this work, a multilabel classification algorithm given in Algorithm 2 is proposed which assigns multiple labels to an instance. Here, the instances are the nodes of the social network, and the labels assigned to the nodes show the social dimensions of the node. The clusters formed by the edge clustering algorithm are used by the classification algorithm to predict the social dimensions of the new node joining the network.

3.3 Dataset

Here the network considered is a blog site, called BlogCatalog [7]. It is a blogging community where the users write blog posts under different categories of their interest. The dataset consists of information about the different bloggers in the network and their tags [8, 9]. A subset of the dataset with 500 nodes was considered for this work. The bloggers or the nodes of the network were assigned with ids starting from 1 up to 500. The tags and categories which were missing were suitably assumed. The network considered had 1358 edges and 340 tags.

Algorithm 1 Edge Clustering Algorithm

Input: Set of edges of the social network in the form (x, y) where x and y are nodes in the network which are numbered from 1 upto 500, tags of nodes in the network, tags to category mapping table M.

Output: n number of clusters with edges where n is the number of categories relevant to the network.

```
1: Read the list of edges in the network
2: ne = number of edges in the network
3: Arrange the edges in the increasing order of the node id
4: k=1;
5: Group_1=(x_1,y_1)
6: for i = 1 upto ne do
7:     if (Node_id(x_i) == Node_id(x_{i+1})) then
8:         check the tags of the nodes x_i and x_{i+1}.
9:         Refer M
10:        if x_i and x_{i+1} belong to same category then
11:            place the edge (x_{i+1},y_{i+1}) in Group_k
12:        else
13:            place the edge (x_{i+1},y_{i+1}) in Group_{k+1}
14:            k = k + 1
15:        end if
16:    else
17:        place the edge (x_{i+1},y_{i+1}) in Group_{k+1}
18:        k = k + 1
19:    end if
20: end for
21: for i = 1 upto number_of_groups do
22:     Take any edge from Group_i, find the category of edge using M
23:     for j = i + 1 upto number_of_groups do
24:         if Group_j ≠ empty then
25:             Take any edge from Group_j and find category of edge using M
26:         end if
27:         if (category(Group_i) == category(Group_j)) then
28:             Merge Group_j into Group_i
29:             Group_j = empty
30:         end if
31:     end for
32: end for
33: Display the n clusters
```

4 Implementation and Results

4.1 *Edge Clustering*

When the edge clustering algorithm is applied to the dataset, 82 clusters are formed. Two sample clusters among the 82 clusters obtained are shown below.

Cluster: Travel

Algorithm 2 Multilabel Classification Algorithm

Input: Set of edge clusters, the node *nd* which connect to the nodes in the network and the connection information
Output: Set of labels for the node *nd* which connects to the network.

1: Read the clusters.
2: num = number of clusters;
3: n= total number of nodes in the network;
4: **for** i=1 upto num **do**
5:     **for** j=1 upto n **do**
6:         $frequency_{ji}$ = number of occurences of node j in $cluster_i$
7:     **end for**
8: **end for**
9: **for** j=1 upto n **do**
10:     **for** i=1 upto num **do**
11:         $highest$ = highest value of $frequency_{ji}$
12:         $secondhighest$ = second highest value of $frequecy_{ji}$
13:         **if** $highest - secondhighest > 1$ **then**
14:             $threshold_{ji} = highest$
15:         **else**
16:             $threshold_{ji} = secondhighest$
17:         **end if**
18:     **end for**
19: **end for**
20: Read the new node *nd* along with its connections.
21: nc = number of connections
22: **for** i=1 upto nc **do**
23:     Output the labels of clusters corresponding to the threshold value of node at the end of *i*th connection.
24: **end for**
25: Display the final list of labels for node *nd*

Nodes: 5, 113, 176, 219, 233, 283, 446, 449, 176, 233, 446, 124, 195, 248, 311, 417, 430, 482, 131, 195, 328, 330, 362, 397, 472, 219, 229, 269, 411, 340, 449, 474

Cluster: Beauty

Nodes: 7, 40, 176, 283, 364, 269, 306, 85, 86, 394, 263, 295, 301, 343, 353, 449, 454, 87, 159, 231, 239, 279, 296, 285, 302, 291, 490, 359, 393, 399, 471

4.2 Multilabel Classification

A multilabel classification model is built and is validated by using 88 random nodes from the dataset whose social dimensions were already known, and the classification performance was analyzed. The model was validated using accuracy as a metric. The accuracy was determined using true positive (tp), true negative (tn), false positive (fp), and false negative (fn) values as given in Eq. 1. The accuracy was found out to be 84.9%.

$$Accuracy = tp + tn/(tp + fn + fp + tn) \quad (1)$$

The classifier performance was again tested for 50 new nodes which were independent of the dataset. The new nodes were labeled with ids as 501 to 550, and these nodes connect to already existing nodes in the network. The prediction of social dimensions was satisfactory for all 50 nodes. For example, node with id 510 is considered which connects to the nodes with ids 6, 7, 17, 36, 40, 129, and 167. Based on the proposed multilabel classification algorithm, it was predicted that the social dimensions of the node 510 were Fashion, Beauty, and Flowers, which are also the common interests of the node to which node 510 is connected.

5 Conclusion and Future Work

This work was set out to predict the social dimensions or the interests of new nodes which connects to the heterogeneous social network. Social dimensions can be very useful for the advertising companies to market their products to the interested people in the social network. In this work, two algorithms were proposed and implemented. One is an edge clustering algorithm, which is used to partition the edges of the network into disjoint clusters, based on the social dimensions of the node and the other algorithm is a multilabel classification algorithm, which is used to predict the social dimensions of a new node which joins the existing network. The algorithms were applied on a blog network of 500 nodes, and the clustering and the prediction performance were found to be satisfactory.

As a future work, the algorithms can be tried on a different social networking site dataset and the clustering and prediction performance can be analyzed. The proposed multilabel classification algorithm can be further extended to enhance the prediction performance.

References

1. Gundecha, P., Liu, H.: Mining Social Media: a brief introduction. Tutorials in Operations Research, Informs (2012)
2. Jensen, D., Neville, J., Gallagher, B.: Why collective inference improves relational classification. In: Proceeding of the 10th ACM SIGKKD International Conference on Knowledge Discovery and Data Mining, pp. 593–598. Seattle, WA, USA, August (2004)
3. Tang, L., Liu, H.: Leveraging social media networks for classification. Data Min. Knowl. Discov. **33**(3), 447–478 (2011)
4. Tang, L., Liu, H.: Relational learning via latent social dimensions. In: Proceedings of the 15th ACM SIGKDD International Conference on Knowledge discovery and Data Mining, pp. 817–826. New York, USA (2009)
5. Tang, L., Wang, X., Liu, H.: Scalable learning of collective behavior. IEEE Trans. Knowl. Data Eng. **24**(4), 1080–1091 (2012)

6. Tang, L., Liu, H.: Toward collective behavior prediction via social dimension extraction. IEEE Int. Syst. **25**(4), 19–25 (2010)
7. BlogCatalog. (2013). http://www.blogcatalog.com
8. Zafarani, R., Liu, H.: Social Computing Data Repository at ASU, Tempe, AZ: Arizona State University, School of Computing, Informatics and Decision Systems Engineering
9. BlogCatalogDataset. (2013). http://dmml.asu.edu/users/xufei/datasets.html

Game Theory-Based Defense Mechanisms of Cyber Warfare

Monica Ravishankar, D. Vijay Rao and C. R. S. Kumar

Abstract Threat faced by wireless network users is not only dependant on their own security stance but is also affected by the security-related actions of their opponents. As this interdependence continues to grow in scope, the need to devise an efficient security solution has become challenging to the security researchers and practitioners. We aim to explore the potential applicability of game theory to model the strategic interactions between these agents. In this paper, the interaction between the attacker and the defender is modeled as both static and dynamic game and the optimal strategies for the players are obtained by computing the Nash equilibrium. Our goal is to refine the key insights to illustrate the current state of game theory, concentrating on areas relevant to security analysis in cyber warfare.

Keywords Game theory · Static game · Dynamic game · Nash equilibrium

1 Introduction

Cyber technology, in contrast to making communication in the wireless network less obtrusive, has made privacy the most "often-cited criticism" [1]. Even though security systems are designed against the attacks of the highly skilled adversaries, they are still vulnerable to cyber threats [2]. Accordingly, advancements in technology have paved way for the growing risk of security concerns that are well exemplified by recent incidents. This list of security incidents is certainly inexhaustive; [3] gives a perception of this growing risk of cybercrimes. Recent studies reveal that defending against sophisticated antagonists is a challenging task which requires not only high technical skills, but also a keen understanding of incentives behind their attacks and different strategies used by them. Thus, being able to

M. Ravishankar (✉) · C. R. S. Kumar
Defence Institute of Advanced Technology, Pune 411 025, India
e-mail: monica_pcse14@diat.ac.in

D. Vijay Rao
Institute of Systems Studies & Analyses, Delhi 110054, India

D. Reddy Edla et al. (eds.), *Advances in Machine Learning and Data Science*,
Advances in Intelligent Systems and Computing 705,
https://doi.org/10.1007/978-981-10-8569-7_16

defend against and survive cyber attacks is still a great concern. Very aware of that, security researchers have analyzed a wide range of mechanisms for successful deterrence [4].

Of late, security decisions have been scrutinized analytically in a more meticulous way. Decisions made analytically are well grounded and persistent since it can be numerically implemented and checked experimentally with further improvements. Many mathematical models like Decision Theory, Machine Learning, Control Theory, Fuzzy Logic, and Pattern Recognition have been used to model, analyze, and solve the security decision problems. But among all the available approaches, game theory seems very effective whose models pave way for capturing the nature of adversaries related to security problem. Since game-theoretic methods stand out for their obstinacy, they have a striking virtue to anticipate and design defense against a sophisticated attacker, rather than responding randomly to a specific attack [5]. Furthermore, game theory can model issues of risk, trust, and other externalities (such as, beliefs) that arise in security systems.

2 Game Model

Our work focuses on mitigating cyber attacks using game-theoretic approach and validating the game models using network simulator for monitoring the network traffic and mitigating malicious flow. For illustration purpose, DoS/DDoS attacks are considered where the attacking nodes attempt to disrupt the network services by flooding with malicious traffic. The attack scenario is considered with an assumption on the network setting that the defender is uncertain about the normal flow and attack flow. The work presents a game model for the DoS attacks, in the form of interaction between the attacker and defender. In an attack scenario, the network traffic flow rate is given by

$$T = n \times r_n + a \times r_a \tag{1}$$

where r_n signifies normal traffic rate for the chosen n legitimate nodes and r_a signifies attack flow rate for the chosen number of a attack nodes. In case there is no defense mechanism in place, it is assumed that θ fraction of traffic pass the firewall to reach the destination and $(1-\theta)$ fraction of flow will be dropped without passing through the firewall. For the rate of each packet, θr, the average number of normal packets, which are able to reach the server, is given by

$$n_{avg} = \frac{n \times r_n}{n \times r_n + a \times r_a} \tag{2}$$

and the average of legitimate nodes deprived of the network services is estimated as

$$n_l = \frac{n - n_{avg}}{n} \tag{3}$$

The attacker's objective is to increase n_l, which will incur him some cost proportional to a. Accordingly, the attacker's net expected payoff is given by:

$$E_a = n_l - a \tag{4}$$

while the defender's expected payoff is defined as:

$$E_d = -n_l + a \tag{5}$$

Now assume a case where the network is configured with an appropriate defense mechanism such as firewall, which filters the incoming packets depending upon the flow rate X. The rate of filtering is given by fast sigmoid function as:

$$F(x) = 0.5 \times \left((x - X) \times \left[\frac{\delta}{1 + abs(x + \delta)} \right] \right) + 0.5 \tag{6}$$

Thus for the expected rate of normal traffic, the average rate of legitimate packets reaching the server through the firewall is given by

$$r_n^{'} = r_n \times (1 - F(r_n)) \tag{7}$$

while the average rate of attack flow reaching the server through the firewall is given by

$$r_a^{'} = r_a \times (1 - F(r_a)) \tag{8}$$

We then compute the attacker's and defender's payoff by replacing r_n by $r_n^{'}$ and r_a by $r_a^{'}$ in Eqs. (2) and (3). The attacker has to set optimal values for a and r_a, and the defender has to set optimal value for X in the fast sigmoid function used by the firewall, in order to maximize their payoffs. The notion of Nash equilibrium is used to determine the equilibrium state of the game which defines the best response strategies of the two players. For the given strategy profile of the two players, (r_a^*, a^*, X^*), the Nash equilibrium is defined to satisfy the following two relations simultaneously.

$$\begin{aligned} E^a_{(r_a^*, a^*, X^*)} &\geq E^A_{(r_a, a, X^*)} \quad \forall\, r_a, a \\ E^d_{(r_a^*, a^*, X^*)} &\geq E^D_{(r_a^*, a^*, X)} \quad \forall\, X \end{aligned} \tag{9}$$

The discussed model is made dynamic, which allows the players to change their strategies based on his/her anticipation of the opponent's behavior. Assuming the game duration as the sequence of k time steps, attacker's and defender's total expected payoff, over the entire game, is given by, $E_a = \sum_{t=1}^{k} E_a^t$ and $E_d = \sum_{t=1}^{k} E_d^t$ and denoted by the strategy profile (r_a^t, a^t, X^t) at the tth step ∀t = 1, . . ., k. For the given strategy profile, $\left(r_{a_t}^*, a_t^*, X_t^*\right)$, the Nash equilibrium is defined to satisfy the following two relations simultaneously.

$$\begin{aligned} E^a_{\left(r_{a_t}^*, a_t^*, X_t^*, \quad t=1,\ldots,k\right)} &\geq E^a_{\left(r_{a_t}, a_t, X_t^*, \quad t=1,\ldots,k\right)} & \forall\, r_a, a \\ E^d_{\left(r_{a_t}^*, a_t^*, X_t^*, \quad t=1,\ldots,k\right)} &\geq E^d_{\left(r_{a_t}^*, a_t^*, X_t, \quad t=1,\ldots,k\right)} & \forall\, X \end{aligned} \tag{10}$$

3 Discussions

Game theory is not about the prescription for the clever strategy but the search for effective decision. What game theory can elucidate is how an interaction proceeds, representation of these interactions as mathematical models that allow a meticulous analysis of the problem, and to help analysts to predict each other's behavior for real-world attacks and defenses. Attackers have their own selection criteria over their targets and are sound enough to alter their attack strategies based on the available defensive schemes. But traditional security approaches which uses heuristic solutions fail to capture this fact in their decision model and prefer the strategy of the attacker alone as an input to the model. Whereas in game-theoretic model, both the defense strategies and the hacker's actions are endogenously realized. This signifies that there is the potential for game theory to play a significant role in cyber warfare.

References

1. Kotenko, I., Chechulin, A.: A cyber attack modeling and impact assessment framework. In: Podins, K., Stinissen, J., Maybaum, M. (Eds.) Proceedings of the Fifth International Conference on Cyber Conflict. 2013 © NATO CCD COE Publications, Tallinn
2. Nagurney, A., Nagurney, L.S., Shukla, S.: A supply chain same theory framework for cyber security investments under network vulnerability in computation, cryptography, and network security, pp 381–398. Springer International Publishing Switzerland (2015)
3. US-CERT: Technical Cyber Security Alerts. http://www.uscert.gov/cas/techalerts/index.html. Dec 2010
4. Lim, S.-H., Yun, S., Kim, J.-H.: Prediction model for botnet-based cyber threats. In International Conference on Convergence, pp. 340–341. IEEE Press, Jeju Island (2012)
5. Bier, V.M., Naceur Azaiez, M. (eds.): Game Theoretic Risk Analysis of Security Threats. Springer, New York (2009)

Challenges Inherent in Building an Intelligent Paradigm for Tumor Detection Using Machine Learning Algorithms

A. S. Shinde, V. V. Desai and M. N. Chavan

Abstract Machine learning is at the heart of the big data rebellion sweeping the world today. It is the science of getting the computers to learn without being explicitly programmed as most of the technological systems are in an insurrection to be operated by intelligent machines capable to make the human like verdict to automatically solve human task with perfect results. Artificial intelligence is the heart of every major technological system in the world today. This paper presents the challenges faced to develop a model to acquiesce excellent results and the different techniques of machine learning; here, we also present the broad view of the current techniques used for detection of Brain tumor in computer-aided diagnosis and an innovative method for detection of Brain tumor by artificial intelligence using the algorithm of k-nearest neighbor which is established on the training a model with different values of k and the appropriate distance metrics are used for the distance calculation between pixels.

Keywords Machine learning · Artificial intelligence · Supervised learning Classification · Regression · Unsupervised learning · K-nearest neighbor

1 Introduction

Brain is the most respective part of the human body. The Brain and the spinal cord comprise of the body's nervous system. The Brain controls the action of the body through this nervous system and this involves receiving information from the sensory organs of the body, interprets the information, and then guides the body's response. Any infection, damage, abnormality to the Brain can cause threat to the normal functioning of the Brain and even human life. Advances in computer-aided

A. S. Shinde (✉) · V. V. Desai
GIT Belgaum, Belgaum, Karnataka, India
e-mail: ashsshinde@gmail.com

M. N. Chavan
ADCET, Ashta, Sangli, India

D. Reddy Edla et al. (eds.), *Advances in Machine Learning and Data Science*,
Advances in Intelligent Systems and Computing 705,
https://doi.org/10.1007/978-981-10-8569-7_17

diagnosis have simplified the task of a radiologist for manual detection of the tumors. Advancement in technology like artificial intelligence and machine learning has made diagnosis more accurate, time efficient, and cost-effective. In this paper, we discuss some of the machine learning techniques used for Brain tumor detection.

1.1 Challenges in Machine Learning

The kind of learning where a machine improves its performance based on experience is called as inductive learning. An agent is said to learn from experience with respect to some task and is measured by performance say ‘p,’ then ‘p’ improves with experience, and this is the actual concept of machine learning (a canonical definition stated by Tom Mitchell in 1997) [1]. There are several challenges to build a machine learning solution. These are: How good the model is? How one does choose a model? Does one have enough data? Is the data of sufficient quality? Is there any error in the data? Is one describing the data correctly? How confident can one be the result? The above questions have to be solved correctly in order to build a machine learning solution using some of the important machine learning assumptions/rules.

Supervised Learning: In the supervised learning, there are labeled data and the goal here is to come up with a function where input is given to get the output. The process for the supervised learning algorithm is shown in Fig. 1.

In the supervised learning, the training algorithm gets the input and produces the output and compares the output with the target, and if there is difference between the target and the output, error occurs which is sent to the agent for updating the

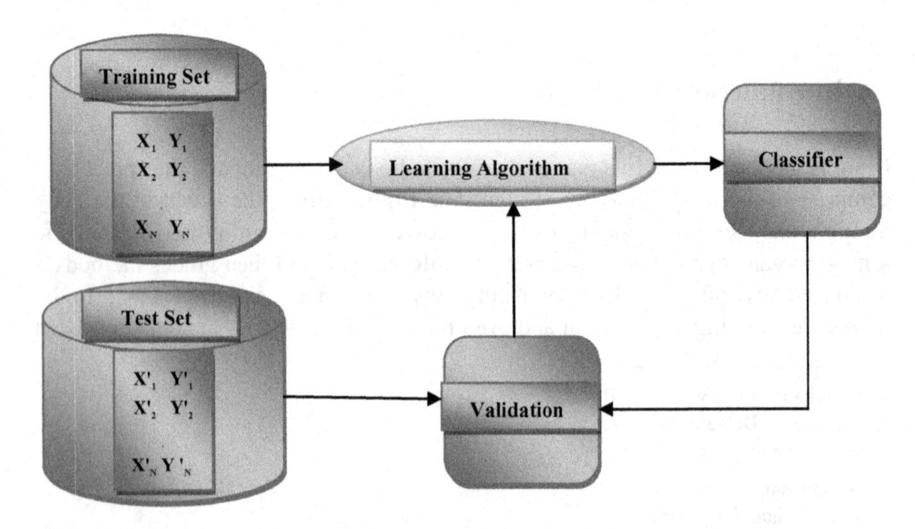

Fig. 1 Block diagram for process of supervised learning

weights [2]. Another problem in supervised learning is prediction or regression where the output predicted is not a discrete value and is a continuous value where most of the error occurs and the solution to fit the noise/error in the data is to over fit the data; here, care must be taken to avoid the over fit of the training data and so we can go for linear regression to minimize the sum of square errors where the aim is to minimize the prediction error by taking the square of the errors.

Unsupervised Learning: Here, there is an unlabeled training data set where clustering is done with groups of coherent or cohesive data points in the input space. The analysis is based on how spread out the points is that belong to the single group. In image processing, it can be with respect to image pixels depending on the similar pixels. The association rules are mining or learning frequent patterns or rules which are conditional dependency and these patterns are with respect to sequences, transactions, and graphs [2] (Fig. 2).

Reinforcement Learning: It is the learning of a machine to control a system through trial and error and minimal feedback (Fig. 3).

1.2 A Broad View on Different Techniques Used for Brain Tumor Detection

Brain tumor is detected from images acquired from Brain Magnetic Resonance Imaging (MRI). The image thus captured is preprocessed and then the region of interest is extracted; this process is called as segmentation of the image. The region of interest (ROI), i.e., the tumor region, is obtained by segmenting the input image which is a machine learning process.

The approach of the segmentation depends upon classifying the similar category of intensity values in an image and to form group of such similar intensity values (subjects). Here, the approach is based on two standards; i.e., similarity of intensity

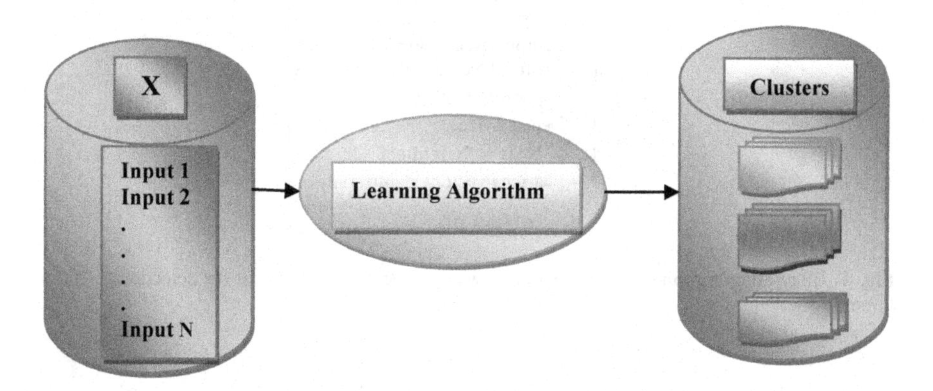

Fig. 2 Block diagram for process of unsupervised learning [1]

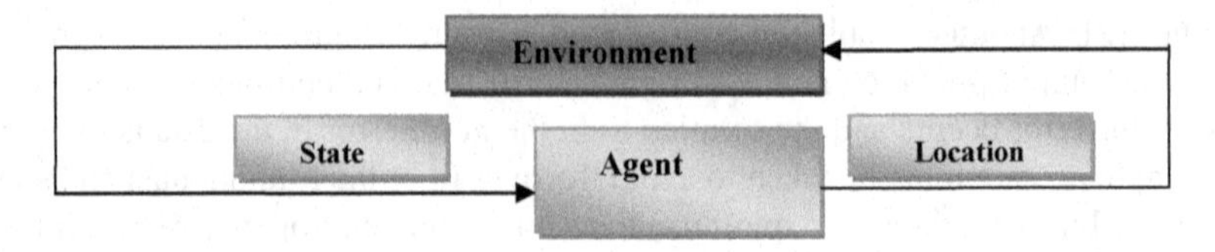

Fig. 3 Block diagram for process of reinforcement learning [1]

values is the dissimilarity between them and the task is either to find the similarity measure between the subjects and group them, and these subjects will have most similar index grouped into clusters or the dissimilarity among the objects separate to give the most dissimilar subjects in the space [3]. Figure 4 shows the different methods used for segmentation [4].

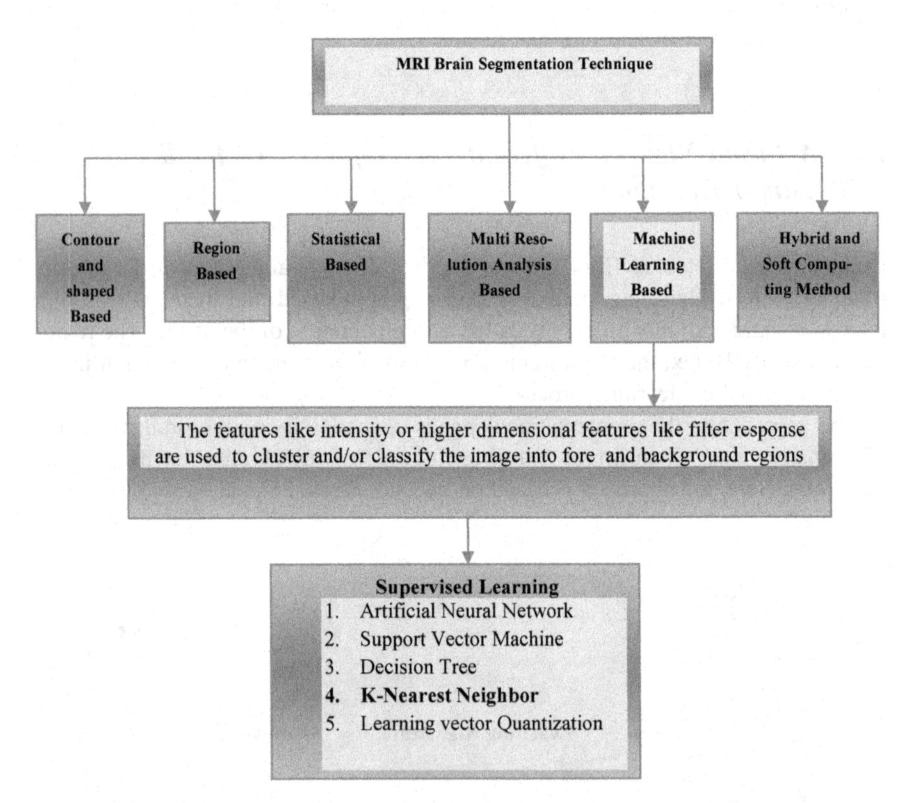

Fig. 4 Block diagram to represent a broad view on the techniques present for detection of Brain tumor

2 Related Work

In this paper, the segmented tumor is obtained by machine learning algorithm. The flow for the algorithm is as shown in the shaded region of Fig. 4. The Brain MRI images obtained are corrupt with noise so the initial step is to preprocess the image to reduce the unwanted noise and the high-frequency components. For a machine learning algorithm, the data is given as some kind of a probability distribution, where the input is drawn from a P-dimensional real space and the output is also a real space which can be given as

$$\text{Input } X = R^P \tag{1}$$

$$\text{Output } Y = R \tag{2}$$

The probability distribution is obtained from a set of samples that are drawn from $\{(x_1, y_1)\ (x_2, y_2), (x_3, y_3), \ldots, (x_n, y_n)\}$, and these can also be called as training data set.

Learn a function

$$f(x)R^P = R \tag{3}$$

Then, the function produces a predicted output which can be given as

$$\widehat{\text{Y}} = \text{f(x)} = \beta_0 + \beta_1 x_1 + \beta_2 x_2 + \cdots + \beta_p x_p \tag{4}$$

$$f(x) = \beta_0 + \sum_{j=1}^{p} (\beta_j, x_j) \tag{5}$$

Set $Xo = 1$ then

$$f(x) = \sum_{j=1}^{p} (x_j, \beta_j) \tag{6}$$

Equation (6) is of linear regression.

For the nearest neighbor classifier, we get

$$\widehat{Y}(x) = \frac{1}{k_0} \sum_{xi = Nx}^{\infty} (Y) \tag{7}$$

The performance measure is to compare the true output with the predicted output where the loss function is given by

$$L = (Y, f(x)) \tag{8}$$

Find $f(x)$ to minimize the loss function where the loss function is the expected predicted error. Minimal error is obtained on conditioning a specific value of x

$$f(x) = \arg\min E_Y |x([y - c])^2 |x \tag{9}$$

The value of C is the error and should be as small as possible, and this is called as the conditioning on a point.

$$\therefore (y - x)^2 = EPE(x) \tag{10}$$

One of the multi-clustering algorithms that are used for extracting tumor in MR images is the k-means clustering. This is a region-based segmentation which divides the region into number of clusters. The required number of clusters is represented by 'K' for particular application. Here, the technique is to find the nearest distance of every pixel to each cluster. Compared with the traditional algorithms like hierarchical algorithms, one of the advantages of using k-means algorithm is that for huge sample space on smaller value of 'k' the algorithm runs much faster. The other limitations that are associated with the algorithm are: 1. For a certain application, it is pretty difficult to guess the ideal value of 'k'; 2. Bounding of the initial partitioning varies these results in different clusters; 3. One of the main factors is the density of the clusters which affects the performance of the algorithm as the clusters with different density algorithm does not work well enough [5].

Similarly K-Nearest Neighbor (K-NN) is a clustering algorithm, which is proficient in integration the nearby pixels that have similar intensity values by measuring the Euclidean distance among the classes of K-samples of the Brain MRI. The algorithm is shown in Fig. 5.

The MRI images are first preprocessed using some of the image enhancement techniques like filtering, mostly the apostrophic filtering or the median filtering techniques are used. The main component as an input to the K-NN algorithm is the

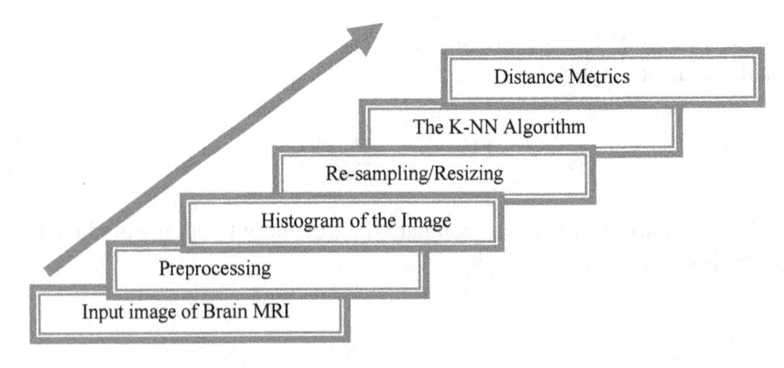

Fig. 5 Block diagram to represent of the K-NN algorithm the flow shows the bottom-up approach

feature values of various intensity levels obtained by the histogram of the input. The next step is re-sampling where the proper geometrical representation of the image is obtained [6]. The K-NN is a non-parametric algorithm which is trained with data, as the number of trained samples increases the output is more efficient or accurate. If S is the number of samples and C is the number of classes where S > C, the distance between the two pixels is found out by the Euclidean distance measure which is given as [7]. Let X = (x_1, x_2, x_3, …, x_n) and U = (u_1, u_2, u_3, …, u_n) be the two points then Y, the distance of order is defined as

$$D(x,u) = \sum_{i=1}^{n} \left(|x_i - u_i|^y\right)^{\frac{1}{y}} \tag{11}$$

- For y = 1: Manhattan Distance
- y = 2: Euclidean Distance
- y = ∞: Infinity Distance

Thus, the distance metrics are obtained and the results for some of the input images are as given in Table 2.

3 Experimental Results

Table 1 shows the results of the segmented tumor, and the corrupted input MRI images are preprocessed to initially enhance and eliminate the unwanted noise. High-frequency components are extracted from the enhanced image and the segmented output is obtained by clustering using the k-means algorithm. Manhattan distance metric is used for distance measurement. The segmented output is as shown in the last column of Table 1.

One of the supervised learning algorithms is the k-nearest neighbor. Researcher (Sudharani) [7] worked on a set of images to train as well as test the K-NN algorithm. The results obtained are shown in Tables 2 and 3 where Table 2 gives the distance measure of the data of tumors and Table 3 gives classification and identification score.

Table 1 Results for different input images using k-means algorithm

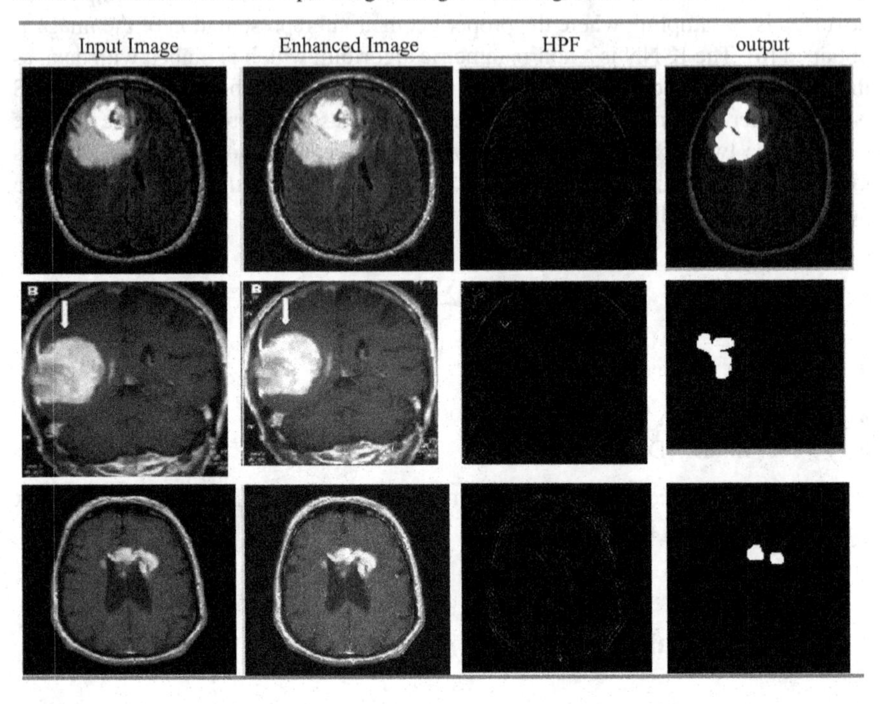

Table 2 Results for distance metrics

S. No.	Tumor 1	Tumor 2	Tumor 3	Tumor 4
Tumor1	0	53.57	312.43	47.67
Tumor2	53.57	0	274.71	73.22
Tumor3	312.43	274.71	0	328.11
Tumor4	47.67	73.22	328.11	0

Table 3 Results for different input images using K-NN algorithm. The classification and the identification score

S. No.	Actual label	Classification score	Identification score
Patient 1	Tumor 1	252.76	992.29
Patient 2	Tumor 2	178.78	883.99
Patient 3	Tumor 3	506.34	977.06
Patient 4	Tumor 4	446.7	969.29

4 Conclusion

This research is piloted to identify Brain tumor using medical imaging techniques. Here, more emphasis is given on the machine learning techniques and the results obtained are better than the results got form morphological operations. We observe that machine learning techniques have the learning capability which adapts itself for new data sets, which helps in minimizing errors by updating weight. Taking into account these powerful techniques of machine learning we are looking forward to find the volume of the tumor which is even more challenging which may help the radiologist for planning the therapy of diagnosis.

References

1. Chapelle, O., Schölkopf, B., Zien, A. (eds.): Semi-Supervised Learning, pp. 508. MIT Press, London, U.K (2006). ISBN:978-0-262-03358-9
2. Duch, W., Mańdziuk, J. (eds.): Challenges for computational intelligence. In: Series on Studies in Computational Intelligence, Vol. 63, pp. 488. Springer, New York (2007). ISBN:978-3-540-71983-0
3. Soleimani, V., Vincheh, F.: Improving ant colony optimization for brain MR image segmentation and brain tumor diagnosis. In: First Iranian Conference on Pattern Recognition and Image Analysis (PRIA). IEEE (2013)
4. El-Dahshan, E.-S.A., Mohsen, H.M., Revett, K., Salem, A.-B.M.: Computer-aided diagnosis of human brain tumor through mri: a survey and a new algorithm. Expert Syst. Appl. **4**, 5526–5545 (2014), Contents lists available at Science Direct. www.elsevier.com/locate/eswa. https://doi.org/10.1016/j.eswa.2014.01.021
5. Khare, S., Gupta, N., Srivastava, V.: Optimization technique, curve fitting and machine learning used to detect brain tumor in MRI. In: 2014 IEEE International Conference on Computer Communication and Systems (ICCCS 114), 20–21 Feb 2014, Chennai, India. https://doi.org/10.1109/iccs.2014.7068202
6. Selvakumar, J.: Brain tumor segmentation and its area calculation in brain MR images using K-mean clustering and Fuzzy C-mean algorithm. In: IEEE Conference on ICAMSE, pp. 186–190 Mar 2012
7. Sudharanil, K., Sarma, T.C., Satya Rasad, K.: Intelligent brain tumor lesion classification and identification from MRI images using kNN technique. In: 2015 International Conference on Control, Instrumentation, Communication and Computational Technologies (lCCICCT). 978-1-4673-9825-1/15/$3 1.00 ©2015 IEEE. https://doi.org/10.1109/iccicct.2015.7475384

Segmentation Techniques for Computer-Aided Diagnosis of Glaucoma: A Review

Sumaiya Pathan, Preetham Kumar and Radhika M. Pai

Abstract Glaucoma is an eye disease in which the optic nerve head (ONH) is damaged, leading to irreversible loss of vision. Vision loss due to glaucoma can be prevented only if it is detected at an early stage. Early diagnosis of glaucoma is possible by measuring the level of intra-ocular pressure (IOP) and the amount of neuro-retinal rim (NRR) area loss. The diagnosis accuracy depends on the experience and domain knowledge of the ophthalmologist. Hence, automated extraction of features from the retinal fundus images can play a major role for screening of glaucoma. The main aim of this paper is to review the different segmentation algorithms used to develop a computer-aided diagnostic (CAD) system for the detection of glaucoma from fundus images, and additionally, the future work is also highlighted.

Keywords Cup-to-disk ratio · Glaucoma · Neuro-retinal rim area
Optic disk · Optic cup · Segmentation

1 Introduction

The most important sense of a human body is the eye, because most of the information is perceived by the eye. The retina is the innermost surface of the eye, and it is made up of several transparent tissue layers, which absorb and convert the light signal into neural signal. These neural signals are transmitted to the brain through the optic nerve. The bright and circular shape region in the fundus images

S. Pathan · P. Kumar (✉) · R. M. Pai
Department of Information and Communication Technology, Manipal Institute of Technology, Manipal Academy of Higher Education, Manipal 576104, India
e-mail: preetham.kumar@manipal.edu

S. Pathan
e-mail: sumaiyakhanpathan45@gmail.com

R. M. Pai
e-mail: radhika.pai@manipal.edu

D. Reddy Edla et al. (eds.), *Advances in Machine Learning and Data Science*, Advances in Intelligent Systems and Computing 705,
https://doi.org/10.1007/978-981-10-8569-7_18

is called as ONH. There are several reasons that cause blindness. The World Health Organization (WHO) ranks glaucoma as the second leading cause of visual impairment [1]. Glaucoma is caused when the eye fails to maintain a balance between the amount of aqueous humor produced by the pupil and the amount of aqueous humor that drains out of the pupil [2]. If glaucoma is not detected at an early stage, it damages the optic nerve and leads to permanent blindness.

There are two types of glaucoma, namely open-angle and angle-closure glaucoma. The most common form of glaucoma is open-angle, which is painless, and vision loss is noticed only after a significant progress in condition. In angle-closure glaucoma, the patient experiences pain and the rate of vision loss is high. Conventional techniques for identification of retinal diseases are subjective and are based on visual perception of an individual along with the domain knowledge. The retina of a patient is periodically imaged using an optical coherence tomography (OCT) or a fundus camera [3]. Figure 1 shows the retinal fundus image captured using a fundus camera. These diagnostic modalities are invasive and consume time. Hence, CAD system for automated analysis of retinal images is found be more consistent, for detecting glaucoma at an early stage, by overcoming the subjective variability in a time efficient way. CAD system can also be used as a decision support system by clinicians [4].

In this paper, we have reviewed the state-of-art techniques for detection of glaucoma from fundus images. The main contributions of this paper are to give key ideas that will help researchers to judge the importance of various methods that aid in diagnosis of glaucoma and give an insight into future research.

1.1 Symptoms of Glaucoma

The major changes that take place while developing glaucoma are the variance in the structure of ONH, the loss of NRR area, defect in the retinal nerve fiber layer (RNFL), and the development of peripapillary atrophy (PPA).

The ONH is the place where the optic nerve originates and is distinguished as optic cup (OC) area and optic disk (OD) area. Due to glaucoma, the optic nerve

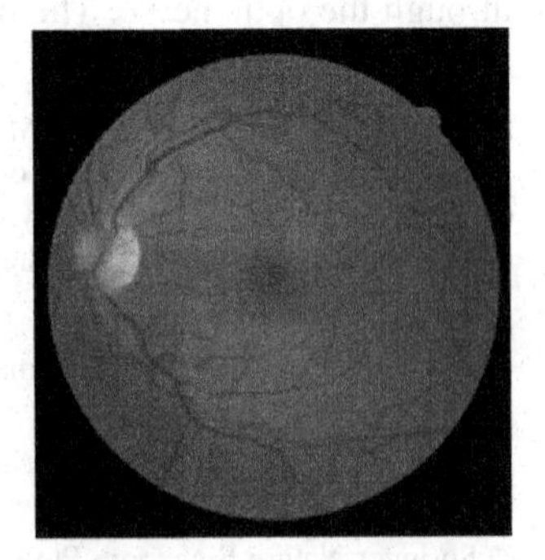

Fig. 1 Retinal fundus image [3]

fibers are lost, which in turn increases the OC area with respect to the OD area. Thus, cup-to-disk ratio (CDR) is an important criterion for detecting glaucoma. CDR can be defined as a ratio with respect to area, and horizontal and vertical length of both OC and OD. CDR value less than 0.3 indicates normal eye. For a person suffering from mild glaucoma, CDR is between 0.4 and 0.7, and for a person suffering from severe glaucoma, the CDR is usually more than 0.7 [5].

There is a circular area between OC and OD called as neuro-retinal rim (NRR) area. For a healthy person, the NRR configuration is in the decreasing order of thickness along the inferior, superior, nasal, and temporal regions, respectively. Due to glaucoma, the NRR area is lost. This is measured by the inferior, superior, nasal, temporal (ISNT) rule [5]. Figure 2 shows the OD, OC, NRR area and the ISNT regions.

In healthy eye, the RNFL is bright striations which are visible in temporal inferior region, followed by temporal superior, nasal superior, and nasal inferior regions. Due to glaucoma, the thickness of RNFL decreases as the nerve fibers are lost [7].

The development of PPA is closely related to open-angle glaucoma [8]. PPA has α zone and β zone. For a healthy person, both the zones are located in the temporal region followed by inferior and superior region. For a person suffering from glaucoma, β zone occurs in temporal region [9].

The CDR and ISNT rule analysis in the NRR area are preferred symptoms used by the ophthalmologist for diagnosis of glaucoma [5].

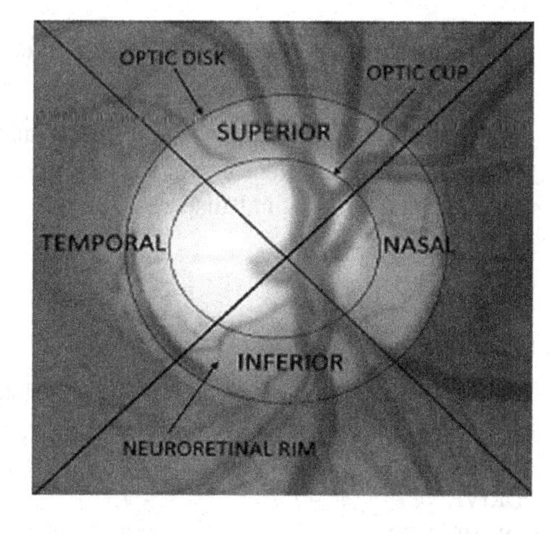

Fig. 2 Fundus image with OD, OC, NRR, and ISNT regions [6]

1.2 Imaging Modalities

Clinicians diagnose glaucoma by measuring the IOP, by analyzing the visual area, and by analyzing the appearance of ONH. IOP is measured using Goldman applanation tonometer, visual field is analyzed using Humphrey Analyzer, and the appearance of the ONH is analyzed using fundus images. For mass screening of glaucoma, the appearance of ONH is the most efficient way [3]. The other imaging modalities used for diagnosing retinal diseases are confocal scanning laser ophthalmoscopy (CSLO), scanning laser polarimetry (SLP), and optical coherence tomography (OCT) [3].

The review paper is organized as follows. Section 2 gives the segmentation approaches. Section 3 gives discussion regarding the segmentation methods. The paper concludes in Sect. 4.

2 Segmentation Approaches

The defects in the appearance ONH can help in detection of glaucoma at an early stage [3]. Segmentation techniques help in better analysis of ONH, by segmenting out various anatomical structures of the retina. This section reviews the segmentation methods required to develop a CAD system for glaucoma detection. The reviews are summarized in the form of a table. This section also gives the available retinal databases and evaluation parameters used for segmentation methods.

2.1 Publically Available Databases

To develop a CAD system for automated analysis of retinal images, there are some publically available databases. Table 1 gives an overview of the publically available databases for retinal diseases.

Table 1 Publically available databases

Database	Total no. of images	No. of retinal diseased images	No. of healthy images	References
STARE	81	50	31	[10]
DRIVE	40	7	33	[11]
DIARETDB0	130	110	20	[12]
DIARETDB1	89	84	5	[13]
MESSIDOR	1200	660	540	[14]
RIM-ONE	169	39	85	[15]

2.2 Segmentation Methods

The main feature of the fundus image is the OD. The OD is vertically oval and is present toward the nasal side. For a healthy retinal image, the OD is circular yellowish object. Any change in the color, shape, or depth of OD can be used as an indication of glaucoma [16]. In the center of OD, there is a small, white, cup-like area of variable dimensions called as the OC.

Most of the reviewed papers have tested their OD and OC segmentation methodologies on publically available online datasets such as RIM-ONE, STARE, DRIVE, ORIGA, and MESSIDOR. The literature divides the segmentation into three main parts: First is OD segmentation, which is a primary approach for detecting retinal pathologies including glaucoma. Second is methods proposed for OC segmentation. Third is segmentation approaches proposed for OD and OC, for the estimation of CDR and ISNT rule analysis in the NRR area.

OD Segmentation Methods

In automated diagnosis of glaucoma, the segmentation of OD is an important step, since most of the pathological changes reside inside the OD. Hence, several OD segmentation algorithms have been defined. Processing of OD involves locating the OD center point known as localization and detection of OD boundary known as segmentation. Table 2 illustrates the OD segmentation methods used in the literature.

Lu et al. [17] used circular transform apart from Hough to detect the color variation and circular boundary of the OD. However, the method showed accurate results only for images having clear variation of color across the boundary of OD. A future work is required if the boundary of OD is asymmetric with respect to the center of OD. Hsiao et al. [18] suggested a two-phase approach, i.e., localization and segmentation. For localization, illumination correction is used to extract the intensity distribution features from the green band of the image. For segmentation, supervised classifier along with gradient vector flow (GVF) is used. Bayesian classifier was used for classifying the disk boundary. The accuracy of the proposed algorithm can be improved by using efficient supervised classifier. Roychowdhury et al. [19] stated a three-step approach. In the first step, the bright region is determined using thresholding. In the second step, the bright region is retained using region-based classification, and in the final step, the OD boundary is segmented using best-fit ellipse around the created convex hull. As a future work, the proposed method can be combined with Hough circular transform to detect the boundary of OD. Bharkad [20] used a low-pass filter for blurring the blood vessels, and median filter along with morphological dilation was used for OD segmentation. However, the proposed algorithm fails to detect the OD center for fundus images with light color OD. Hence, future work could be to rectify this drawback.

Mary et al. [21] used circular Hough transform (CHT) for OD localization and active contours using GVF for OD segmentation. The OD segmentation accuracy is 94%. As a future work, the segmentation accuracy can be further improved by

Table 2 OD segmentation methods

Method	Image dataset	Accuracy	No. of test images	Year and references
Circular transformation	MESSIDOR, ARIA, STARE	Average segmentation accuracy is 99% (overlap score)	1401	2011 [17]
Supervised gradient vector flow snake (SGVF)	DRIVE, STARE	Average segmentation accuracy is 91% (absolute mean difference)	60	2012 [18]
Best-fit circle	DRIVE, DIARETDB1, DIARETDB0, CHASE_DB1, MESSIDOR, STARE	Average segmentation accuracy is 98% (overlap score)	81	2016 [19]
Grayscale morphological and median filtering operation	DRIVE, DIRATEDB0, DIRATEDB1, DRIONS	100% for DRIVE, 96% for DIRATEDB0, 98% for DIRATEDB1, and 100% for DRIONS (absolute mean difference)	369	2017 [20]
Active contour using gradient vector flow (GVF)	RIM-ONE	Success rate of 94% (absolute mean difference)	169	2015 [21]
Hough circle cloud	DRIVE, DIARETDB1	Average accuracy of 99% (absolute mean difference)	129	2016 [22]
Variation model along with multiple energies	MESSIDOR, ONHSD, DRIONS	99% for MESSIDOR, 100% for ONHSD, 100% for DRIONS (overlap score)	90	2017 [23]
Based on edge information	DRIVE, STARE, DIARETDB0, DIARETDB1	Average success rate of 98%. (overlap score)	340	2016 [24]

overcoming the interface caused due to atrophy which is located around the OD. Diaz Pernil et al. [22] used graphics processing unit (GPU) technology. Edges of the image are extracted using AGP-color segmentation. Hough circle cloud is used for OD detection. The proposed technique does not depend on color space. Hence, the method proves efficient for analyzing pathological retinal images.

Dai et al. [23] localized OD using sparse coding method, and OD is segmented by integrating energy with a variation model. The method can be further developed for segmenting OC for glaucoma detection. Xiong and Li [24] stated a method to locate the center of OD based on edge extraction and size of the bright object. It efficiently locates the OD for retinal images with less contrast and images with pathological changes. Hence, we have reviewed some OD localization and segmentation methods with their preliminary research and their future work.

OC Segmentation Methods

The segmentation of OC aids in determining CDR and NRR area loss. Table 3 illustrates the OC segmentation methods. Joshi et al. [25] used spline interpolation and kinks of blood vessels to determine the OC boundary. During OC segmentation, it was observed that segmentation error occurs due to lack of depth information. Hence, future work directs cup segmentation in 3D images. Damon et al. [26] suggested a method to detect the OC by using vessel kinks. A multi-scale shifting window is used to localize and detect kinks. The pallor information is used along with the kinks to segment OC. The proposed method can be further studied to use kinks in detection of glaucoma. Fondon et al. [27] used JCh color space for OC segmentation. The position, shape, and color with respect to OD were used to train the random forest classifier to detect the pixel of OC edge. At the end, the pixels are joined in the form of a circle. This technique can be further used in developing a CAD system for glaucoma screening. Hu et al. [28] used interpolation of vessel bend along with color difference information to segment OC. This method segments OC even for fundus images with small color and vessel bend information. The estimated OC boundaries are consistent. Hence, it can be further combined with other features for glaucoma screening.

From Table 3, it can be seen that there are limited algorithms which deal only with the segmentation of OC and their results are preliminary.

OD and OC Segmentation Methods

Estimation of CDR and ISNT rule require segmentation of both OD and OC. CDR is the main clinical indicator for detecting glaucoma. The ISNT rule is analyzed in the NRR area for accurate evaluation of glaucoma along with CDR. Table 4 illustrates the OD and OC segmentation methods for calculating CDR and ISNT.

Narsimhan et al. [29] stated a method to segment OD and OC using K-mean clustering. CDR is then calculated using elliptical fitting, and ISNT is computed using entropy threshold approach. SVM, KNN, and NB classifiers were used. The

Table 3 OC segmentation methods

Method	Image dataset	Accuracy	No. of test images	Year and references
Interpolation of vessel kink	Aravind eye hospital, India.	82% (overlap score)	138	2011 [25]
Vessel kinks	SERI	82% (overlap score)	67	2012 [26]
Random forest classifier	Glaucoma Repo	95% (absolute mean difference)	55	2015 [27]
Vessel bend and color information	DRIVE	82% (overlap score)	40	2016 [28]

Table 4 OD and OC segmentation methods

Method	Image dataset	Accuracy	No. of test images	Year and references
K-means clustering	AEH	Average accuracy of 95% (coefficients of correlation)	36	2012 [29]
Morphological methods	DMED FAU MESSIDOR	Average accuracy of 94% (overlap score)	50	2013 [30]
Active contour model and morphological methods	KMC hospital, Manipal	Average accuracy of 96% (overlap score)	200	2015 [31]
Adaptive threshold	Local hospital database	Average accuracy of 94% (overlap score)	34	2015 [32]
Active contour model and structural properties	RIM-ONE DIARETDB0	OD segmentation is 97% and for OC segmentation is 89% (coefficient of correlation)	59	2016 [33]

efficiency of classification can be further improved by increasing the training dataset. Khan et al. [30] utilized morphological methods to estimate CDR and to analyze ISNT rule in the NRR area. Classification was done manually. As a future work, classifiers can be used for automated classification.

Lotankar et al. [31] calculated CDR, by segmenting the OD using geodesic ACM and OC are segmented using morphological methods. Along with CDR, the ratio of blood vessels in the ISNT quadrant was measured. SVM was then used for classification. As a future work, the OD damage scale can be measured, to improve the glaucoma detection rate. Issac et al. [32] used adaptive threshold for segmenting OD and OC. The accuracy of the method is improved since the threshold is adaptive in nature. The NRR area and the blood vessels area in ISNT region are estimated along with CDR. Classification is done using SVM and ANN classifiers. The performance can be further improved by extracting different features and using alternate classifiers. Mittapalli and Kande [33] proposed two segmentation methods. Active contour model (ACM) is used for OD segmentation, and thresholding technique based on clustering is used for OC segmentation. CDR and ISNT rule are estimated along with disk damage likelihood scale (DDLS) to estimate the OD damage. In general, accurate OD and OC segmentation will capture the disease effect by computing CDR and ISNT rule.

2.3 Evaluation Parameters for Segmentation

The evaluation parameters used in the literature for evaluating the accuracy of segmentation techniques for ONH are given in Table 5.

Table 5 Evaluation parameters for ONH segmentation

Parameter	Definition
Overlap score	It is defined as $S(E, F) = \|E \cap F\|/\|E \cup F\|$ where, E is segmented image, F is annotation made by physicians, ∩ denotes intersection, and ∪ denotes union. The value is between zero and one. Greater value means large degree of overlap
Absolute mean difference	It is the difference between the boundary extracted and the average obtained boundary from the physicians Absolute mean difference $= \frac{1}{n}\sum_{i=1}^{n}\sqrt{\left(S_{x_i}^2 + S_{y_i}^2\right)}$ where S_{x_i} and S_{y_i} are the value of error at a given point
Coefficients of correlation	$\rho_{A,B} = \frac{E(AB) - E(A)E(B)}{\sqrt{E(A^2) - E(B^2)}\sqrt{E(B^2) - E(A^2)}}$ where A and B are the datasets and E () is dataset mean

3 Discussion

Glaucoma is irreversible. Early detection can help the ophthalmologist to prevent the person from blindness. In order to diagnose glaucoma, ophthalmologist evaluates the changes in the retinal structure around the ONH. Examination of ONH is based on measuring the CDR and analysis of ISNT rule in the NRR area. Hence, approaches that segment OD and OC for computing CDR and ISNT are found to be more promising for detection of glaucoma. Automated analysis of retinal images can be used to provide a second opinion to the ophthalmologist. Robust segmentation methods are required for analyzing the retinal images. The presence of atrophy hinders the segmentation of OD, because the atrophy has similar intensity as that of OD. Algorithms that will segment the OD accurately even in the presence of atrophy are required. For pathological retinal images, extraction of main features along with vascular convergence algorithm can improve the accuracy of OD detection. In fundus images, the edge between OD and OC is not clearly visible in normal contrast. Hence, pre-processing step which includes selection of appropriate image channel and overcoming the variations caused during image acquisition can be employed for accurate segmentation of OD and OC. The blood vessels are denser in OC compared to OD. Hence, segmentation of OC is difficult compared to segmentation of OD. Blood vessel extraction algorithm can be used for segmentation of OC.

4 Conclusion

Vision loss due to glaucoma can be prevented only if it is detected at an early stage. Detection of glaucoma depends on individual perception of vision and clinical experience of ophthalmologist. With the use of 2D retinal scan images, automated analysis of retinal images for detection of glaucoma can be used in cost- and time-effective manner. This paper reviews the recent segmentation methods proposed for automated diagnosis of glaucoma, along with the research direction. The techniques summarized in the related work are promising, but there is still a scope for improved and accurate segmentation techniques as discussed in above sections.

References

1. Quigley, H.A., Broman, A.T.: The number of people with glaucoma worldwide in 2010 and 2020. Brit. J. Opthalmol **90**(5), 262–267 (2006)
2. Lim, R., Golberg, I.: The glaucoma book, 2nd edn. Springer Science Business Media, New York (2010)
3. Acharya, R., Yun, W.L., Ng, E.Y.K., Yu, W., Suri, J.S.: Imaging systems of human eye: a review. J. Med. Syst **32**(2), 301–315 (2008)
4. Nishikawa, R.M., Giger, M.L., Vyborny, C.J., Schmidt, R.A.: Computer-aided detection of clustered micro classifiations on digital mammograms. J Comput. Methods Progr. Biomed. **116**(3), 226–235 (2014)
5. Cheng, J., Liu, J., Xu, Y., et al.: Optic disk segmentation based on variational model with multiple energies. IEEE Trans. Med. Imaging **32**(6), 1019–1032 (2013)
6. Das, Nirmala, S.R., Medhi, et al.: Diagnosis of glaucoma using CDR and NRR area in retina images. Netw. Model Anal Health Inf. Bioinform. **5**(1), 91–96 (2015)
7. Jonas, J., Budde, W., Jonas, S.: Opthalmoscopic evaluation of optic nerve head. Surv. Ophthalmol. **43**(5), 293–320 (1999)
8. Jonas, J.: Clinical implication of peripapillary attropy in glaucoma. Curr. Opin. Ophthalmol. **16**(3), 84–88 (2005)
9. Ehrlich, J.R., Radcliffe, N.M.: The role of clinical parapaillary atrophy evaluation in the diagnosis of open angle glaucoma. Clin. Ophthalmol. **4**(3), 971–976 (2010)
10. Structured Analysis of retina (Stare). http://cecas.clemson.edu/~ahoover/stare/. Accessed 18 July 2017
11. Stall, J., Abramoff, M., Niemeijer, M., Viergever, M., et al.: Ridge based vessel segmentation in color images of the retina. IEEE Trans. Med. Imaging **23**(4), 501–509 (2004)
12. Diaretdb0: Evaluation database and methodology for diabetic retinopathy algorithms. http://www.it.lut.fi/project/imageret/diaretdb0/. Accessed 18 July 2017
13. Diaretdb1: Diabetic retinopathy evaluation protocol. http://www.it.lut.fi/project/imageret/diaretdb1/. Accessed 18 July 2017
14. Messidor. http://www.adcis.net/en/Download-Third-Party/Messidor.html
15. RIM-ONE: http://medimrg.webs.ull.es/research/retinal-imaging/rim-one/. Accessed 19 July 2017
16. Dai, B., Wu, X., Bu, W.: Superpixel classification based optic disk and optic cup segmentation for glaucoma screening. J. Pattern Recognit. **64**(7), 226–235 (2017)
17. Lu, S., et al.: Accurate and efficient optic disc detection and segmentation by a circular transformation. IEEE Trans. Med. Imaging **30**(12), 2126–2133 (2011)
18. Hsiao, H.-K., Liu, C.-C., Yu, C.-Y., Kuo, S.-W., Yu, S.-S.: A novel optic disc detection scheme on retinal images. J Expert Syst. Appl. **39**(12), 10600–11066 (2012)

19. Roychowdhury, S., Koozekanani, D.D., Kuchinka, S.N., Parhi, K.K.: Optic disc boundary and vessel origin segmentation of fundus images. IEEE J. Biomed. Health Inform. **20**(6), 1562–1574 (2016)
20. Bharkad, S.: Automatic segmentation of optic disk in retinal images. J. Biomed. Signal Process Control **31**(5), 483–491 (2017)
21. Mary, M.C.V.S., Rajsingh, E.B., Jacob, J.K.K., Anandhi, D., Amato, U., Selvan, S.E.: An empirical study on optic disc segmentation using an active contour model. J. Biomed. Signal Process Control **18**(5), 19–29 (2015)
22. Díaz-Pernil, D., Fondón, I., Peña-Cantillana, F., Gutiérrez-Naranjo, M.A.: Fully automatized parallel segmentation of the optic disc in retinal fundus images. J. Pattern Recognit. Lett. **83** (3), 99–107 (2016)
23. Dai, B., Wu, X., Bu, W.: Optic disc segmentation based on variational model with multiple energies. J. Pattern Recognit. **64**(6), 226–235 (2017)
24. Xiong, L., Li, H.: An approach to locate optic disc in retinal images with pathological changes. J. Comput. Med. Imaging Graph. **47**, 40–50 (2016)
25. Joshi, G.D., Sivaswamy, J., Krishnadas, S.R.: Optic disk and cup segmentation from monocular color retinal images for glaucoma assessment. IEEE Trans. Med. Imaging **30**(6), 1192–1205 (2011)
26. Damon, W.W.K., Liu, J., Meng, T.N., Fengshou, Y., Yin, W.T.: Automatic detection of the optic cup using vessel kinking in digital retinal fundus images. In: 9th IEEE International Symposium on Biomedical Imaging (ISBI) (2012)
27. Fondon, I., Valverde, J.F., Sarmiento, A., Abbas, Q., Jimenez, S., Alemany, P.: Automatic optic cup segmentation algorithm for retinal fundus images based on random forest classifier. In: International Conference on Computer as a Tool IEEE EUROCON 2015 (2015)
28. Hu, M., Zhu, C., Li, X., Xu, Y.: Optic cup segmentation from fundus images for glaucoma diagnosis. J. Bioeng. **8**(1), 21–28 (2016)
29. Narasimhan, K., Vijayarekha, K., Jogi Narayan, K.A., Siva Prasad, P., Satish Kumar, V.: An efficient automated system for glaucoma detection using fundus images. Res. J. Appl. Sci. Eng. Technol. (2012)
30. Khan, F., Khan, S.A., Yasin, U.U., Haq, I.U., Qamar, U.: Detection of glaucoma using retinal fundus images. In: The 6th 2013 Biomedical Engineering International Conference (2013)
31. Lotankar, M., Noronha, K., Koti, J.: Glaucoma screening using digital fundus images through optic disk and optic cup segmentation. Int. J. Comput. Appl. 9975–8887 (2015)
32. Issac, A., Sarathi, M.P., Dutta, M.K.: An adaptive threshold based image processing technique for improved glaucoma detection and classification. J Comput. Methods Programs Biomed. **122**(2), 229–244 (2015)
33. Mittapalli, P.S., Kande, G.B.: Segmentation of optic disk and optic cup from digital fundus images for the assessment of glaucoma. J. Biomed. Signal Process. Control **24**, 34–46 (2016)

Performance Analysis of Information Retrieval Models on Word Pair Index Structure

N. Karthika and B. Janet

Abstract This paper analyzes the performance of word pair index structure for various information retrieval models. Word pair index structure is the most efficient for solving contextual queries and it is a precision-enhancing structure. The selection of information retrieval model is very important as it precisely influences the outcome of information retrieval system. This paper analyzes the performance of different information retrieval models using word pair index structure. It is found that there is an increase in precision of 18% when compared with traditional inverted index structure, and recall is 8% in the inverted word pair index structure.The mean average precision is increased by 26%, and R-precision is increased by 20%.

1 Introduction

Information retrieval is a process of retrieving a set of documents which are relevant to the query [1]. The user indicates his/her information need which is a description of what the user is expecting as a query. It may be a list of words or a phrase. Document collection/corpus is a group of documents. Each document contains an information that the user needs.

Indexing is a process of creating a document representation of information content which makes querying faster. There are various indexing structures available. The most popular one is inverted index structure. The Fig. 1 gives the modules of the information retrieval process using the inverted index.

Preprocessing [2] module has tokenization, stop word removal, and stemming [3]. Tokenization fragments the document into word pairs [4]. Stop word removal is an optional step which is applied to the result of tokenization where the words whose

N. Karthika (✉) · B. Janet
Department of Computer Applications, National Institute of Technology, Tiruchirappalli, Tamil Nadu, India
e-mail: bharathikarthika@gmail.com

B. Janet
e-mail: janet@nitt.edu

D. Reddy Edla et al. (eds.), *Advances in Machine Learning and Data Science*, Advances in Intelligent Systems and Computing 705,
https://doi.org/10.1007/978-981-10-8569-7_19

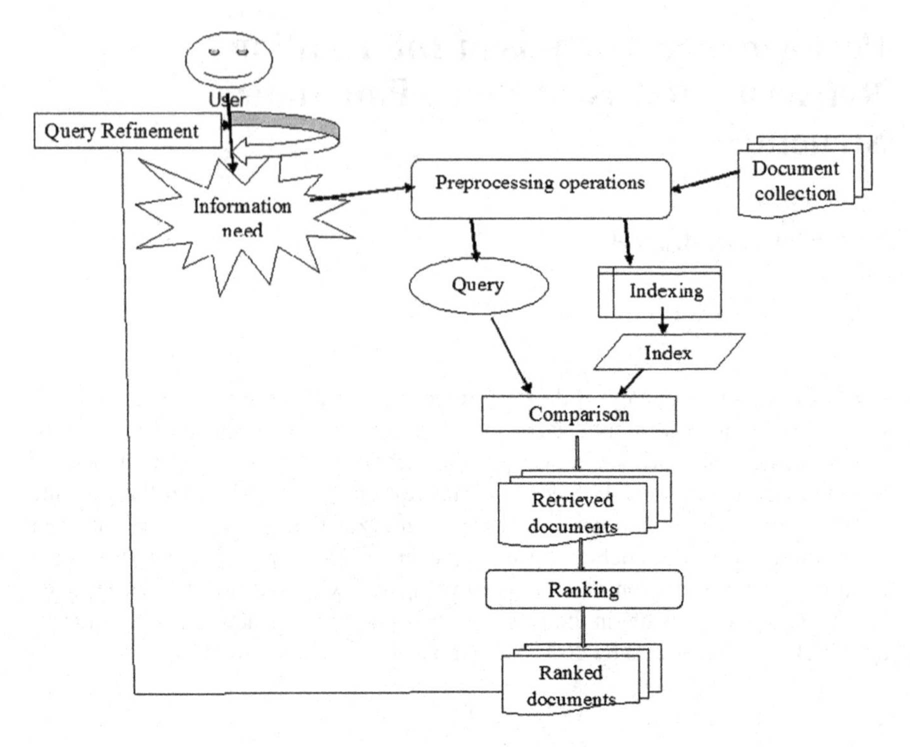

Fig. 1 Information retrieval process

contribution is very less to the meaning of the document and occurs very frequently in the document is eliminated. Then, stemming is used to normalize the word, i.e., it gives the grammatical root of word and reduces the number of words.

Comparison module fetches the documents that matches with given query. Ranking module grades the documents which are retrieved corresponding to a retrieval model based on relevance.

If the user is not satisfied with the retrieved results of documents, then he/she will refine the query accordingly to improve the retrieval performance [2].

This paper is organized as follows. The former works are summarized in Sect. 2. Various text indexing structures are explained in Sect. 3. Variety of information models are discussed in Sect. 4. The measures used in performance evaluation are described in Sect. 5. Experimental setup and results are discussed in Sect. 6. This paper analyzes the performance evaluation measures for various information retrieval models on the word pair index structure implemented with Terrier 3.5. The following are the contributions of the paper.

- For inverted word pair index structure BM25,IFB2, and TF–IDF retrieval models performed better than other retrieval models.

- It is found that there is an increase in precision of 18% when compared with traditional inverted index structure and the recall is 8% in the inverted word pair index structure.
- The mean average precision is increased by 26% and the R-precision is increased by 20%.

2 Related Works

Many statistical approaches for retrieval models have been developed, including [5–7]. According to Harter, terms are of different types [8]. They are speciality terms and non-speciality terms. Speciality terms occur frequently in documents and they contribute toward the meaning of information content. Non-speciality terms are ones whose contribution is less and they are not contributing toward the content of the documents. Both of them follow a Poisson distribution. Then it was fine-tuned by Robertson et.al and promoted by Robertson and Walker as a chain of productive implementations named BMs (eg., BM25) [9]. Later, a matured probabilistic model termed divergence-from-randomness (DFR) was used [5]. The general issues like efficiency, computational cost, and benefits of the information retrieval models were discussed in [10]. The partial and exact matching models are interdependent, and they are joined together to get better results [11]. According to the needs of the user or the application, information retrieval models have to be selected [12].

Inverted index structure is the best for evaluating Boolean queries that are ranked on huge collections. The inverted files outperform in many aspects like space, speed, and functionality in full text indexing [13]. Instead of a term indexing, phrase-based indexing achieved improved retrieval results. The phrase-based systems are precision-enhancing systems in terms of both indexing and retrieval of documents in a collection [14].

Word pair index was implemented in small document collection [15]. Word pair index was implemented on FIRE data set and found that the retrieval speed is increased [4]. This word pair index structure has not been analyzed on various retrieval models. In this paper, the analysis over various information retrieval models on word pair index structure is done using Terrier 3.5.

3 Text Indexing Structures

3.1 Inverted Index Structure

Inverted index is used in most of the search engines [13]. It has two parts. They are vocabulary and postings list. Vocabulary consists of all the distinct terms which are being indexed. Each indexed term is associated with the posting list, which has identifier of the documents in the collection. The structure used to implement the inverted index is arrays, B-tree, B+tree, hash tables, suffix tree, and trie.

3.2 Word Pair Index Structure

Retrieval systems have queries which are of words or of a list of words [16]. In word pair indexing structure, instead of the term, every consecutive pair of terms is treated as term for representation. Compared to inverted index structure, word pair indexing contains small postings list due to the pair of words [15]. It also captures the semantic meanings of text. The queries can be resolved quicker than conventional inverted index structure since it directly captures the postings for the corresponding word pair itself.

4 Information Retrieval Models

The objective of an information retrieval process is to acquire and rank the collection of documents which are relevant to the information needs. The IR system not only finds a single document in the collection but instead identifies multiple documents with various degrees of relevance to the query. To recognize the query, information retrieval system employs a lot of models to allocate/compute a numeric score to the documents. Accordingly, it credits the ranks to the documents and retrieves the top relevant documents from the collection of documents.

4.1 Boolean Model

This model is a very classic traditional model. It is grounded in set theory and Boolean algebra [11]. This model finds the documents which are matched with the query terms exactly. Here, index term weights are considered to be binary, i.e., (0, 1). The query is framed by three logical operators (AND, OR, NOT) linked with index term. This is also known as exact match model.

It is very efficient and precise model. Due to exact matching, it retrieves either too many or very few results. Not all the queries can be easily translated into Boolean expression.

4.2 Vector Space Model

The retrieval of documents is done through partial matches unlike the Boolean match model. Compared to Boolean model, the vector space model gives better results. This model expresses queries and documents as vectors. To find out numeric score between a query and a document, this model uses cosine similarity between query vector and document vector.

$$\cos(\mathbf{q},\mathbf{d}) = sim(\mathbf{q},\mathbf{d}) = \frac{\mathbf{q}\cdot\mathbf{d}}{||\mathbf{q}||\cdot||\mathbf{d}||} \tag{1}$$

where $\mathbf{q}$ is the query vector. $\mathbf{d}$ is the document vector. $||\mathbf{q}||$ and $||\mathbf{d}||$ are length of the query vector and document vector.

In the vector space model, computing weight of the terms that are existing in the query and the documents plays a vital role [10]. Three important elements are to be considered to compute the weight of the terms. They are term frequency t_f (frequency of the term in the document), document frequency d_f (frequency of the term in the collection), the length of the document which contains the term d_l, the number of documents n_d [17].

$$TF\_IDF = Roberston\_tf * idf * K_f \tag{2}$$

where

$\text{Roberston}\_t_f = k_1 * \left[\frac{t_f}{t_f + k_1 * \left(1 - b + \frac{b*d}{dl_{avg}}\right)}\right]$

$idf = \log\left(\frac{n_d}{d_f+1}\right)$

d_l = The length of the document which contains the term

t_f = The frequency of the term in the document

d_f = The frequency of the term in the collection

K_f = The term frequency in the query

n_d = The number of documents

$b = 0.75 \quad k_1 = 1.2$

It produces better recall compared to the Boolean model due to the partial matching and term weighting. The ranking of documents is done with the help of relevance. Semantic sensitivity, i.e., terms with the similar contextual meaning cannot be identified.

4.3 Probabilistic Model

The probability retrieval model is established on the probability ranking principle which ranks the documents based on the probability of relevance to the query [18]. Common models are binary independence model, language models, divergence-from-randomness model [19], and Latent Dirichlet allocation [12]. It provides the partial matching with relevance ranking and queries can be expressed in easier language but the score computation is expensive.

Table 1 gives the formulae for scoring computation of different information retrieval models. The notations used in the probabilistic models are
t = The term.
d = The document in the collection.
q = The query.
t_f = Within document frequency of t in d.
N = The total number of documents in the collection.
df_t = The frequency of the term in the collection.
qt_f = The term frequency in the query.
l = The length of the document which contains the term.
l_{avg} = Average number of terms in a document.
n_t = The document frequency of t.
tfn = The normalized term frequency.
c, b, k_1 = Constants.
F = The term frequency of t in the whole collection.
λ = The variance and mean of a Poisson distribution.

5 Performance Evaluation Measures

The evaluation of an information retrieval system is the mechanism for estimating how robust or healthy a system is to meet the information needs of the users. Information retrieval systems are not only interested in quality of retrieval results (effectiveness) but also interested in quickness of results retrieved (efficiency). The methods which increase effectiveness will have a complementary effect on efficiency because of their interdependent behavior. Naturally, there is an inverse relationship between precision and recall. If the user needs to increment precision, then they have to submit small confined queries in the system, whereas if he wants to increase the recall then the expanded queries may be submitted.

The common parameters employed to compute the efficiency of the retrieval process are recall and precision.

Table 1 Probabilistic models

Retrieval model	Score computation
BM25	$w(t,d) = \sum_{t \in q \cap d} \left(\frac{t_f}{t_f + k_1 \cdot n_b} \cdot log\left(\frac{N - df_t + 0.5}{df_t + 0.5} \right) \cdot qt_f \right)$ $n_b = (1-b) + b \cdot \frac{l}{l_{avg}}$ $k_1 = 1.2 \quad b = 0.75$
BB2	$w(t,d) = \frac{F+1}{n_t\left(tfn+1\right)}\left(-log_2(N-1) - log_2(e) + f(N+F-1, N+F-tfn-2) - f(F, F-tfn)\right)$ $f = \frac{F}{N}$ $tfn = tf \cdot log_2\left(1 + c \cdot \frac{l_{avg}}{l}\right)$
PL2	$w(t,d) = \frac{1}{tfn+1}\left(tfn \cdot log_2 \frac{tfn}{\lambda} + \lambda + \frac{1}{tfn} - tfn\right) \cdot log_2 e + 0.5 + log_2(2\pi \cdot tfn)$ $\lambda = \frac{F}{N}$ $tfn = tf \cdot log_2\left(1 + c \cdot \frac{l_{avg}}{l}\right)$ $c = 1, ByDefault$
InL2	$w(t,d) = \frac{1}{tfn+1}\left(tfn \cdot \log_2 \frac{N+1}{n_t+0.5}\right)$ $tfn = tf \cdot log_2\left(1 + c \cdot \frac{l_{avg}}{l}\right)$ $c = 1$
IFB2	$w(t,d) = \frac{F+1}{n_t \cdot (tfn+1)}\left(tfn \cdot \log_2 \frac{N+1}{F+0.5}\right)$ $tfn = tf \cdot log_2\left(1 + c \cdot \frac{l_{avg}}{l}\right)$ $c = 1$
In(exp)B2	$w(t,d) = \frac{F+1}{n_t \cdot (tfn+1)}\left(tfn \cdot \log_2 \frac{N+1}{n_e+0.5}\right)$ $tfn = tf \cdot log_2\left(1 + c \cdot \frac{l_{avg}}{l}\right)$ $c = 1$
In(exp)C2	$w(t,d) = \frac{F+1}{n_t \cdot (tfn_e+1)}\left(tfn_e \cdot \log_2 \frac{N+1}{n_e+0.5}\right)$ $n_e = N \cdot \left(1 - \left(1 - \frac{nt}{N}\right)F\right)$ $tfn_e = tf \cdot log_e\left(1 + c \cdot \frac{l_{avg}}{l}\right)$ $c = 1$
DPH	$w(t,d) = K_f * N_b * \left(tf * Idf \cdot log\left(\left(\frac{tf * l_{avg}}{l}\right) * \left(\frac{n_d}{d_f}\right)\right) + 0.5 * \mathrm{Idf} \cdot log\left(2\pi * tf(1-f)\right)\right)$ $f = \frac{tf}{l}$ $Idf = log\left(\frac{n_d}{d_f+1}\right)$ $N_b = (1-f) * \left(\frac{1-f}{tf+1}\right)$
Lemur TF-IDF	$w(t,d) = \left(tf_d * tf_q * idf^2\right)$ $tf_d = \frac{k_1 * tf}{tf + k_1 * \left(1 - b + b * \frac{dl}{l_{avg}}\right)}$ $idf = log\left(\frac{n}{N} + 1\right)$ $k_1 = 1.2 \quad b = 0.75$
DLH	$w(t,d) = \frac{1}{tf+0.5} \cdot \left(\log_2\left(\frac{tf \cdot l_{avg}}{l} \cdot \frac{N}{F}\right) + (l - tf)\log_2(1-f) + 0.5\log_2\left(2\pi * tf(1-f)\right)\right)$ $f = \frac{tf}{l}$

5.1 Recall

Recall is the ratio of the number of relevant documents retrieved to the total number of documents relevant. It is an indicator of the exhaustivity of the indexing [1].

$$Recall = \frac{|\,relevant \quad documents \cap retrieved \quad documents\,|}{|\,relevant \quad documents\,|}$$

5.2 Precision

Precision is the ratio of the number of relevant documents retrieved to the total number of documents retrieved. It is thus an indicator of the specificity of the indexing [1].

$$Precision = \frac{|\,relevant \quad documents \cap retrieved \quad documents\,|}{|\,retrieved \quad documents\,|}$$

5.3 Mean Average Precision

In recent days, mean average precision (MAP) is highly recommended to show the quality of retrieval results in a single measure [1]. Average precision is the average of precision value acquired for the collection of top k documents existing after each relevant document is retrieved, and this is average over the queries to gain MAP. Simply MAP for a collection of queries is the mean of average precision scores for each query.

$$MAP = \frac{\sum_{q=1}^{Q} avgP(q)}{Q}$$

where Q is the total number of queries.

6 Results and Discussion

The experiment has been carried out on HP Workstation Z640 which has Intel Xeon E5-2620V3/2.4 GHz processor, 1 TB Hard Drive Capacity, 8 GB RAM for Windows 7 Professional and Java Runtime Environment of version 1.7.0.51 using Terrier 3.5. The Terrier is a suitable platform for carrying out and testing information retrieval tasks. This is an open source available at [20]. Terrier's properties are modified according to our need.

Table 2 Inverted index versus Inverted word pair index

Particulars	Inverted index	Inverted word pair index
Number of tokens	73689988	62157864
Number of terms	330737	7824526
Number of pointers	46945079	58264918
Size of index (GB)	0.22	0.99
Time to index	11 min 8 s	68 min 37 s
Retrieval time (s)	94	5
Query	50 queries	50 queries

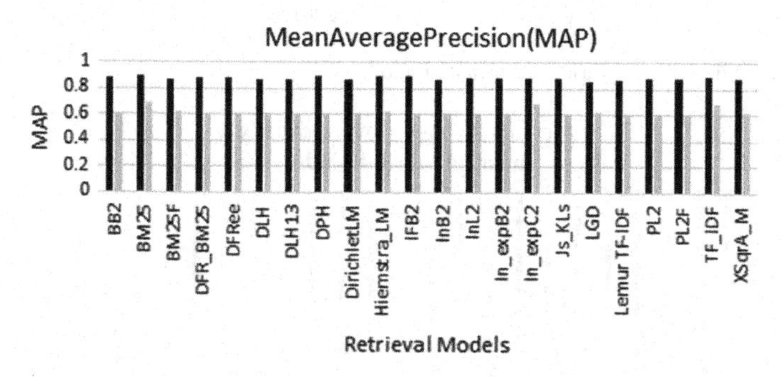

Fig. 2 Comparison of mean average precision between inverted indexing structure and inverted word pair index structure

FIRE data set has been used for testing. FIRE stands Forum for Information Retrieval and Evaluation. FIRE collection maintains the same classic representation of TREC collection. The FIRE 2011 comprises of various documents from newspapers and Websites. It is available and has been downloaded from [21]. Table. 2 shows the comparison between traditional inverted index and inverted word pair index structure for FIRE data set with time to index, retrieval time, and query size.

Figure 2 depicts the mean average precision (MAP) between inverted index structure and inverted word pair index structure.The modified inverted word pair index structure gave better precision value than traditional inverted index structure.The MAP is increased by 26% in inverted word pair index structure. For inverted word pair index structure, BM25,IFB2, and TF–IDF retrieval models performed better than other retrieval models.

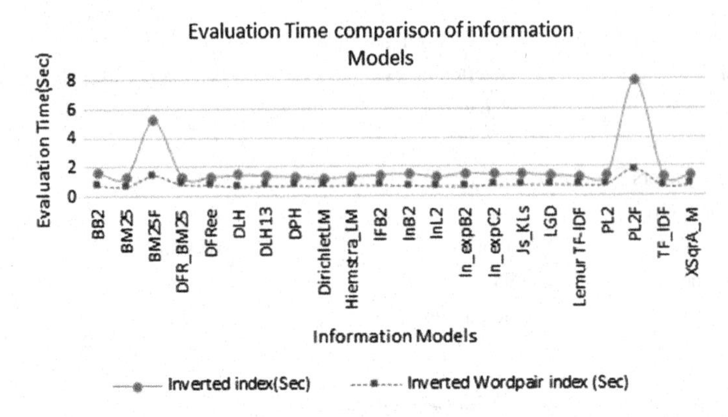

Fig. 3 Evaluation time of various information models

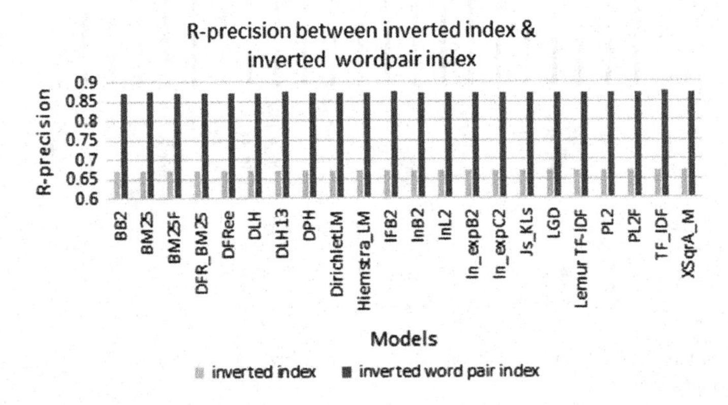

Fig. 4 R-precision comparison of inverted index and inverted word pair indexing structure

Figure 3 shows the time taken by various information models to evaluate the performance. In both the structures, PL2F model takes more time than other models. BM25F takes second largest time than others.

Figure 4 illustrates the R-precision comparison of inverted index and inverted word pair index structure. For inverted index structure, BM25 and IFB2 information retrieval models perform better than others. For inverted word pair index structure, BM25, DLH13, and IFB2 models gave better results than others.

Figures 5 and 6 show the precision–recall curve for BM25 and TF–IDF information retrieval models. The precision–recall curve for inverted word pair index structure appears smoother than the inverted index structure.

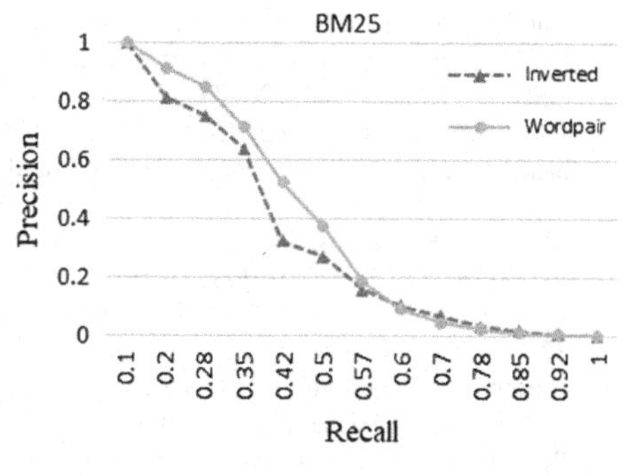

Fig. 5 Precision–recall curve for BM25 Model

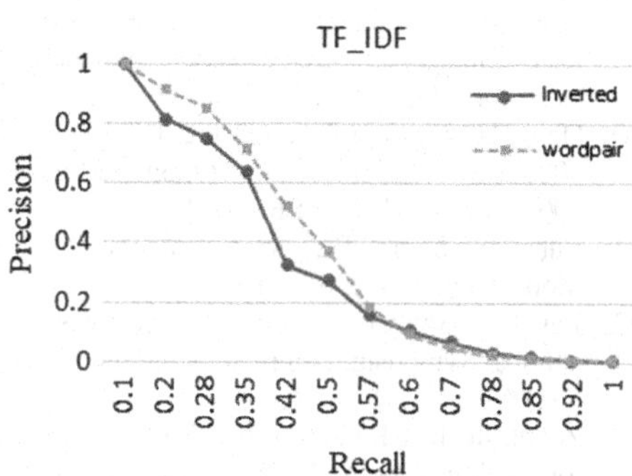

Fig. 6 Precision–recall curve for TF–IDF Model

7 Conclusion

From the results, we come to a conclusion that for inverted word pair index structure BM25,IFB2, and TF–IDF retrieval models performed better than other retrieval models. In the same way, for traditional inverted index structure BM25, In-expC2, TF–IDF retrieval models have better performance than others. It is found that there is an increase in precision of 18% when compared with traditional inverted index structure, and recall is reduced by 8% in the inverted word pair index structure.The mean average precision is increased by 26%, and R-precision is increased by 20%. In future, it can be implemented on big data and the results can be analyzed for various applications.

References

1. Salton, G., McGill, M.J.: Introduction to modern information retrieval. In: McGraw-Hill Computer Science Series. McGraw-Hill (1983)
2. Manning, C.D., Raghavan, P., Schütze, H.: Introduction to Information Retrieval. Cambridge University Press, New York, NY, USA (2008)
3. Porter, M.F.: Readings in information retrieval. In: Chapter An Algorithm for Suffix Stripping, pp. 313–316. Morgan Kaufmann Publishers Inc., San Francisco, CA, USA (1997)

4. Karthika, N., Janet, B.: Word pair index structure for information retrieval using terrier 3.5. In: IEEE Technically Sponsored International Conference on Computational Intelligence on Data Science (ICCIDS) June, 2017 (In Press)
5. Amati, Gianni, Rijsbergen, Cornelis Joost Van: Probabilistic models of information retrieval based on measuring the divergence from randomness. ACM Trans. Inf. Syst. (TOIS) **20**(4), 357–389 (2002)
6. Maron, M.E., Kuhns, J.L.: On relevance, probabilistic indexing and information retrieval. J. ACM (JACM), **7**(3), 216–244 (1960)
7. Robertson, S.E., Jones, K.S.: Relevance weighting of search terms. J. Assoc. Inf. Sci. Technol. **27**(3), 129–146 (1976)
8. Harter, S.P.: A probabilistic approach to automatic keyword indexing. Part ii. An algorithm for probabilistic indexing. J. Assoc. Inf. Sci. Technol. **26**(5), 280–289 (1975)
9. Robertson, S.E., Walker, S.: Some simple effective approximations to the 2-poisson model for probabilistic weighted retrieval. In: Proceedings of the 17th Annual International ACM SIGIR Conference on Research and Development in Information Retrieval, pages 232–241. Springer, New York, Inc. (1994)
10. Dong, H., Hussain, F.K., Chang, E.: A survey in traditional information retrieval models. In: 2008 2nd IEEE International Conference on Digital Ecosystems and Technologies, DEST 2008, pp. 397–402. IEEE (2008)
11. Ruban, S., Sam, S.B., Serrao, L.V., Harshitha.: A Study and Analysis of Information Retrieval Models, pp. 230–236 (2015)
12. Jiang, H.: Study on the performance measure of information retrieval models. In: 2009 International Symposium on Intelligent Ubiquitous Computing and Education, pp. 436–439. IEEE (2009)
13. Zobel, Justin, Moffat, Alistair, Ramamohanarao, Kotagiri: Inverted files versus signature files for text indexing. ACM Trans. Database Syst. (TODS) **23**(4), 453–490 (1998)
14. Mitra, Mandar, Chaudhuri, B.B.: Information retrieval from documents: a survey. Inf. Retr. **2**(2–3), 141–163 (2000)
15. Janet, B., Reddy, A.V.: Wordpair index: a nextword index structure for phrase retrieval. Int. J. Recent Trends Eng. Technol. **3**(2) (2010)
16. Bahle, D., Williams, H.E., Zobel. J.: Efficient phrase querying with an auxiliary index. In: Proceedings of the 25th Annual International ACM SIGIR conference on Research and Development in Information Retrieval, pp. 215–221. ACM (2002)
17. Paik, J.H.: A novel tf-idf weighting scheme for effective ranking. In: Proceedings of the 36th International ACM SIGIR Conference on Research and Development in Information Retrieval, pp. 343–352. ACM (2013)
18. Gasmi, K., Khemakhem, M.T., Jemaa, M.B.: Word indexing versus conceptual indexing in medical image retrieval. In: CLEF (Online Working Notes/Labs/Workshop) (2012)
19. Singh, A., Dey, N., Ashour, A.S., Santhi, V.: Web Semantics for Textual and Visual Information Retrieval, 01 (2017)
20. Retrieval platform. http://ir.dcs.gla.ac.uk/terrier/
21. Fire dataset. http://storm.cis.fordham.edu/~gweiss/data-mining/datasets.html

Fast Fingerprint Retrieval Using Minutiae Neighbor Structure

Ilaiah Kavati, G. Kiran Kumar and Koppula Srinivas Rao

Abstract This paper proposes a novel fingerprint identification system using minutiae neighborhood structure. First, we construct the nearest neighborhood for each minutia in the fingerprint. In the next step, we extract the features such as rotation invariant distances and orientation differences from the neighborhood structure. Then, we use these features to compute the index keys for each fingerprint. During identification of a query, a nearest neighbor algorithm is used to retrieve the best matches. Further, this approach enrolls the new fingerprints dynamically. This approach has been experimented on different benchmark Fingerprint Verification Competition (FVC) databases and the results are promising.

Keywords Minutiae · Quadruplet · Nearest neighbors · Indexing
Identification · Retrieval

1 Introduction

Biometrics is the study of automatic identification of an individual using his/her biological data. Biological data generally refers to human body characteristics such as fingerprint, iris, voice pattern especially for identification purposes [14]. One of the most widely used biometric for identification purpose is fingerprint [10]. The ridge pattern found in fingerprint for every individual is unique. These ridge patterns do not alter in whole lifetime unless some severe accident.

I. Kavati (✉)
Department of CSE, Anurag Group of Institutions, Hyderabad 500038, India
e-mail: kavati089@gmail.com

G. Kiran Kumar · K. Srinivas Rao
Department of CSE, MLR Institute of Technology, Hyderabad 500043, India
e-mail: ganipalli.kiran@gmail.com

K. Srinivas Rao
e-mail: ksreenu2k@gmail.com

D. Reddy Edla et al. (eds.), *Advances in Machine Learning and Data Science*,
Advances in Intelligent Systems and Computing 705,
https://doi.org/10.1007/978-981-10-8569-7_20

Fingerprint recognition is either verification or identification [4]. In verification, the user identifies himself by giving his fingerprint data on the system, and only matching is done with the acquired template with the stored template of his/her claimed identity. It is a 1:1 matching process. In identification, a probe fingerprint image is given and the system has to determine the identity of the candidate from database of stored fingerprint image. A stored fingerprint image, best matching the probe image provides the identity of the input probe. A naive method is to match each fingerprint image with the probe, and the user with highest matching score is considered as identity. A typical fingerprint matching algorithm can perform a matching of two fingerprint image in real time. But the identification for larger database for example, dataset containing million of fingerprint can require a large amount of time. Therefore, this exhaustive approach is impractical and hence requires a fast identification approaches.

This manuscript is organized as follows: The next section explores the recent developments in fast identification of the fingerprints. The proposed methodology is discussed in Sect. 3. Experimental results are explored in Sect. 4. Conclusions are given in Sect. 5.

2 Related Work

One of the solutions is classification of fingerprint by Henry classes (left loop, right loop, arch, tended arch and whorl) [1]. This pattern divides the fingerprint dataset into five different clusters. So, the search is reduced to the relevant class of cluster. But the problem with this approach is that there are only five clusters. Mostly, 90% of the fingerprints are of three classes. Therefore, the search space reduction of possible candidate is not efficient in this approach.

Another solution is fingerprint indexing as explained in [7, 9], which gives better accuracy than classification method. It also gives better search space reduction. In this approach, a preprocessing of the fingerprint is done which includes extraction of minutiae points. These minutiae points are used to get feature vectors that identifies each fingerprint image. Feature vectors are used to form index which are stored in an index table. In retrieving stage, the query probe is used to get the query index and the correspondence between the query index and stored index is used to get the candidate list, reducing the search space to number of image ids in candidate list.

There are several methods [3, 5, 6] and many more for indexing of fingerprint. These diversity is because of the type of feature selected for indexing. Some of them use minutiae triplet [6] for feature extraction. Others use minutiae quadruplet [3]. Few of them use local features or global features. Iloanusi et al. used minutiae quadruplet for feature extraction and k-means algorithm for clustering [2]. In an other approach, Illoanusi [3] used similar method for feature extraction and clustering but they used fusion of different fingerprint type for improving matching accuracy. In an approach, Umarani et al. have used minutiae points but for storing and retrieval strategy they used geometric hashing technique [5].

Praveer et al. proposed an indexing approach in which they have taken minutiae points and stored it in minutiae tree using k-plet approach [11]. Kavati et al. proposed an indexing technique in which they used leader algorithm for clustering which helped in reducing the overload for reclustering whenever a new enrollment is done [8]. Mehrotra and Majhi discussed various tree like k-d-b tree and r-tree for indexing [12]. Somnath et al. proposed three different indexing methods (linear, clustered, and clustered k-d tree) based on minutiae triplets [13].

In this paper, we propose an indexing algorithm that consists of two-stage indexing and retrieving. In indexing, we need to extract features with the help of minutiae quadruplets. Lookup table is formed with the help of these features. In retrieving stage, the query image is used to get the query index and these query indices are used to get the candidate list from the lookup table.

3 Methodology

This method follows these steps: minutiae extraction form fingerprint, neighborhood structure formation, feature extraction and fingerprint enrollment, and finally query retrieval.

3.1 Minutiae Extraction

The minutiae points (ridge bifurcation and ridge end points) extracted for a fingerprint are shown in Fig. 1. Ridge bifurcation is a position where finger ridge line is separated into two or more lines, whereas an ridge end is a point where finger ridge ends abruptly. Let A be a fingerprint image and $MIN = \{m_1, m_2, \dots, m_i, \dots, m_n\}$ be its minutiae set, where m_i is the ith minutiae point of the fingerprint and n is the total number of minutiae points in the fingerprint. Each $m_i = (x_i, y_i, \theta_i)$ contains the detail of a minutiae point where x_i, y_i are the minutiae coordinates, θ_i is the minutiae orientation. We used Nuerotechnology VeriFinger SDK for fingerprint minutiae extraction.

3.2 Minutiae Neighbor Structure

After getting the minutia points, neighborhood structure is calculated for each minutia. In this work, we considered three nearest neighbors for each minutia and form a quadruplet (shown in Fig. 2). In three closest point quadruplet formation, let m_i be a reference minutiae and m_j, m_k, m_l be its three nearest neighbor minutia's in Euclidean space and d_{ij}, d_{ik}, d_{il} are distances between m_i and m_j, m_k, m_l. We consider

Fig. 1 Sample fingerprint and its minutiae

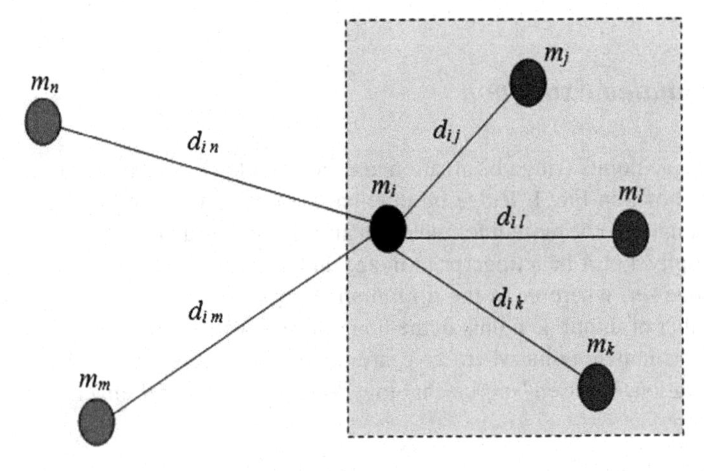

Fig. 2 Three closest minutiae points m_j, m_k, m_l (blue color) for a reference minutiae m_i (black color)

these minutiae for formation of quadruplet $q_i = \{m_i, m_j, m_k, m_l\}$ as shown in Fig. 3. Like this, a quadruplet is computed for each minutia of the fingerprint. Finally, a quadruplet set $Q = \{q_1, q_2, \ldots, q_i, \ldots, q_n\}$ is formed for each fingerprint, where n is total minutiae points of fingerprint image.

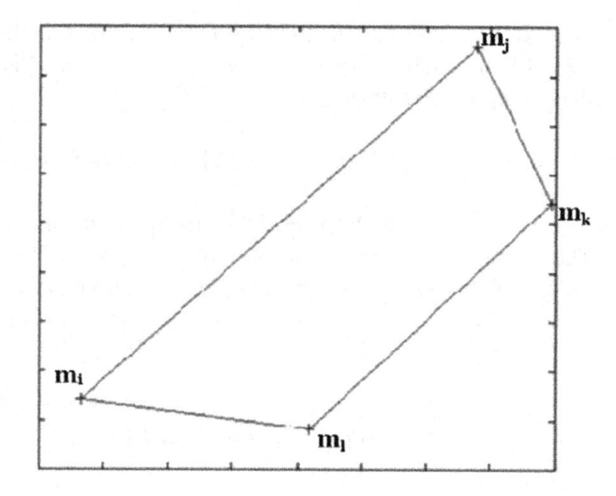

Fig. 3 Quadruplet formed by three closest minutiae points m_j, m_k, m_l and reference minutiae m_i

3.3 Feature Extraction and Fingerprint Enrollment

Now from the computed quadruplet, we extract the geometric features. This is shown in Fig. 4. Let q_i be one of the quadruplet of a fingerprint, and we compute its index $X = \{\phi_1, \phi_2, \delta\}$, where ϕ_1, ϕ_2 are the difference between the opposite angles of the quadruplet, i.e., $\phi_1 = \theta_1 - \theta_3$ and $\phi_2 = \theta_2 - \theta_4$; δ is the average length of two diagonals of the quadruplet, i.e., $\delta = \frac{\delta_1 + \delta_2}{2}$.

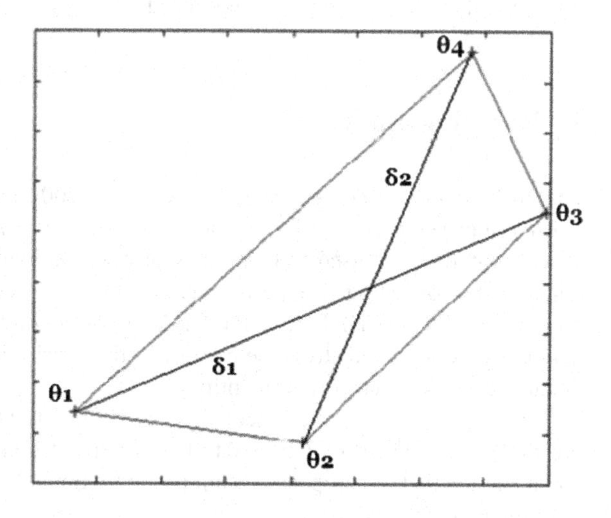

Fig. 4 Quadruplet features

Next, we enroll the quadruplet into a 3D lookup table based on its index I. The size of lookup table is $max(\phi_1) \times max(\phi_2) \times max(\delta)$. Enrolling the quadruplet into the lookup table is shown in Eq. 1.

$$Table[X] = Fingerprint_{id} \tag{1}$$

Likewise other quadruplets of the fingerprint are also mapped to the lookup table. The remaining fingerprints also enrolled likewise. This results many quadruplets may enrolled to the same location of the lookup table as different quadruplets may have similar index and therefore every location may occupy a list of fingerprint ids.

3.4 Query Fingerprint Identification

Query fingerprint identification is the method to identify the suitable candidate from the large dataset. To obtain suitable candidate, we need to generate an index for each quadruplet of the query fingerprint as discussed in Sect. 3.3. Let $X^1 = \{\phi_1, \phi_2, \delta\}$ be the index of a query quadruplet. We use this index X^1 of the query quadruplet to access the lookup table and select the list of ids found there as similar quadruplets. In other words, two quadruplets are said to be similar, if their indexes are same. The retrieved fingerprint ids are stored into a list L. We repeat this process to every quadruplet of the query and retrieve the ids to a list L. Next, we count the occurrences (i.e., Votes) of every id in list L to form a set $\{fingerprint_{id}, Votes\}$. Finally, the fingerprints whose votes are more than a preselected threshold T as similar fingerprints (i.e., candidate set C) to the test fingerprint.

4 Experimental Results

This technique has been experimented on FVC 2002 DB1, DB2, DB3 databases. These databases each consist of 800 fingerprints of 100 different individuals. In experimentation, we used four samples of every individual for training and other four samples for testing. The performance of this technique is evaluated using Penetration Rate (PR) and Hit Rate (HR). The PR is the average search space in the dataset to identify a query; and HR is the fraction of the queries for which the retrieved candidate set contains the correct identity.

The reported results of this technique on different databases are given in Fig. 5. We changed the threshold T from 0 to 100 in steps of 1 and observed the PR and HR. It can be seen that the proposed technique approximately achieves 10% PR at an HR of more than 85% for all the datasets.

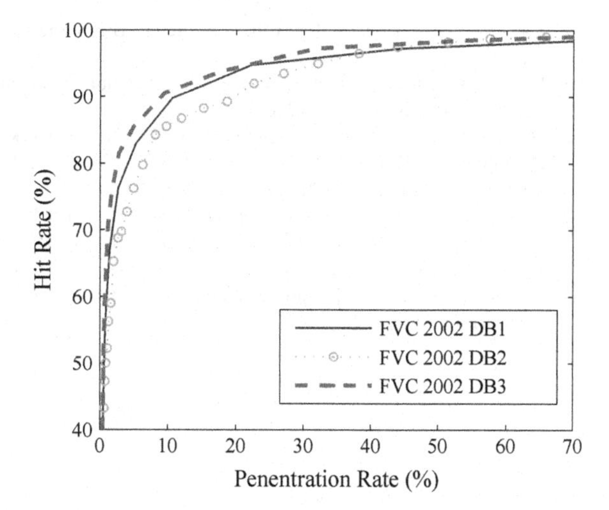

Fig. 5 Retrieval performance of on FVC 2002 databases

5 Conclusion

In this paper, we proposed a novel technique to identify a fingerprint quickly. This method is based on the minutiae neighbor structure called quadruplets and drastically reduced the search space. The use of neighbor minutiae information along with the minutiae made the system more accurate and increases the hit rate. From this, we can conclude that this indexing technique is suitable for fast fingerprint retrieval. In future, we will improve this work to make the system more robust.

References

1. Cappelli, R., Lumini, A., Maio, D., Maltoni, D.: Fingerprint classification by directional image partitioning. IEEE Trans. Pattern Anal. Mach. Intell. **21**(5), 402–421 (1999)
2. Iloanusi, O., Gyaourova, A., Ross, A.: Indexing fingerprints using minutiae quadruplets. In: 2011 IEEE Computer Society Conference on Computer Vision and Pattern Recognition Workshops (CVPRW), pp. 127–133. IEEE (2011)
3. Iloanusi, O.N.: Fusion of finger types for fingerprint indexing using minutiae quadruplets. Pattern Recognit. Lett. **38**, 8–14 (2014)
4. Jain, A.K., Ross, A.A., Nandakumar, K.: Fingerprint recognition. Introd. Biom. 51–96 (2011)
5. Jayaraman, U., Gupta, A.K., Gupta, P.: An efficient minutiae based geometric hashing for fingerprint database. Neurocomputing **137**, 115–126 (2014)
6. Kavati, I., Chenna, V., Prasad, M.V.N.K., Bhagvati, C.: Classification of extended delaunay triangulation for fingerprint indexing. In: 8th Asia Modelling Symposium (AMS), pp. 153–158. IEEE (2014)
7. Kavati, I., Prasad, M.V., Bhagvati, C.: Search space reduction in biometric databases: a review. In: Developing Next-Generation Countermeasures for Homeland Security Threat Prevention p. 236 (2016)
8. Kavati, I., Prasad, M.V., Bhagvati, C.: A clustering-based indexing approach for biometric databases using decision-level fusion. Int. J. Biom. **9**(1), 17–43 (2017)

9. Kavati, I., Prasad, M.V., Bhagvati, C.: Efficient Biometric Indexing and Retrieval Techniques for Large-Scale Systems. Springer (2017)
10. Maltoni, D., Maio, D., Jain, A., Prabhakar, S.: Handbook of Fingerprint Recognition. Springer Science & Business Media (2009)
11. Mansukhani, P., Tulyakov, S., Govindaraju, V.: A framework for efficient fingerprint identification using a minutiae tree. IEEE Syst. J. **4**(2), 126–137 (2010)
12. Mehrotra, H., Majhi, B.: An efficient indexing scheme for iris biometric using kdb trees. In: International Conference on Intelligent Computing, pp. 475–484. Springer (2013)
13. Singh, O.P., Dey, S., Samanta, D.: Fingerprint indexing using minutiae-based invariable set of multidimensional features. Int. J. Biom. **6**(3), 272–303 (2014)
14. Wayman, J., Jain, A., Maltoni, D., Maio, D.: An introduction to biometric authentication systems. Biom. Syst. 1–20 (2005)

Key Leader Analysis in Scientific Collaboration Network Using H-Type Hybrid Measures

Anand Bihari and Sudhakar Tripathi

Abstract In research community, most of the research work is done by the group of researchers and the evaluation of scientific impact of individual is based on either citation-based metrics or centrality measures of social network. But both type of measures have its own impact in scientific evaluation, and the centrality measures are based on number of collaborators and their impact, whereas the citation-based metrics are based on the citation count of articles published by individual. For the evaluation of scientific impact of individual required a hybrid approach of citation-based index and centrality measure of social network analysis. In this article, we have discussed some of the h-type hybrid measures which is the combination of citation-based index and the centrality-based measures for scientific evaluation and find out the prominent leader in scientific collaboration network.

Keywords Social network · Research collaboration · Centrality · h-type hybrid centrality

1 Introduction

In research community, generally evaluation of scientific impact of individual is based on their published articles and their citation count. h-index [1] gives a break-through to evaluate the scientific impact of individual in community based on their articles' citation count, and also, social network analysis metrics plays an important role to evaluate the scientific impact of individual [2] because most of the research work is done by the group of researchers. So the evaluation of scientific impact of individual required a hybrid approach which is the combination of number of collab-

A. Bihari (✉) · S. Tripathi
Department of Computer Science and Engineering, National Institute of Technology Patna, Bihar, India
e-mail: anand.cse15@nitp.ac.in

S. Tripathi
e-mail: stripathi.cse@nitp.ac.in

D. Reddy Edla et al. (eds.), *Advances in Machine Learning and Data Science*, Advances in Intelligent Systems and Computing 705,
https://doi.org/10.1007/978-981-10-8569-7_21

orators as well as citation count of articles published together. Some of the eminent researchers proposed the hybrid approach of citation-based index and social network analysis metrics which is categorized in the following categories: (i) lobby-index [3], al-index [4] and gl-index [4] consider only the degree of neighbour nodes; (ii) h-degree [5], a-degree [4] and g-degree [4] consider the weight of the edge between nodes; (iii) HW-degree, AW-degree and GW-degree [4] consider both neighbour's degree and edge weight.

In this article, first we form scientific collaboration network of researchers and find out prominent leader using h-type hybrid centrality measures. To validate the analysis of h-type hybrid measures and for finding prominent leader in scientific collaboration network, an experimental analysis has been made on 91,546 author's collaboration network.

2 Background

Newman [6] recommended the idea of weighted collaboration network based on the quantity of co-authors. The co-authorship weight between two collaborators is the ratio of total number of publication, which is published together. Then, they use power law to estimate the overall co-authorship weight among researchers. Abbasia et al. [7, 8] build a weighted co-authorship network and used social network analysis metrics for the evaluation of performance of individual researchers. Here, weight is the total number of collaboration between researchers. Bihari et al. [9] discussed the undirected weighted collaboration network of researcher and discovered the prominent researcher based on centrality measures (degree, closeness, betweenness and eigenvector centrality) of social network and citation-based indicators (frequency, citation count, h-index, g-index and i10-index). In this article, author used total citation count earn by the collaborators. Arnaboldi et al. [10] deliberate the co-authorship ego network and study the performance of respectively every node. In this paper, the author used h-index and average amount of authors to evaluate the scientific influence of researchers and furthermore discover the relation among h-index and average number of authors. Bihari et al. [11] in this paper used the maximum spanning tree to remove weak edges from a collaborative network of researchers and used social network centrality measures to appraise the scientific impact of individual and find out the prominent actor in the network of a given data set.

3 Methodology

3.1 Social Network Analysis (SNA)

It maps the social relationship between nodes in the graph and in terms of mathematics and graph theory. SNA is the combination of nodes (e.g. individuals, organization, information) and edges (e.g. friends, trade, financial, sexual) [2, 12, 13]. In collaboration network, the node and edge represent individual researcher and the collaboration, respectively. It provides both visual and mathematical analyses of networks [14, 15].

3.2 h-Type Hybrid Centrality Measures

Lobby-index: The lobby-index [3] of node k is largest integer one such that the node k has at least l neighbour with at least l-degree and the rest of the nodes may have one or less degree.

The lobby-index does not give any extra credit to all those nodes which have more than one degree. To overcome this limitations, [4] proposed the al-degree and gl-index.

al-degree: The al-degree of node k is the average degree of top n neighbours which have at least n degree each.

gl-index: The gl-index of node k is the highest integer g such that the top g nodes have at least g^2 connection (edge) together.

h-degree: The h-degree of a node k is the largest number h such that the node k has at least h neighbours and the strength of the edge between k and neighbour node is at least h and rest of the node or edge may h or less.

But the h-degree does not give any extra credit to all those edges which have more than h-edge weight. To overcome this limitations of h-degree, [4] proposed a-degree and g-degree.

a-degree: The a-degree of node k is the average of top h-edge weights between nodes which have minimum h-degree.

g-degree: The g-degree of node k is the largest number g such that the top g edges have at least g^2 edge weight together.

But all those measures consider either only neighbours' degree or edge weight for the evaluation of nodes. Abbasi [4] combine both degree of the neighbours and edge weight and proposed HW-degree, AW-degree and gW-degree for the evaluation of a node in weighted network. To calculate the Hw-degree, Aw-degree and Gw-degree, first calculate the neighbours' weighted degree. The neighbour's weighted degree is the product of degree of neighbours and the edge weights.

Hw-degree: The Hw-degree of node k is largest integer k such that their k neighbours have at least k neighbour's weighted degree each.

Aw-degree: The Aw-degree of node k is the average of neighbour's weighted degree of top k neighbours which have at least k neighbour's weighted degree.

Gw-degree: The Gw-degree of node k is the largest integer g such that the top g neighbours have at least g^2 neighbour's weighted degree together.

4 Collaboration Network of Research Professionals' an Example

In order to evaluate the scientific impact of individual researcher, the collaboration network has been constructed by using the following technique. Lets an author published an article with 3 other co-authors has collaborated with each other means first author links with second, third and fourth authors, second author links with also first author and with third and fourth author and so on [16, 17]. In this collaboration network, nodes represent the individual author and the edge between nodes represents the collaboration between nodes. This network is considered as a weighted undirected network, where weight is the total number of citation counts earned from all publications published together. Here, the weight of edge shows the collaboration impact of both authors. The network is constructed by using Python and NetworkX [18]. The network is like Fig. 1 (Table 1).

5 Dataset

To create a data set of author's publication, we have used IEEE Xplore with the computer science keyword for the period of Jan-2000 to July-2016. The extracted publication data do not represent the whole world computer science community.

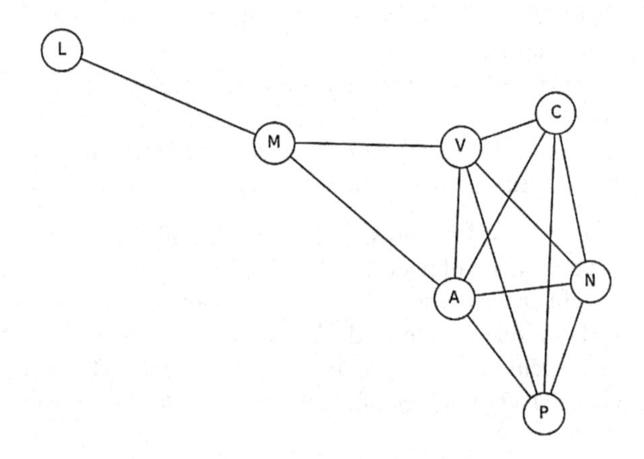

Fig. 1 Research professionals' Collaboration network an example

Table 1 Example of publication and their authors

Paper ID	Author's name	Citation
Pub1	L, M	112
Pub2	M, A, V	92
Pub3	P, A, V, C, N	113

The extracted publication data are in comma separated value (CSV), and it contains several fields such as document title, authors, author's affiliation, journal or conference name, publication year, citation count etc. After that cleaning is performed on raw data and found 126,802 articles and 91,546 authors to conduct our experiments. After analysis of raw data, it seems that 17% publications are written by single author, whereas 41% publications are written by two authors.

6 Analysis and Result

In this section, our primary goal is to evaluate the scientific impact of individual based on h-type hybrid measures such as l-index, al-index, gl-index, h-degree, a-degree and g-degree, Hw-degree, Aw-degree and Gw-degree and try to discover the prominent leader in the network. For this first, we construct the collaboration network of all 91,546 authors and then calculate l-index, al-index, gl-index, h-degree, a-degree and g-degree, Hw-degree, Aw-degree and Gw-degree for all authors'. The l-degree, al-degree and gl-degree consider the collaborators' strength, and h-degree, a-degree and g-degree consider the collaboration strength as well as impact of collaborators. The Hw-degree, Aw-degree and Gw-degree also consider the collaborators' strength as well as impact of collaborator with base node. To analyse the impact of author's, we select top 10 authors from each measures.

The top 10 authors from l-degree, al-degree and gl-degree are based on neighbours' degree shown in Table 2. In this result, it seems that the seven authors (Boehm, D. H.; Nobe, J.; Takahashi, T.; Tanaka, T.; Hatayama, M.; Brunner, B.; Groger, Martin) present in top 10 in all three measures. So we can say that these seven authors are prominent leader in the community based on these three measures.

The results of top 10 authors from Hw-degree, Aw-degree and Gw-degree are shown in Table 3 which are based on the product of edge weight and the degree of the neighbours' node. Here, results show that the seven authors (Bochm, D. H.; Takahashi, T.; Borst, C.; Nobe, J.; Bauml, B.; Seitz, N.; Ortmaier, T.) are present in all three measures in top 10. So, we can say that these seven authors are key actor in the network.

Table 2 Results of top 10 researcher from l-degree, al-degree and gl-degree

Sl. No.	Name of author	l-degree	al-degree	gl-degree
1	Boehm, D.H.	63	80	89
2	Nobe, J.	55	68	68
3	Takahashi, T.	48	44	48
4	Borst, C.	46	22.25	46
5	Bauml, B.	44	20	45
6	Tanaka, T.	42	40.25	42
7	Wolf, S.	35	26.13	43
8	Hatayama, M.	33	36.97	35
9	Seitz, N.	33	28	35
10	Groger, Martin	33	32.25	39
11	Ortmaier, T.	33	25	34
12	Grebenstein, M.	32	19.23	36
13	Friedl, W.	23	52.1	25
14	Konietschke, R.	23	50	25
15	Albu-Schaffer, A.	22	42.40909091	25
16	Brunner, B.	21	63.78	22
17	Ott, C.	10	42.23	12

Table 3 Results of top 10 authors from Hw-degree, Aw-degree and Gw-degree

Sl. No	Name of author	Hw-degree	Aw-degree	Gw-degree
1	Boehm, D.H.	66	62.12	69
2	Takahashi, T.	58	58.13	63
3	Tanaka, T.	49	25.12	50
4	Borst, C.	49	40.25	49
5	Nobe, J.	49	45.96	49
6	Bauml, B.	44	45.26	40
7	Seitz, N.	39	36.89	40
8	Groger, Martin	39	22.36	41
9	Ortmaier, T.	39	41.2	36
10	Wolf, S.	39	39	40
11	Grebenstein, M.	33	36.79	36
12	Hagn, U.	28	36	30

The results of top 10 authors from h-degree, a-degree and g-degree are shown in Table 4 which are based on the edge weight and the degree of the node. Here, results show that the all authors are present in all three measures in top 10.

Table 4 Results of top 10 Researcher from h-degree, a-degree and g-degree

Sl. No.	Name of author	h-degree	a-degree	g-degree
1	Boehm, D.H.	58	45.14	59
2	Nobe, J.	45	40	45
3	Takahashi, T.	44	40	50
4	Bauml, B.	41	30.78	39
5	Borst, C.	40	29	41
6	Tanaka, T.	32	25.25	33
7	Wolf, S.	30	25.14	39
8	Hatayama, M.	28	19	33
9	Seitz, N.	28	19	30
10	Ortmaier, T.	28	18.23	36

Table 5 A comparative results of top 10 authors from all measures

Sl. No.	Name of author	l-degree	al-degree	gl-degree	h-degree	a-degree	g-degree	Hw-degree	Aw-degree	Gw-degree
1	Takahashi, T.	48	44	48	44	40	50	58	58.13	63
2	Boehm, D.H.	63	80	89	58	45.14	59	66	62.12	69
3	Tanaka, T.	42	40.25	42	32	25.25	33	49	25.12	50
4	Borst, C.	46	22.25	46	40	29	41	49	40.25	49

Finally, we made a comparative analysis on top 10 authors from all there different types of h-type hybrid measures and found that only four authors are present in all measures as shown in Table 5. So, we can say that these four authors are key leader in the network.

7 Conclusions

In this paper, we have investigated the research community of research professionals from the h-type hybrid centrality point of view and analyse the impact of author in nine different measures. The experimental analysis shows that the same type of centrality measures gives almost similar results for all authors with little variations. In this contribution, we have formed a collaboration network of 91,546 authors and compute the impact of all authors individually based on nine different types of h-type hybrid centrality measures. To analyse the impact of author, we consider top 10 authors from each measure and found that only four authors are present in all measure. These four authors are called key leaders in the research community.

References

1. Hirsch, J.E.: An index to quantify an individual's scientific research output. Proc. Natl. Acad. Sci. USA **102**(46), 16569–16572 (2005)
2. Wasserman, S., Faust, K.: Social Network Analysis: Methods and Applications, vol. 8. Cambridge University Press (1994)
3. Korn, A., Schubert, A., Telcs, A.: Lobby index in networks. Phys. A: Stat. Mech. Appl. **388**(11), 2221–2226 (2009)
4. Abbasi, A.: h-type hybrid centrality measures for weighted networks. Scientometrics **96**(2), 633–640 (2013). https://doi.org/10.1007/s11192-013-0959-y
5. Zhao, S.X., Rousseau, R., Fred, Y.Y.: h-degree as a basic measure in weighted networks. J. Inform. **5**(4), 668–677 (2011)
6. Newman, M.E.: Scientific collaboration networks. I. Network construction and fundamental results. Phys. Rev. E **64**(1), 016131 (2001)
7. Abbasi, A., Altmann, J.: On the correlation between research performance and social network analysis measures applied to research collaboration networks. In: 44th Hawaii International Conference on System Sciences (HICSS), 2011, pp. 1–10. IEEE (2011)
8. Abbasi, A., Hossain, L., Uddin, S., Rasmussen, K.J.: Evolutionary dynamics of scientific collaboration networks: multi-levels and cross-time analysis. Scientometrics **89**(2), 687–710 (2011)
9. Bihari, A., Pandia, M.K.: Key author analysis in research professionals relationship network using citation indices and centrality. Proc. Comput. Sci. **57**, 606–613 (2015)
10. Arnaboldi, V., Dunbar, R.I., Passarella, A., Conti, M.: Analysis of co-authorship ego networks. In: Advances in Network Science, pp. 82–96. Springer (2016)
11. Bihari, A., Tripathi, S., Pandia, M.K.: Key author analysis in research professionals' collaboration network based on MST using centrality measures. In: Proceedings of the Second International Conference on Information and Communication Technology for Competitive Strategies, p. 118. ACM (2016)
12. Umadevi, V.: Automatic co-authorship network extraction and discovery of central authors. Int. J. Comput. Appl. **74**(4), 1–6 (2013)
13. Jin, J., Xu, K., Xiong, N., Liu, Y., Li, G.: Multi-index evaluation algorithm based on principal component analysis for node importance in complex networks. IET Netw. **1**(3), 108–115 (2012)
14. Liu, B.: Web Data Mining. Springer (2007)
15. Said, Y.H., Wegman, E.J., Sharabati, W.K., Rigsby, J.T.: Retracted: social networks of author-coauthor relationships. Comput. Stat. Data Anal. **52**(4), 2177–2184 (2008)
16. Wang, B., Yao, X.: To form a smaller world in the research realm of hierarchical decision models. In: 2011 IEEE International Conference on Industrial Engineering and Engineering Management (IEEM), pp. 1784–1788. IEEE (2011)
17. Wang, B., Yang, J.: To form a smaller world in the research realm of hierarchical decision models. In: Proceedings of PICMET'11, PICMET (2011)
18. Swart, P.J., Schult, D.A., Hagberg, A.A.: Exploring network structure, dynamics, and function using NetworkX. In: Proceedings of the 7th Python in Science Conference (SciPy 2008)

A Graph-Based Method for Clustering of Gene Expression Data with Detection of Functionally Inactive Genes and Noise

Girish Chandra, Akshay Deepak and Sudhakar Tripathi

Abstract Noise that presents in gene expression data creates trouble in clustering for many clustering algorithms, and it is also observed that some non-functional genes may be present in the gene expression data that should not be the part of any cluster. A solution of this problem first removes the functionally inactive genes or noise and then clusters the remaining genes. Based on this solution, a graph-based clustering algorithm is proposed in this article which first identified the functionally inactive genes or noise and after that clustered the remaining genes of gene expression data. The proposed method is applied to a cell cycle data of yeast, and the results show that it performs well in identification of highly co-expressed gene clusters in the presence of functionally inactive genes and noise.

Keywords Clustering · Gene expression data · Data mining

1 Introduction

Gene expression data contain expression levels of thousands of genes measured simultaneously in different experiment conditions. Clustering is the primary task for the analysis and identification of co-expressed genes of the gene expression data that have a high degree chance of co-regulation [1, 2]. Clustering is the task of grouping objects in such a way that similar objects are put into the same group. The similarity or dissimilarity between objects is measured by the distance between them [3].

G. Chandra (✉) · A. Deepak · S. Tripathi
Department of Computer Science and Engineering, National Institute of Technology Patna, Bihar, India
e-mail: gcchandra440@gmail.com

A. Deepak
e-mail: akshayd@nitp.ac.in

S. Tripathi
e-mail: stripathi.cse@nitp.ac.in

D. Reddy Edla et al. (eds.), *Advances in Machine Learning and Data Science*, Advances in Intelligent Systems and Computing 705,
https://doi.org/10.1007/978-981-10-8569-7_22

Due to the complex procedure of experiments to measure gene expression level, inherent noise comes with these data. This noise creates trouble for many clustering algorithms. To improve the performance of clustering algorithms, these noises are removed from the data set. For the task, filtering is an important step in removing the noise before clustering the data [4]. However, it increases the time complexity of the overall process. In addition, it has been also observed that few of the genes are not involved in any biological functions [1, 2]. That is, these genes are not a part of any clusters. Therefore, it is necessary that few of the genes must be segregated to not being the part of clusters. Many existing clustering algorithms such as k-means [5], fuzzy c-means [6], SOM [7], hierarchical agglomerative clustering algorithm [8], CLICK [9], CAST [10] and SiMM-TS [11] ignore this problem and focus only to assign all the genes to the clusters.

Analysis of the gene expression data suggests that the distance between a noisy gene with the other genes is greater compared to distance between genes of a cluster. In addition, intracluster gene distances are smaller in comparison of intercluster gene distances. Based on these observations, a distance threshold t is defined which can play an important role in the separation of clusters and detection of noise or non-functional genes.

To overcome the aforementioned issue, a graph-based clustering method is proposed in this article, which clusters the gene expression data and also detects the functionally inactive genes and noise. First a graph is created from the data, and the edges having weight greater than threshold t are removed. After that, a vertex with degree 0 is marked as functionally inactive or noise, and remaining genes are clustered by assigning highest degree vertex as cluster centre. To validate the proposed method, we implemented on a real gene expression data and validate using four different validation indices and also compared with k-means, SOM and hierarchical algorithms.

The paper is divided into four sections. After the Introduction, Sect. 2 contains a detailed description of the proposed method. Next Sect. 3 discussed implementation of the proposed method and result of the method on a data set. After that, Sect. 4 contains the conclusion of the whole discussion

2 Proposed Approach

The proposed method has three steps: the first step is creating a graph from gene expression data, the second step is removing the scattered noise and functionally inactive genes from the data set and the third step is clustering of the remaining genes.

Let a gene expression data $GE = \{W_{ij} | 1 \leq i \leq n; 1 \leq j \leq m\}$ contains n genes and m experiment conditions. The graph of the gene expression data is G(V,E), where $V = \{g_1, g_2, \ldots, g_n\}$ and $E = \{d(g_i, g_j) | 1 \leq i \leq n; 1 \leq j \leq n; i \neq j\}$. The vertices of this graph are the genes of the gene expression data set, and the weight of an edge, $d(g_i, g_j)$, between a pair of genes g_i and g_j is the Euclidean distance between these

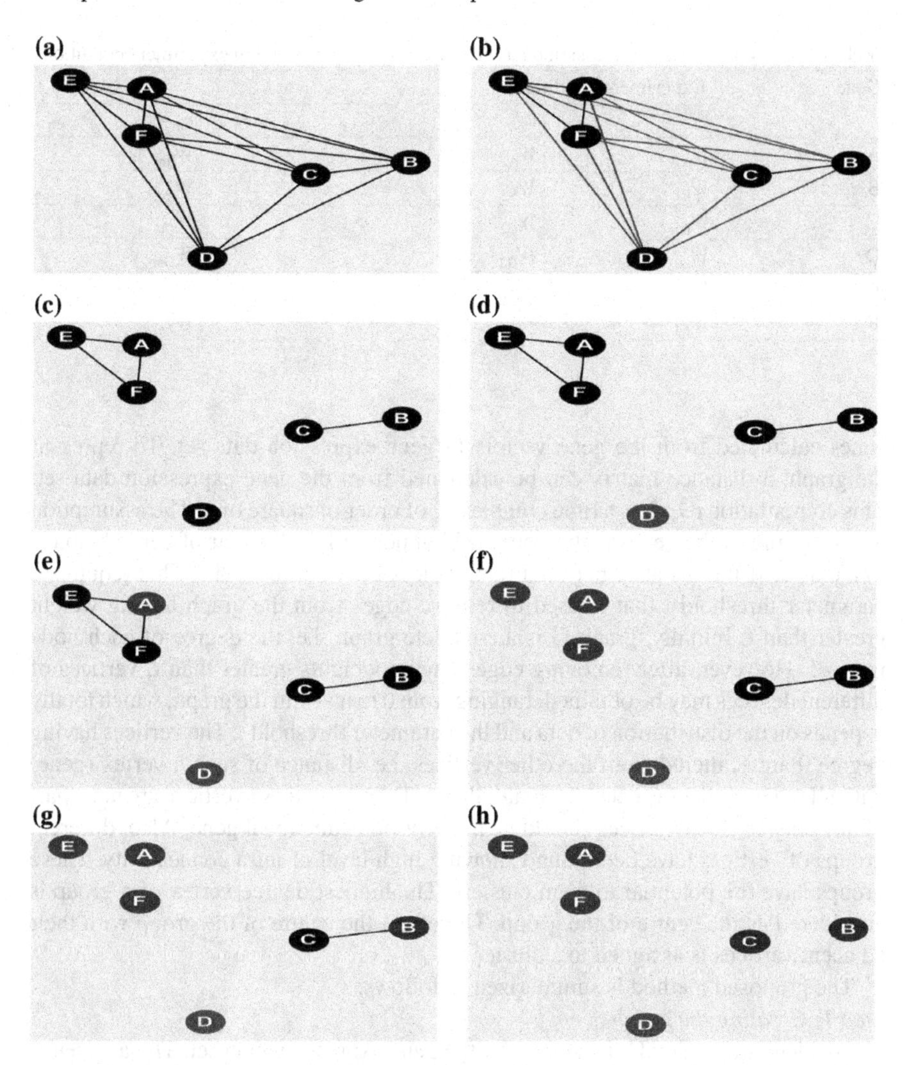

Fig. 1 Proposed approach applied on a six sample genes of Table 1. **a** Initial complete graph. **b** Edges having weights greater than threshold t are shown in red colour. **c** After removing edges having weights greater than threshold t. **d** gene D has degree 0 and is marked as noise or functionally inactive (shown in red colour). **e** Highest degree gene, i.e. gene A (degree is 2) is selected. **f** All the genes adjacent to gene A are put in a cluster with A (A, E, F are the members of the cluster) and edges attached to these genes are removed. **g** Next, select highest degree gene, i.e. B (degree is 1). **h** Genes adjacent to gene B are put in a new cluster with B and remove the edges. Finally, after completing the process we get two clusters and a noise

Table 1 A sample of gene expression data set containing six genes and m experiment conditions

Gene	Experiment condition			
	e_1	e_2	...	e_m
A	W_{A1}	W_{A2}	...	W_{Am}
B	W_{B1}	W_{B2}	...	W_{Bm}
C	W_{C1}	W_{C2}	...	W_{Cm}
D	W_{D1}	W_{D2}	...	W_{Dm}
E	W_{E1}	W_{E2}	...	W_{Em}
F	W_{F1}	W_{F2}	...	W_{Fm}

genes calculated from the gene vectors of gene expression data set. To represent the graph, a distance matrix can be calculated from the gene expression data set. This computation requires a time complexity of order of square of n. The assumption for clustering of the genes is that there is high possibility of a pair of genes belongs to a cluster if the distance between them is less than a threshold t. This requires a parameter threshold t that is used to remove edges from the graph having weight greater than t. Initially, graph G is a complete graph, i.e. the degree of each node is n – 1. However, after removing edges having weights greater than t, vertices of different degrees may be obtained, ranging from 0 to n – 1 in the graph, which totally depends on the distribution of data and the parameter threshold t. The vertices having degree 0 are scattered from the other vertices, i.e. distance of such a vertex (gene) with all other vertices (genes) is greater than t. Therefore, these vertices are not a part of any cluster. Such vertices should be noise or non-functional gene. Also, different groups of vertices have been found showing high level of intra connectivity. These groups have the potential to form clusters. The highest degree vertex of a group is considered as the centre of the group. Therefore, the centre of the group with their adjacent vertices is assigned to a cluster.

The proposed method is summarized as follows:

Step 1: Creating the graph

A complete graph G(V,E) is created for the gene expression data set. The adjacency matrix of this graph is represented by a $n \times n$ distance matrix, computed from the gene expression data set, in which an element (the weight of an edge) at row i and column j is the Euclidean distance $d(g_i, g_j)$ between genes g_i and g_j.

Step 2: Removing the scattered noise and functionally inactive genes

The edges having weight greater than a threshold t are removed from the graph. After that, the vertices having degree 0 are marked as noise or functionally inactive.

Step 3: Clustering of genes

The highest degree vertex vmax , considering the centre, with their adjacent vertices is assigned to a new cluster. After that, the edges connected to these clustered vertices are removed from the graph. Therefore, the degree of these clustered vertices becomes zero. These aforementioned steps are repeated till each vertex of the graph having degree greater than 0. At the end of the iteration, the degree of each vertex of the graph becomes zero and different clusters and noise are obtained.

An example of a data set having six genes is given in Fig. 1. The step-by-step implementation and output of the proposed approach after applying on the data set are shown in the pictures.

3 Implementation and Analysis

This section contains detailed description of the gene expression data and cluster validation indices. The result of the proposed approach applied to the data is evaluated, and the performance of the proposed method is compared with k-means, hierarchical and SOM.

3.1 Data Set

The proposed method is applied to a gene expression data which is subset of cell cycle data set of yeast that contains 384 genes [12]. Originally, the data contain expression levels of 6220 genes. The expression levels of genes are observed and measured at 17-time points, with an interval of 10 min. Only a set of 416 genes is identified that shows significant change during the experiment [13, 14]. The expression levels of the data are normalized such that mean and variance become 0 and 1, respectively [7]. The yeast cell cycle data containing 384 genes is available on the link http://faculty.washington.edu/kayee/cluster.

3.2 Cluster Validation Index

For evaluation and validation of proposed method four cluster validation indices, Dunn Index [15, 16], Dynamic Validity Index (DVIndex) [17, 18], Silhouette Index [16, 19] and Connectivity [16] have been used.

Dunn Index: It is the ratio of the minimum separation in the centre of two clusters and maximum separation in two objects of the same cluster. It is defined as

$$Dunn = min_{1 \leq i \leq n, 1 \leq j \leq n, i \nleq j} \left(\frac{d(i,j)}{max_{1 \leq k \leq n} d'(k)} \right) \tag{1}$$

where n indicates the total number of clusters, d(i, j) is the separation in the centre of clusters i and cluster j and d′(k) is the maximum separation distance between two elements of cluster k.

Dynamic Validity Index (DVIndex): It is defined as

$$min_{k=1,\ldots,K} IntraRatio(k) + y \times InterRatio(k) \tag{2}$$

where IntraRatio represents the overall scaled compactness of clusters and InterRatio represents the overall scaled separateness of clusters. Here, K is the upper bound assumption of number of clusters and y is the modulating parameter that is used to balance the importance between the IntraRatio and InterRatio. The value of DVInex should be minimized.
Silhouette: Silhouette is the average Silhouette value of each object. Silhouette is defined as

$$s(i) = \frac{b_i - a_i}{max(b_i, a_i)} \tag{3}$$

where a_i is the average distance of all objects of the cluster, in which object i belongs, with object i and b_i is the average distance of object i with objects of the nearest neighbour clusters.
Connectivity: Connectivity indicates how the neighbours of each object are clustered. It is defined as

$$conn = \sum_{i=1}^{n} \sum_{j=1}^{k} Z_{ij} \tag{4}$$

where n is the number of objects of the data set. If jth nearest neighbour of object i is the member of the cluster in which object i belongs, then z_{ij} is 0, otherwise 1/j. k is the nearest neighbour parameter.

3.3 Implementation Platform

To implement the proposed method, R package is used. For calculation of cluster validation indices, two R packages, clValid [16] and clv [17], are used.

3.4 Result

The proposed approach is applied to gene expression data of yeast cell cycle with different threshold parameter t from 0.1 to 8 with an interval of 0.1. However, it only identifies clusters from 0.8 to 7. The values of Dunn, Connectivity and Silhouette indices are shown in Table 2.

The observation from Table 2 is as follows:

- Dunn index value is greater for smaller threshold t in comparison to large threshold t. This indicates that cluster quality is higher at smaller t. Dunn index at

Table 2 Result of proposed approach on cell cycle data of yeast

Threshold t	Dunn	DVIndex	Connectivity	Silhouette	Number of clusters	Identified genes
0.8	1.252174	0.694638	27.43452	0.578104	7	15
0.9	0.860789	1.006744	40.11627	0.468389	10	23
1	0.535894	1.009343	58.125	0.376179	14	36
1.1	0.331553	0.827565	66.20119	0.313003	14	44
1.2	0.360941	0.728225	90.41984	0.265537	18	61
1.3	0.3592	0.616198	113.5353	0.232523	21	80
1.4	0.288298	0.630968	162.7056	0.189627	30	115
1.5	0.288298	0.646462	189.554	0.192762	35	141
1.6	0.331712	0.638321	192.4754	0.183077	35	162
1.7	0.315199	0.57713	230.9552	0.156417	42	192
1.8	0.205216	0.611229	256.1718	0.132744	41	213
1.9	0.188099	0.616078	265.1492	0.123468	40	241
2	0.35	0.586568	257.673	0.140994	41	261
2.1	0.343388	0.61282	291.1738	0.105229	41	283
2.2	0.295224	0.611579	262.1302	0.123818	34	285
2.3	0.226595	0.670567	269.3119	0.116486	34	300
2.4	0.269775	0.64032	239.6317	0.121789	31	308
2.5	0.219216	0.711549	276.0063	0.090563	35	329
2.6	0.252841	0.700745	233.0421	0.130163	25	335
2.7	0.199611	0.689685	244.2746	0.109662	29	344
2.8	0.194197	0.702204	228.1893	0.118723	24	348
2.9	0.187897	0.675215	221.8067	0.129448	24	354
3	0.22898	0.736265	210.6333	0.12396	20	353
3.1	0.182377	0.8125	254.5393	0.070089	26	370
3.2	0.176245	0.800789	227.7627	0.097515	21	373
3.3	0.225102	0.908551	189.5639	0.123752	18	370
3.4	0.223016	0.801958	156.5905	0.140225	16	371
3.5	0.155368	0.802181	183.5675	0.106357	14	373
3.6	0.151567	0.800307	179.2004	0.141286	12	377
3.7	0.151567	0.829722	197.8817	0.135456	11	378
3.8	0.148584	0.829412	204.873	0.071495	14	381
3.9	0.172898	0.881599	178.1417	0.053904	8	378
4	0.121996	0.822822	160.7349	0.178207	7	378
4.1	0.209045	0.86933	156.2694	0.124317	7	377
4.2	0.199526	0.844268	146.3373	0.11611	6	377

(continued)

Table 2 (continued)

Threshold t	Dunn	DVIndex	Connectivity	Silhouette	Number of clusters	Identified genes
4.3	0.197169	0.830101	129.5294	0.205077	5	379
4.4	0.195311	0.858611	151.6298	0.177224	5	380
4.5	0.133541	0.907983	146.3603	0.100818	6	383
4.6	0.156092	0.928689	125.9488	0.1053	6	384
4.7	0.214571	0.95575	114.3111	0.097556	6	384
4.8	0.167744	0.89237	134.1099	0.168967	5	384
4.9	0.182598	0.924505	106.3456	0.12978	5	383
5	0.202297	0.915762	99.57302	0.18744	4	384
5.1	0.179697	0.975714	108.8667	0.07145	5	384
5.2	0.191415	1.005459	99.00159	0.04334	5	384
5.3	0.081456	0.980374	105.1353	0.218711	3	382
5.4	0.080207	0.990626	98.74365	0.212818	3	383
5.5	0.189745	1.013886	109.0381	0.082138	4	384
5.6	0.157459	1.08174	128.3607	0.149505	3	384
5.7	0.151576	1.123154	150.2738	0.105666	3	384
5.8	0.150824	1.096008	153.0893	0.115131	3	384
5.9	0.14394	1.103633	154.2452	0.085285	3	384
6	0.157818	1.149132	114.4377	0.158357	2	383
6.1	0.133609	1.173131	101.0087	0.144285	2	383
6.2	0.193753	1.185812	71.96825	0.138866	2	384
6.3	0.193753	1.191782	65.04246	0.136945	2	384
6.4	0.193753	1.194994	55.76429	0.134822	2	384
6.5	0.142378	1.230938	59.93849	0.066488	2	384
6.6	0.142378	1.250312	52.92421	0.05885	2	384
6.7	0.246495	1.188541	24.49246	0.137392	2	384
6.8	0.20858	1.228704	26.025	0.070837	2	384
6.9	0.282311	1.21935	12.93651	0.037813	2	384
7	0.313201	1.197755	6.780159	0.071815	2	384

t = 0.8 is 1.252174 which is highest, and it shows best cluster is obtained. However, the number of genes identified in clusters, i.e. 15, is very small which is not desirable. Therefore, we consider some more clusters of moderate value of Dunn index where number of genes identified in clusters is significant. One of the Dunn index values is 0.35 at threshold t = 2 where 216 genes are identified.

- According to DVIndex, the best clusters are obtained at threshold t = 1.7 where its value, that is 0.57713, is minimum and 42 clusters are found containing all 192 genes.

Table 3 Result of k-means, hierarchical and SOM on cell cycle data of yeast

	k-means		Hierarchical		SOM	
	Value	Cluster	Value	Cluster	Value	Cluster
Dunn	0.2431213	9	0.2770405	30	0.2117015	5
Connectivity	51.646428	3	28.403571	2	40.983310	3
Silhouette	0.3261063	5	0.3080457	5	0.2901607	5
DVIndex	1.00292	3	1.070709	4	1.092031	4

- According to Connectivity, the best clusters are obtained at threshold $t = 7$ where its value, that is 6.780159, is minimum and two clusters are found containing all 384 genes.
- Silhouette index shows similar behaviour to Dunn index. Its value, i.e. 6.78015, is highest at $t = 0.8$ and indicates that best clusters are obtained. Similar to Dunn index, we can consider some more clusters with moderate Silhouette value with significant number of genes obtained in clusters.

The k-means, SOM and hierarchical algorithms are also applied on cell cycle data of yeast. The best clusters obtained by these algorithms that are indicated by an optimal score of Dunn, Connectivity and Silhouette indices shown in Table 3.

Comparison of the proposed method with k-means, SOM and hierarchical algorithms is as follows:

- The Dunn index values of the proposed method are greater for threshold t at 0.8 to 1.7, 2.2, 6.9 and 7 in comparison to optimal Dunn index values of k-means, SOM and hierarchical.
- The DVIndex values of the proposed method are smaller for threshold t at 0.8, 1.1 to 5.1, 5.3 and 5.4 in comparison to optimal DVIndex values of k-means, SOM and hierarchical.
- The Connectivity values of proposed method are smaller for threshold t at 0.8, 0.9, 6.7, 6.8, 6.9 and 7 in comparison to optimal Connectivity values of k-means, SOM and hierarchical.
- The Silhouette values of proposed method are greater for threshold t at 0.8 ,0.9 and 1 in comparison to optimal Silhouette values of k-means, SOM and hierarchical.

Thus, all four indices indicate that proposed method performs better than these algorithms.

4 Conclusions

In this paper, a graph-based method is proposed for clustering of gene expression data in the presence of functionally inactive genes and noise. The proposed method is applied to a gene expression data, and result shows it identifies highly co-expressed gene clusters. To evaluate and validate the results, four cluster validation indices are used. The proposed method is compared with k-means, hierarchical and SOM. The validation indices show that the proposed method performs significantly better than these algorithms.

References

1. Jiang, D., Tang, C., Zhang, A.: Cluster analysis for gene expression data: a survey. IEEE Trans. Knowl. Data Eng. **16**(11), 1370–1386 (2004)
2. Kerr, G., Ruskin, H.J., Crane, M., Doolan, P.: Techniques for clustering gene expression data. Comput. Biol. Med. **38**(3), 283–293 (2008)
3. Young, W.C., Yeung, K.Y., Raftery, A.E.: Model-based clustering with data correction for removing artifacts in gene expression data. arXiv:1602.06316
4. Yun, T., Hwang, T., Cha, K., Yi, G.-S.: Clic: clustering analysis of large microarray datasets with individual dimension-based clustering. Nucleic Acids Res. **38**(suppl 2), W246–W253 (2010)
5. Tavazoie, S., Hughes, J.D., Campbell, M.J., Cho, R.J., Church, G.M.: Systematic determination of genetic network architecture. Nat. Genet. **22**(3), 281–285 (1999)
6. Dembélé, D., Kastner, P.: Fuzzy c-means method for clustering microarray data. Bioinformatics **19**(8), 973–980 (2003)
7. Tamayo, P., Slonim, D., Mesirov, J., Zhu, Q., Kitareewan, S., Dmitrovsky, E., Lander, E.S., Golub, T.R.: Interpreting patterns of gene expression with self-organizing maps: methods and application to hematopoietic differentiation. Proc. Natl. Acad. Sci. **96**(6), 2907–2912 (1999)
8. Eisen, M.B., Spellman, P.T., Brown, P.O., Botstein, D.: Cluster analysis and display of genome-wide expression patterns. Proc. Natl. Acad. Sci. **95**(25), 14863–14868 (1998)
9. Sharan, R., Shamir, R.: Click: a clustering algorithm with applications to gene expression analysis. In: Proceedings of International Conference on Intelligent Systems for Molecular Biology, vol. 8, p. 16 (2000)
10. Ben-Dor, A., Shamir, R., Yakhini, Z.: Clustering gene expression patterns. J. Comput. Biol. **6**(3–4), 281–297 (1999)
11. Bandyopadhyay, S., Mukhopadhyay, A., Maulik, U.: An improved algorithm for clustering gene expression data. Bioinformatics **23**(21), 2859–2865 (2007)
12. Ma, P.C., Chan, K.C.: A novel approach for discovering overlapping clusters in gene expression data. IEEE Trans. Biomed. Eng. **56**(7), 1803–1809 (2009)
13. Yeung, K.Y., Haynor, D.R., Ruzzo, W.L.: Validating clustering for gene expression data. Bioinformatics **17**(4), 309–318 (2001)
14. Cho, R.J., Campbell, M.J., Winzeler, E.A., Steinmetz, L., Conway, A., Wodicka, L., Wolfsberg, T.G., Gabrielian, A.E., Landsman, D., Lockhart, D.J., et al.: A genome-wide transcriptional analysis of the mitotic cell cycle. Mol. Cell **2**(1), 65–73 (1998)
15. Dunn, J.C.: Well-separated clusters and optimal fuzzy partitions. J. Cybern. **4**(1), 95–104 (1974)
16. Brock, G., Pihur, V., Datta, S., Datta, S., et al.: clvalid, an r package for cluster validation. J. Stat. Softw. (2008)
17. Shen, J., Chang, S.I., Lee, E.S., Deng, Y., Brown, S.J.: Determination of cluster number in clustering microarray data. Appl. Math. Comput. **169**(2), 1172–1185 (2005)
18. Hosseininasab, S.M.E., Ershadi, M.J.: Optimization of the number of clusters: a case study on multivariate quality control results of segment installation. Int. J. Adv. Manuf. Technol. 1–7 (2013)
19. Rousseeuw, P.J.: Silhouettes: a graphical aid to the interpretation and validation of cluster analysis. J. Comput. Appl. Math. **20**, 53–65 (1987)

OTAWE-Optimized Topic-Adaptive Word Expansion for Cross Domain Sentiment Classification on Tweets

Savitha Mathapati, Ayesha Nafeesa, S. H. Manjula and K. R. Venugopal

Abstract The enormous growth of Internet usage, number of social interactions, and activities in social networking sites results in users adding their opinions on the products. An automated system, called sentiment classifier, is required to extract the sentiments and opinions from social media data. Classifier that is trained using the labeled tweets of one domain may not efficiently classify the tweets from another domain. This is a basic problem with the tweets as twitter data is very diverse. Therefore, Cross Domain Sentiment Classification is required. In this paper, we propose a semi-supervised domain-adaptive sentiment classifier with Optimized Topic-Adaptive Word Expansion (*OTAWE*) model on tweets. Initially, the classifier is trained on common sentiment words and mixed labeled tweets from various topics. Then, *OTAWE* algorithm selects more reliable unlabeled tweets from a particular domain and updates domain-adaptive words in every iteration. *OTAWE* outperforms existing domain-adaptive algorithms as it saves the feature weights after every iteration. This ensures that moderate sentiment words are not missed out and avoids the inclusion of weak sentiment words.

Keywords Cross Domain Sentiment Classification · Opinion mining and sentiment analysis · SVM classifier · Topic-adaptive features · Tweets

1 Introduction

Social media like Facebook, Twitter, Microblogs is a platform where people build social relations and share their opinion on various topics. People post their views on products to guide others in deciding whether they want to buy or not to buy the product. These reviews help business in many ways. The tweets or reviews

S. Mathapati (✉) · A. Nafeesa · S. H. Manjula · K. R. Venugopal
University Visvesvaraya College of Engineering, Bengaluru, India
e-mail: hiremathsavitha@gmail.com

A. Nafeesa
e-mail: ayeshanafeesa10@gmail.com

D. Reddy Edla et al. (eds.), *Advances in Machine Learning and Data Science*,
Advances in Intelligent Systems and Computing 705,
https://doi.org/10.1007/978-981-10-8569-7_23

posted by the users are voluminous that make it difficult to analyze the complete information. To overcome this problem, we use sentiment analysis, a powerful method of gaining the overview of the public opinion on a particular topic. The different topics can be connected with the same classifier with an assumption that the topics have certain words in common that can effectively be used to compute the overall sentiment. This methodology is more efficient in case of reviews that define the quality of a product. Tweets have more diverse data, and it is sometimes difficult to predict the topics that have been referred in Twitter. Hence, a common classifier built for different tweets may not work efficiently. For example, classifier trained using labeled data or tweets of *Book* domain may not give a good accuracy while classifying *Kitchen* domain tweets. Thus, topic adaptation is required before Sentiment Classification, as the tweets are from mixed topics. This process is called Cross Domain Sentiment Classification.

Motivation: A classifier trained on sentiment data from one topic often performs poorly on the test data from another topic. Shenghua Liu et al. [1] proposed semi-supervised Topic-Adaptive Sentiment Classifier Algorithm. It initially trains a common sentiment classifier for multiple domains and then transforms it into a specific one on an emerging topic or domain in an iterative process. In every iteration, predefined number of topic-adaptive words are added to the list of topic-adaptive words and the remaining are discarded. The drawback of this method is that a few strong topic-adaptive words are missed out if many strong sentiment words are found in that particular iteration. Similarly, weak sentiment words may get selected if not many strong sentiment words are generated in that iteration. Feature weight of each topic-adaptive word that is not selected in an iteration is not saved and when the same words repeat in next iteration, their weight count restarts. This results in missing of few more strong topic-adaptive words. We propose *OTAWE* algorithm that overcomes the problem by carrying the feature weights of the topic-adaptive words to the next iteration and impose threshold on the feature weight of the topic-adaptive words that are selected in every iteration.

Contribution: In this paper, we focus on Optimized Topic-Adaptive Word Expansion *(OTAWE)* algorithm for Cross Domain Sentiment Classification on Tweets. Each sentiment word is assigned weight. In every iteration, sentiment words that pass the predefined threshold weight are added to the topic-adaptive word list. The remaining words are carried to the next iteration along with their weights. This results in saving the strong sentiment words and discarding weak sentiment words for the improvement in classification accuracy.

The rest of the paper is organized as follows. Section 2 analyzes previous related works. Problem definition is stated in Sect. 3. Algorithm for Optimized Topic-Adaptive Word Expansion for Domain Adaptation is explained in Sect. 4. Section 5 gives an overview of the implementation part of the algorithm. Performance analysis with results is analyzed in Sect. 6. Lastly Sect. 7 concludes the paper.

2 Related Work

One of the challenging part of sentiment analysis is the Cross Domain Sentiment Classification. Number of works focused on this topic, and in this section we discuss about it. Spectral feature alignment algorithm is suggested by Pan et al. [2] that used domain-independent words as bridge to align topic-specific words from various domains into unified clusters for domain adaptation. The domain-independent words function as a bridge in this context. The clusters are used to minimize the difference between the domain-specific data that are later used to train the classifier. Li et al. [3] analyzed adaptive bootstrapping algorithm to enhance the sentiment words in target domain by using source domain labeled data.

Twitter data is different from reviews, as it contains data from variety of topics. Twitter data is unpredictable and needs a lot of labeled data to classify each topic. Mejova and Srinivasan [4] have shown that cross media Sentiment Classification can be done by considering news, blogs, and twitter data set. A novel multi-source, multi-topic labeled data set is prepared specifically for comparing sentiment across these social media sources. It is difficult to apply supervised learning on tweets as it lacks sentiment labels. Go et al. [5] introduced a supervised learning applied on tweets with emoticons used for training classifier. Depending on the positive or negative emoticons present in tweets, tweets are labeled and used for training the classifier. Cheng and Pan [6] used semi-supervised domain adaptation for linear transformation from source to target domain. This method can be used in common for all types of loss functions. Semi-supervised Support Vector Machine (SVM) is one of the efficient models that classify data with minimal labeled data and utilize maximum unlabeled data. When features can be easily split into different views, co-training framework achieves good results. Glorot et al. [7] proposed a topic adaptation method that does not require any labeled data. Deep Learning approach is used to extract the topic-adaptive words from the unlabeled data. Bollegala et al. [8] proposed a method of classification in the absence of labeled data in the target domain as well as a small amount of labeled data is available from other domains. This method automatically creates a sentiment sensitive "*thesaurus*" using labeled and unlabeled data from multiple source domains. The constructed "*thesaurus*" is then used to enlarge the feature vectors and to train the classifier.

3 Problem Definition

The objective of the proposed work is to build a Cross Domain Sentiment classifier that adapts itself to classify tweets from different topics or domains. This is achieved using semi-supervised Optimized Topic-Adaptive Word Expansion Algorithm (*OTAWE*). Initially, the classifier is trained using labeled tweets from mixed topics, and then the classifier adapts itself to a particular topic e using the unlabeled tweets iteratively. This algorithm allows only sentiment words that pass the predefined threshold to be added to the topic-adaptive word list. This method avoids weak sentiment words from getting added to the topic-adaptive word list, hence the accuracy of the system increases. In our work, we use only the tweets that pass the confidence threshold and only the text features in the tweets are considered.

4 Optimized Topic-Adaptive Word Expansion *(OTAWE)* for Cross Domain Sentiment Classification for a Topic

In our proposed work, Sentiment Classification is treated as a multi-class classification that classifies tweets into more than two classes (*positive, negative* and *neutral*). The first step of Optimized Topic-Adaptive Word Expansion *(OTAWE)* for Cross Domain Sentiment Classification for a topic e involves building a labeled data set that contributes toward semi-supervised classification. Here, tweets are classified into *positive, negative* and *neutral* classes. The input tweets used as the training set are represented as a pair (x_i, y_i), where x_i is the feature vector for the input data and y_i refers to the sentiment class of x_i. The value of y_i varies between $1 \ldots K$, where K is the maximum number of sentiment classes available in the input data. There are two important steps during the process of creating the labeled data set. The first step involves the creation of multi-class Support Vector Machine (*SVM*) model that classifies the input data set into three sentiment classes (*positive, negative* and *neutral*). The second step is the formulation of the labeled data set as pairs that contain the feature value for each word and its corresponding sentiment class.

4.1 Creation of SVM Model

Support Vector Machine (*SVM*) model is used to classify the input data into its sentiment classes. One strategy of implementing SVM for multi-class problem is by building K *one-versus-rest* classifiers where K is the maximum number of classes considered. This classifier chooses the class that classifies the data with greatest margin. In the *OTAWE* algorithm, the one-versus-rest technique for classification is used. The *SVM* model is implemented using the Eq. (1).

$$\min_{w} \frac{1}{2} \sum_{i=1}^{k} w_i^T w_i + \frac{C}{n} \sum_{i=1}^{n} \max_{y \neq y_i} \{0, 1 - w_{y_i}^T x_i + w_y^T x_i\} \tag{1}$$

$$y_i' = argmax\{w_y^T x_i\} \tag{2}$$

Instead of building multiple binary *SVMs* and processing them to get a single class, we use the result of the Eq. (1) that is a single multi-class *SVM*. Then the class label of any tweet t_i is calculated using the Eq. (2).

4.2 Calculation of Feature Values

The text feature set has two components, (i) common sentiment words and (ii) Topic-Adaptive Sentiment Words. (i) Common sentiment words are the most frequently used sentiment words. Such sentiment word lists are readily downloadable from the Web. The common sentiment words used in the implementation of *OTAWE* algorithm are taken from two sentiment corpuses. *WordNet Affect* is one sentiment corpus that has words labeled as positive and negative. *Public Sentiment Lexicon* is another publicly available corpus that has two separate lists, one labeled as positive and the other negative. Both these corpuses are combined and called as common sentiment word set P. (ii) Topic-Adaptive Sentiment Words: The labeled set also requires some Topic-Adaptive words. Certain tweets are selected from the input corpuses such that they have high confidence and span multiple topics. Part-of-Speech tagging is done for the tweets, and common sentiment words are removed from the resulting set. Only the frequently occurring nouns, verbs, adjectives, and adverbs are selected to be added as topic-adaptive words into the labeled set. This comprises the Topic-Adaptive sentiment word set denoted by Q. The complete labeled set is formed by combining the Common and the Topic-Adaptive Sentiment word sets. The size of the set L containing text-based features $v = |P|$, $u = |Q|$ and labeled set $L = v + u$.

4.3 Adaptive Training

The Adaptive Training transforms the classifier built from the labeled set L to adapt on any given topic e. The classifier starts adapting to the topic, two functions are carried out stepwise on every iteration through the *OTAWE* algorithm. The first step is to train the Classifier. Next, feature extraction and updation is performed. Both these steps help the classifier in drifting toward a single topic.

4.3.1 Training the Classifier

The *OTAWE* algorithm is a semi-supervised learning algorithm. Here, classifier is built using a labeled set and tested using unlabeled data. Addition of unlabeled tweets from topic e into the classifier makes the classifier to adapt to a particular topic e. Therefore, the unlabeled data is added into the *SVM* model of Eq. (1) as an optimization quantity. This term helps the classifier to adapt to the particular topic e. The optimized SVM model after adding the unlabeled tweets is given by Eq. (3). Apart from training, the unlabeled data set U also tests the classifier.

$$\min_{w} \frac{1}{2}\sum_{i=1}^{k} w_i^T w_i + \frac{C}{|L|}\sum_{t_i \epsilon L} \max_{y \neq y_i}\{0, 1 - w_{y_i}^T x_i + w_y^T x_i\} + \frac{C'}{|U|}\sum_{t_j \epsilon U} \max_{y \neq y_j}\{0, 1 - w_{y_j}^T x_j + w_y^T x_j\} \quad (3)$$

The constant co-efficient is taken as C for labeled data and C' for unlabeled data. $|L|$ gives the number of words in the labeled set, and $|U|$ gives the number of words in the unlabeled set. Training the classifier requires unlabeled data. This data is collected by processing the tweets on a topic e. Any random tweets cannot be picked for the initial training of the sentiment classifier as spam or irrelevant input can result the classifier to drift in wrong direction. Hence, the most confident tweets are selected for training the classifier. The confidence of the unlabeled tweets t_j is defined as S_j, and it is obtained by the Eq. (4)

$$S_j = \frac{max_y\{w_y^T x_j\}}{\Sigma_y w_y^T x_j} \tag{4}$$

$$\phi_y(d) = \sum_{y_j=y} f_j(d) \tag{5}$$

Given a confidence threshold τ, number of tweets t_j that satisfy $S_j \geq \tau$ are selected for training the classifier in each iteration. These tweets t_j form the unlabeled data set U.

4.3.2 Feature Extraction and Updation

The set of Topic-Adaptive Sentiment words includes the topic-adaptive sentiment word set Q that is taken as a labeled set as well as the topic related words that are obtained from the unlabeled data. The weight of the topic-adaptive sentiment word d is calculated by computing the frequency of occurrence of that word in unlabeled set as belonging to class y, and it is defined in Eq. (5)

$$x_d = \max_y\{\phi_y(d)\} \tag{6}$$

$$\varpi_d = \frac{\max_y\{\phi_y(d)\}}{\Sigma_y \phi_y(d)} \tag{7}$$

where $f_i(d)$ is the term frequency of the topic-adaptive sentiment word d in the tweet t_j and $\phi_y(d)$ is the summation of the term frequency of the topic-adaptive sentiment word d in the tweet t_j with the predicted class being y. Finally, the feature value of the word d is taken as the maximum of the summation of weight of any word. In Eq. (6), x_d is the feature value for the word d. The size of the labeled set keeps growing with more words being added and in this process the classifier gets trained. Unlike the formation of labeled set, not all words are added during training. The words having a high significance are considered for addition. The significance is computed using the Eq. (7).

$$x_d = \beta_d \cdot \max_y\{\phi_y(d)\} \tag{8}$$

$$\beta_d = I(\varpi_d \geq \theta) \tag{9}$$

In Eq. (7), ϖ_d is the significance of the word d. The words that exceed the threshold of significance are eligible to be added into the labeled set L. Finally, feature values, x_d are computed for the words from the unlabeled training set as given in the Eq. (8). Here β_d is the selection vector and defined in Eq. (9).

In the Eq. (9), I(.) is the indicator function and θ is the significance threshold. When indicator function I(.) returns 1, the word d is selected and d is not selected if I(.) returns 0.

4.4 Word Expansion Using Unlabeled Tweets

The sentiment classifier that is designed using labeled tweets from one domain may not classify the data from other domain with good accuracy. Hence, to train a Topic-Adaptive classifier, we use labeled data set L taken equal proportion from the mixed topics. L is used to train the initial classifier. To adapt this classifier to a topic e, unlabeled tweets U from a topic e are used to convert this weak classifier to a topic-adaptive classifier. The selected unlabeled tweet t_j is predicted to be belong to class $y_j^{'}$ according to Eq. (1). Thus, we can write the Eq. (10) to find the class of tweet t_j.

$$w_{y_j^{'}}^T x_j = \max\{w_y^T x_j\} \tag{10}$$

The complete procedure of adapting the classifier to a specific topic is illustrated in *OTAWE* algorithm. When a classifier is trained, not all unlabeled tweets are selected for training purpose. Only the tweets whose results are of high confidence are selected. This avoid the noise that added into the system. Therefore, confidence score S_j of tweet t_j with the predicted class y_j is defined as in Eq. (4).

Unlabeled tweets t_j that satisfy $S_j > \tau$ are selected for adaptive training, where τ is the confidence threshold. Words that satisfy the condition $\varpi_d > \theta$ are selected for Topic-Adaptive Word Expansion. Topic-Adaptive words are extracted from the confident, significant words that satisfy the condition $w > g$ where w is the weight of the topic-adaptive word and g represents threshold to select the topic-adaptive words from each iteration. These unlabeled tweets U are added to L set to form augmented L set. Topic-Adaptive words are selected, and feature values (weight) are updated in every iteration. Words that are not selected in an iteration carry their weight to the next iteration. When the number of iterations $I > M$, adaptation procedure stops where M indicates maximum number of iterations. Final sentiment classifier is a result obtained by the classifier trained using the augmented labeled data set L.

4.5 Optimized Topic-Adaptive Word Expansion (OTAWE) Algorithm

The complete procedure of adapting the classifier to a specific topic is illustrated in *OTAWE* algorithm. When a classifier is trained, not all unlabeled tweets are selected for training purpose. Only the tweets whose results are of high confidence are selected. This avoids the noise that added into the system. Therefore, confidence score S_j of tweet t_j with the predicted class y_j is defined as in Eq. (4).

Unlabeled tweets t_j that satisfy $S_j > \tau$ are selected for adaptive training, where τ is the confidence threshold. Words that satisfy the condition $\varpi_d > \theta$ are selected for Topic-Adaptive Word Expansion. Topic-Adaptive words are extracted from the confident, significant words that satisfies the condition $w > g$ where w weight of the topic-adaptive word and g represents threshold to select the topic-adaptive words from each iteration. These unlabeled tweets U are added to L set to form augmented L set. Topic-Adaptive words are selected and feature values (weight) are updated in every iteration. Words that are not selected in an iteration, carry their weight to the next iteration. When number of iterations $I > M$, adaptation procedure stops where M indicates maximum number of iterations. Final sentiment classifier is a result obtained by the classifier trained using the augmented labeled data set L. Initially, during the creation of the labeled set, the topic-adaptive features are set to zero. In the first few iterations, number of topic-adaptive words present is less. As the classification progresses, the feature values are adapted to a topic e to give a more precise result. At the end of the algorithm, the updated labeled set L, adapted feature values, and the designed classifier from the text features are obtained. These steps are repeated until the algorithm reaches the maximum number of iterations M.

5 Implementation

The implementation of the *OTAWE* algorithm requires both the labeled and the unlabeled set of data. The corpuses [9] are used as the input data set for training and testing the classifier. Classifier trained using the labeled set L results in a weak classifier. Then the classifier is made to adapt to a particular topic e by training it with the Topic-Adaptive words extracted from the unlabeled tweets U.

Initially, a common sentiment classifier is built using labeled set L. Next step is to adapt this classifier to a particular topic using the unlabeled tweets from that topic or domain. When adaptation process starts, the labeled set L is augmented with words in each iteration. Hence, L becomes augmenting set L_t. After adaptation, classifier does Cross Domain Sentiment Classification or Topic-Adaptive Sentiment Classification. The number of unlabeled tweets selected in each iteration (step length) of the *OTAWE* algorithm for the adaptation process is fixed, and these tweets with a confidence value $S_j > \tau$ are selected and are stored in a separate list. For each selected word of the unlabeled tweet set, we check whether that word exists in the labeled set as well. If

Algorithm OTAWE: Optimized Topic-Adaptive Word Expansion on Topic *e*

1: **Input**
2: L: labeled tweets consisting of K sentiment classes on mixed topics;
3: U: unlabeled tweets from topic e;
4: τ: threshold for selection of most confident topic-adaptive words;
5: θ: threshold for selection of significant topic-adaptive words;
6: l: step length, the maximum number of unlabeled tweets selected in one iteration;
7: g: Threshold to select the topic-adaptive words from each iteration;
8: M: the maximum number of iterations;
9: x: text feature set=0;
10: **Output**
11: L: labeled tweets (augmented) consisting of K sentiment classes on mixed topics;
12: C: SVM classifier trained on features x;
13: Calculate the confidence score S_j of tweets by using Eq. (4);
14: **while** number of iterations $I > M$ **do**
15: Select the l most confident and unlabeled tweets t_j in each sentiment class that satisfy condition $S_j > \tau$;
16: Calculate the weight x_d of the topic-adaptive sentiment word d using (6)
17: Calculate the significance ϖ of topic-adaptive words using Eq. (7);
18: Select the topic-adaptive words whose significance $\varpi > \theta$ from each class;
19: Add the Topic-adaptive words to the labeled set that satisfy the condition $w > g$ from each class;
20: Update feature values
21: Move the tweets with estimated class label from U to L;
22: Increment I;
23: **end while**

the word already exists, its sentiment class and weight are fetched and assigned to the selected word. If it does not exist in the labeled set, the word is simply skipped and process moves to the next word. Finally, when all the words that already exist in the labeled set L are labeled, the overall sentiment of each tweet is calculated. The sentiment class label of the tweet in the corpus can be compared with the result of the *OTAWE* algorithm. In the adaptation process, the feature values or weights of the words are updated. Words with the feature value or weight greater than the threshold g are considered for the addition into the labeled set L. This value of threshold g can be fixed as required and can be changed from time to time. The process is repeated for every iteration, and the weights are updated. Finally, the classifier becomes more robust and adapts to the topic well.

6 Results and Analysis

Simulation of *OTAWE* algorithm is carried out in *Python* language on *Linux* platform. We plot the graph between accuracy of the classifier and the augmenting set L_t for the "*Apple*" tweets with 1% sample ratio with step length $l = 15$ as shown in Fig. 1. Table 1 lists the accuracy values of the classification results for the Apple tweets of 1% sample ratio for 14 iterations with $l = 15$. We observe that the accuracy of classification in the first iteration where $L_t = 120$ tweets is better when compared to the next nine iterations ($L_t = 135$–255), we are using tweets from the labeled set as input in the first iteration. In the second iteration, unlabeled tweets with $l = 15$ are used for training the classifier. Topic-adaptive words from these unlabeled tweets are added to L_t after the completion of iteration and before starting of the next iteration. Compared to the first iteration, Accuracy reduces in second iteration as unlabeled tweets are used for training and at present only few topic-adaptive words present in the labeled set. In *OTAWE* algorithm, from second to seventh iteration (135 to 210 tweets), accuracy of classification remains same. This implies that not many topic-adaptive words cross the threshold but words carried there weights in subsequent iterations. Words having weight more than *g*, i.e., threshold to select the topic-adaptive words from each iteration, are selected as topic-adaptive words. From eight to tenth iteration ($L_t = 225$–255), the number of topic-adaptive words selected increases.

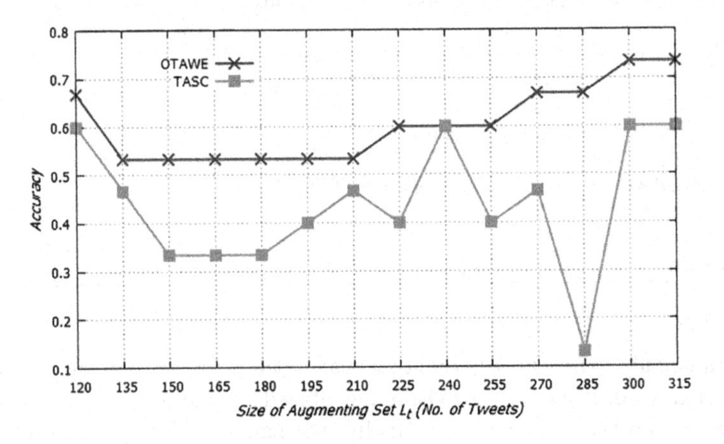

Fig. 1 Sample 1% on Apple tweets

Table 1 Comparison of accuracy values for on Apple tweets of 1% Sample ratio with step length = 15

Iterations	1	2	3	4	5	6	7	8	9	10	11	12	13	14
OTAWE	0.66	0.53	0.53	0.53	0.53	0.53	0.53	0.60	0.60	0.60	0.67	0.67	0.73	0.73
TASC	0.6	0.46	0.33	0.33	0.33	0.40	0.47	0.40	0.60	0.40	0.47	0.14	0.60	0.60

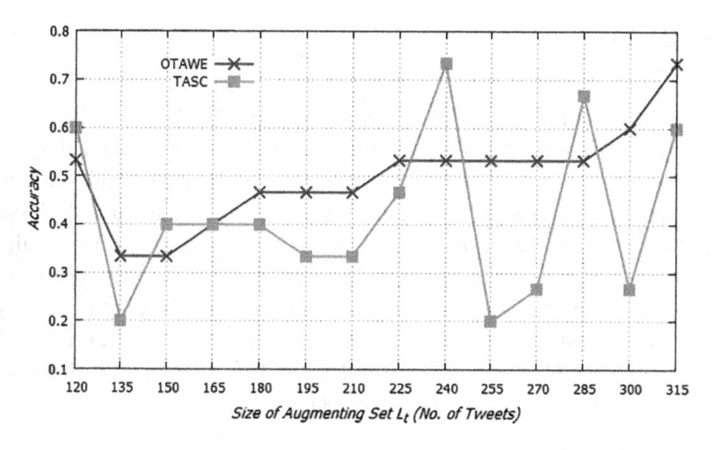

Fig. 2 Sample 1% on Twitter tweets

Table 2 Comparison of accuracy values for on Twitter tweets of 1% sample ratio with step length = 15

Iterations	1	2	3	4	5	6	7	8	9	10	11	12	13	14
OTAWE	0.53	0.33	0.33	0.40	0.47	0.47	0.47	0.53	0.53	0.53	0.53	0.53	0.60	0.73
TASC	0.60	0.20	0.40	0.40	0.40	0.33	0.33	0.46	0.73	0.20	0.26	0.66	0.26	0.60

Table 3 Comparison of average accuracy values

Sample Ratio =	1%	
$\mid L \mid$ =	120	
	OTAWE	TASC
Apple	0.57	0.48
Twitter	0.51	0.39

When the number of iterations increases, strong sentiment words that cross the threshold g increase that result in the increase in the accuracy. Hence, from eleventh to fourteenth iteration, accuracy increases considerably. From Table 3, for the "Apple" tweets, there is a 15% increase in the average accuracy of the *OTAWE* compared to *TASC* [1] algorithm for 1% sample ratio. This performance is the result of selection of strong sentiment words and avoiding the selection of weak sentiment words. From Fig. 2, we say that similar results are observed for "*Twitter*" tweets of 1% sample ratio. Table 2 lists the accuracy values of the classification results for the Twitter tweets of 1% sample ratio for all 14 iterations with $l = 15$. From Table 3, we observe that there is 23% improvement in the average accuracy value of *OTAWE* algorithm compared to *TASC* for these *Twitter* tweets of 1% sample ratio.

7 Conclusions

People tweet their opinion on various topics on a day-to-day basis. A challenging task is to extract the sentiments and opinions in these tweets using a single Topic-Adaptive Classifier instead of having separate classifier for each topic. Extracting the topic-adaptive words plays an important role in adapting a sentiment classifier on variety of topics. Optimized Topic-Adaptive Word Expansion (*OTAWE*) algorithm converts common sentiment classifier to a topic-specific one. Proposed *OTAWE* algorithm achieves promising increase in accuracy along with adapting to any given topic. We observe that there is a 15% increase in the accuracy of the *OTAWE* algorithm compared to *TASC* for 15 tweets step length of 1% sample ratio of *Apple* tweets and 23% increase in accuracy for *Twitter* tweets.

References

1. Liu, S., Cheng, X., Li, F., Li, F.: TASC: topic-adaptive sentiment classification on dynamic tweets. IEEE Trans. Knowl. Data Eng. **27**(6), 1696–1709 (2015)
2. Pan, S.J., Ni, X., Sun, J.-T., Yang, Q., Chen, Z.: Cross-domain sentiment classification via spectral feature alignment. In: Proceedings of the 19th International Conference on World Wide Web, pp. 751–760. ACM (2010)
3. Li, F. Pan, S.J., Jin, O., Yang, Q., Zhu, X.: Cross-domain co-extraction of sentiment and topic lexicons. In: Proceedings of the 50th Annual Meeting of the Association for Computational Linguistics: Long Papers, vol. 1, pp. 410–419 (2012)
4. Mejova, Y., Srinivasan, P.: Crossing media streams with sentiment: domain adaptation in blogs, reviews and twitter. In: Sixth International AAAI Conference on Weblog and Social Media, pp. 234–241 (2012)
5. Go, A., Bhayani, R., Huang, L.: Twitter sentiment classification using distant supervision. CS224N Project Report, vol. 1, 12. Stanford (2009)
6. Cheng, L., Pan, S.J.: Semi-supervised domain adaptation on manifolds. IEEE Trans. Neural Netw. Learn. Syst. **25**(12), 2240–2249 (2014)
7. Glorot, X., Bordes, A., Bengio, Y.: Domain adaptation for large-scale sentiment classification: a deep learning approach. In: Proceedings of the 28th International Conference on Machine Learning (ICML-11), pp. 513–520 (2011)
8. Bollegala, D., Weir, D., Carroll, J.: Using multiple sources to construct a sentiment sensitive thesaurus for cross-domain sentiment classification. In: Proceedings of the 49th Annual Meeting of the Association for Computational Linguistics: Human Language Technologies, vol. 1, pp. 132–141 (2011)
9. Sanders, N.J: Sanders-Twitter Sentiment Corpus. Sanders Analytics LLC (2011)

DCaP—Data Confidentiality and Privacy in Cloud Computing: Strategies and Challenges

Basappa B. Kodada and Demian Antony D'Mello

Abstract Cloud computing is one of the revolutionary technology for individual users due to its availability of on-demand services through the sharing of resources. This technology avoids the in-house infrastructure construction, cost of infrastructure and its maintenance. Hence individual users and organizations are moving towards cloud to outsource their data and applications. But many individuals are worried about the privacy and confidentiality of their data before sharing into cloud. The Cloud service provider has to ensure data security against unauthorized access in and out of the cloud computing infrastructure. Hence this paper identifies and describes the possible threats on Data confidentiality and privacy. The author also presents the anatomy and summary of general observations made on the available mechanisms to counter these threats.

Keywords Cloud computing · Data security · Confidentiality · Privacy · Encryption · Cloud data security

B. B. Kodada (✉)
AJ Institute of Engineering & Technology Mangaluru,
Visvesvaraya Technological University Belagavi, Belgaum, Karnataka, India
e-mail: basappabk@gmail.com

D. A. D'Mello
Canara Engineering College Mangaluru, Visvesvaraya Technological University Belagavi,
Belgaum, Karnataka, India
e-mail: demian.antony@gmail.com

D. Reddy Edla et al. (eds.), *Advances in Machine Learning and Data Science*,
Advances in Intelligent Systems and Computing 705,
https://doi.org/10.1007/978-981-10-8569-7_24

1 Introduction

Cloud computing basically relies on sharing of resources (hardware and software) and provides the flexible way of storing and accessing the data at any time. Cloud is emerging technology, rapidly increasing the usage due to its scalability, availability, flexibility, elasticity, on-demand storage service, and computation power characteristics. The main concept behind the cloud is virtualization technique [11] which allows the cloud services (IaaS, PaaS, SaaS) to be virtualized for the cloud users. Most of the cloud users use the storage services or storage space of cloud at low cost to outsource their data; that is why, the cloud computing is the fastest growing technology for economic benefits for an any organization. Cloud provides the innumerous amount of resources (power, storage etc.) which helps to the small-scale industry where they are not ready to invest for the storage server and its maintenance. They directly buy the storage space for their data at minimal cost compared to buying the new storage server cost and managing cost which avoids the in-house storage system [3, 4].

The cloud computing provides the unlimited storage service to customer, any organizations or to any cloud user. Many organizations outsource their data into the cloud to share with their employees. The user will not have full control on their data once they outsource it and cloud service provider (CSP) can form malicious activity on data even though there is an SLA between both user and provider. The data owner (DO) will be a sufferer and will not be able to fight with CSP on legal issue, because of location of outsourced data might be in different countries and the acts and laws might also be different. So data owner varied about mainly security of data which includes confidentiality, privacy, integrity, access control on their outsourced data along with user authenticity, authorization, its obstacles and in–out threats in cloud. Due to this reason, many organizations have the fearness of adopting the cloud even though many researchers have proposed techniques to counter the data confidentiality and privacy issue. The main mechanism of providing confidentiality and privacy security service by encrypting data/file during transmission and at-rest. But it is really worried to achieve the degree of data security. Achieving the data confidentiality while sharing of data in cloud, Iliana Iankoulova et al. have done a systematic review on security issues and requirements of cloud computing in [10] and found that the less research has been done on privacy and data confidentiality and privacy. It is very essential to provide the security of data-in-motion and at-rest. Hence this paper provides the review of threats on data during its transmission and at-rest in cloud computing. We also classify and present various security issues in cloud computing, types of threats on data confidentiality, and also discuss the general observation made on available mechanisms to counter the threats.

Rest of the paper information is organized as follows: Sect. 2 describes and presents only data confidentiality and privacy threats in cloud storage server (data center) along with detail available counter mechanism and summary of general observations made on these mechanisms. Open challenges and future research directions in cloud computing are also presented in Sect. 3, and finally in Sect. 4, we conclude the paper.

2 DCaP: Data Confidentiality and Privacy Threats in Cloud Computing and Counter Mechanisms

Hiding of user information from the unauthorized access is called data privacy and ensuring cloud service provider will not learn from the user data stored in the cloud called data confidentiality. This issue can be achieved by encrypting the file before uploading into the cloud. The organizations will have many departments and each department will have many employees and would like to outsource the data to cloud. But they are worried about the security of data, because entire data is handed over to CSP and can perform any activity on data even though there is SLA between them. Hence, CSP should ensure the security of infrastructure during data-in-motion and at-rest. There are threats to the data during its transmission where eavesdropper can capture the data, analyze, and modify it. We identified the possible threats on confidentiality and privacy during data in transmission and at-rest are presented in Table 1. The counter mechanisms against these threats during data-in-transmission can be provided by various encryption techniques. But when data-at-rest in cloud data center is very difficult, because once the data is stored in cloud, the CSP will have control on data. The Table 2 present summary of the various work in literature to ensure the confidentiality and privacy of data during data in-motion and at-rest to counter different threats. Table 2 also shows that it is very difficult to provide counter mechanisms for threats when data-at-rest in cloud data center. Hence, this paper gives the possible available encryption mechanisms to ensure the data confidentiality and privacy in cloud computing as shown in Fig. 1 which represents the anatomy of various encryption techniques. Related threats can also be possible when data-at-rest in the cloud data center like, data can be compromised by inside attacker who also make room for outside attacker to gain the access to data center.

2.1 Counter Mechanisms of Threats on Data-in-motion

Saravana et al. have proposed the attribute-based encryption technique [11] for securing the data in the cloud by preventing the data from unauthorized access. This scheme proposes attribute-based encryption (ABE) and hashing technique to address the data confidentiality and data access control, respectively, where each user will be assigned a access structure A and will have the master key, private key, and public key. If user wants to decrypt the data from the cloud, then he needs to authenticate himself by providing the digital signature which is computed $enc(H(A))$. CSP will decrypt the digital signature, match with $H(A)$ which is stored in the access control table. If it matches, then user will be provided with an access permission and can able to decrypt the file by using his private key. So this schema does not show how the confidentiality and integrity of data will be achieved by ABE method and might be high expensive in computational cost due to bilinear pairing in asymmetric cryptosystem.

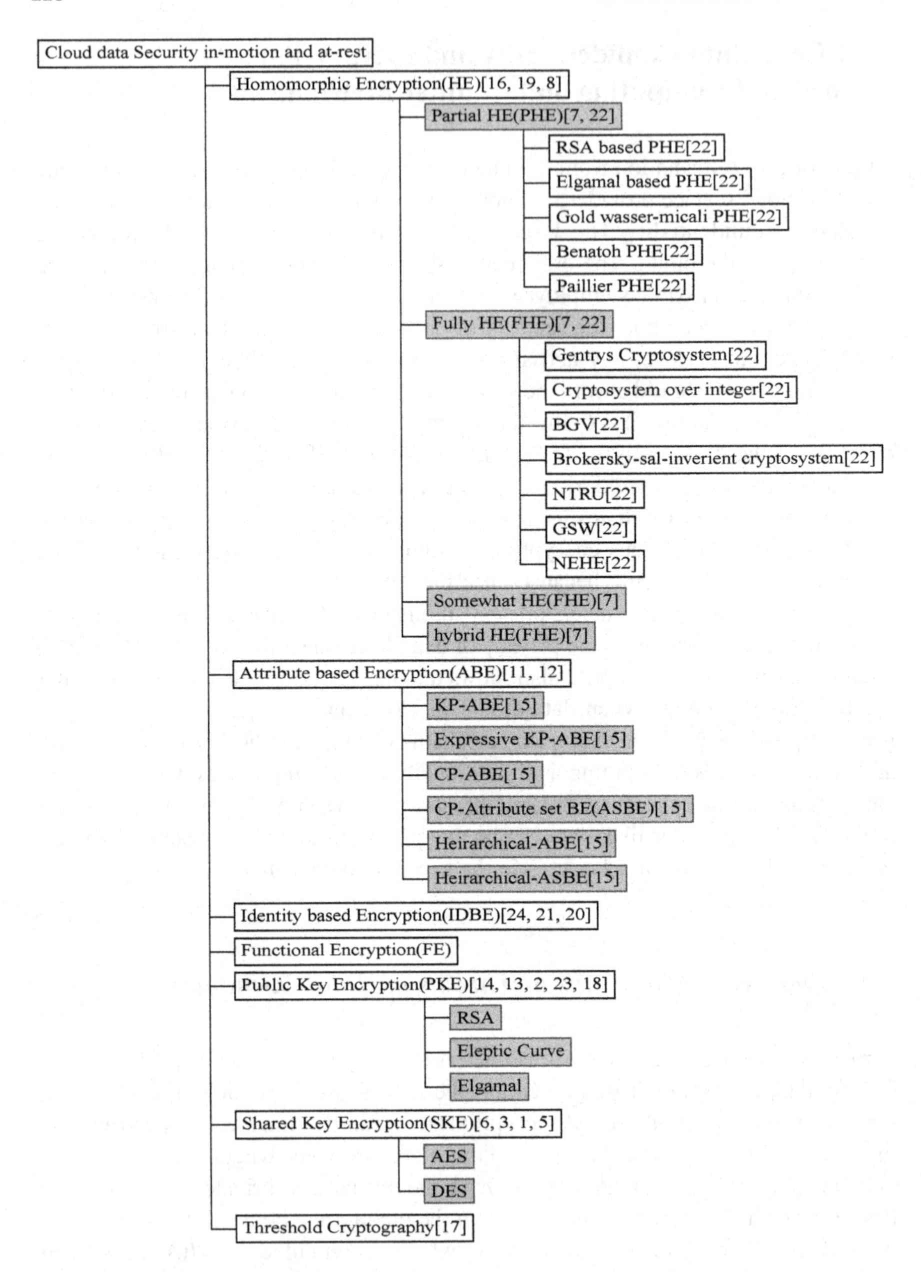

Fig. 1 Anatomy of various encryption techniques

Table 1 Types of threats in data confidentiality and confidentiality

S.No	Data privacy and confidentiality	Possible threats	Threat description
1	Data-in-motion	Spoofing	Activity involves attacker falsifying the data to CSP
		Eavesdropping	Activity involves when attacker gain access to the path of data and able to read data being transferred to CSP
		Man-in-middle	Activity involves when attacker read the data contents being transferred to CSP and modify it
2	Data-at-rest	Data storage	It is an activity where cloud user looses the full control on their own data once the data is outsourced to cloud
		Data recycling	It is an activity where the actual data will be recovered from recycle bin for malicious purpose
		Data location	It is an activity by CSP who hides actual location of user data being stored in cloud data center
		Data relocation	It is an activity where the user data is relocated at different data center to avoid the accidental technical issues
		Data recovery	is an activity where cloud user data is recovered on data loss or disaster

Sushil K. et al. have proposed the threshold cryptography-based data security [17] in which data owner generates the encryption key and shares a part of keys with the group members. That means each member will get a part of key, not complete key, so that user will not be able to decrypt alone due to unavailability of full key. To decrypt the file user needs to perform some levels of decryption at each user in the group by using their own part key; this will be repeated until complete key is found to decrypt it. This will prevent from compromising the insider, but regenerating and deleting keys when users leave or join from/to the group. Marimuthu et al. have proposed secure data sharing for dynamic group in cloud by generating group signature and broadcast encryption technique [13]. In this schema, DO encrypts the data by using broadcast encryption method and uploads it to cloud and also generates the group signature for each group by exchanging the user credentials. All the users in the group can share the data to other users who belong to the group by using group signature and also all can access the file by authenticating themselves with OTP generated by CSP.

The authors of [6] have proposed the data confidentiality by splitting the encrypted file and store into multi-cloud. The metadata will be maintained by private cloud. This method has five entities Cloud A, Cloud B, private cloud, server, and users. Server encrypts the user data, splits it into equal size, and sends it to multi-cloud. Clouds A and B archive their corresponding part and store it in some location; these location will be updated dynamically. Cloud storage service avoids organization from in-house storage system, but sharing of data with group of members in organization is major security issue; Mazhar et al. have proposed secure data sharing in cloud (SeDaSC) [3] among the group of users with the help of cryptographic server. In the proposed schema, the following security issues are addressed among group members (a) confidentiality and integrity (b) access control (c) data sharing without re-encryption (d) insider threats and (e) forward and backward access control. This schema encrypts the file with single key and uploads the encrypted file into the cloud. This encryption key will be partitioned and shared with all group members, and the key management is done by cryptographic server (CS). When user wants to download the file and decrypt it, user has to share the partitioned key to the CS, will download the file, decrypts it, and sends it to user. This schema addresses all issues which are defined in this proposal, but it uses the number theoretical concepts which lead to expensive in computational overhead in the cloud.

The authors of [1] have presented a security architecture model for providing data privacy and integrity by using AES, RSA, and Hash algorithm. The proposed model has four components (a) client application portal—who is a broker between user and provider, (b) key management and storage service—used to store and manage the user keys, (c) integrity check-up service—used to verify the hash code of data and (d) cloud service provider—used to provide the storage service. The data privacy has been provided with help of double encryption. Since the author of this paper used amazon S3 storage service, uploading of large file to S3 leads to high latency. Hence, performance may come down when user tries to upload files in larger size. The paper [16] presents the data privacy and confidentiality model using HE (Paillier algorithm) in cloud computing. Authors have used key distribution center (KDC) for distributing the user keys. In this paper, user needs to trust KDC. The paper [8] also presents the data confidentiality and privacy in cloud computing by encrypting the user data; this also presents the technique to compute and search over the encrypted data. Using this model, the computational cost may be higher due to the bilinear pairing concept in the proposed model. The author of [12] have proposed schema for shared authority-based privacy-preserving protocol and attribute-based access control mechanism. This schema uses the bilinear pairing method and based on number theory concept which causes the high computational cost.

Data confidentiality can also be achieved by public key infrastructure where each user gets the certificate for his identity, public key for encryption, and private key for decryption. Due to bilinear pairing method and complexity of managing the certificate for each user which causes high expensive in computational cost. So the authors of [18] have proposed mediated certificate-less public key encryption (mCL-PKE) which addresses the key screwing, certificate management, certificate revocation, and pairing problems. This schema has Key generation center (KGC) and Security mediator (SEM) for key generation for each user and partial decryption of file respectively resides in public cloud and are not trusted. This schema uses the re-decryption of same file for each user due to different secretes keys of each user, and dynamic updation of group user is not addressed. The cloud user could know the location of data being stored in the cloud data center, the most of the security issues will be resolved. So the authors of [9] have proposed the dual server (Link and Data server)-based secure data storage system in which the link server does not store real data; it only has the address of the data and the data server; on the other hand, it has the real data and access information, and it can only be accessed via the link server. The link server and data server maintains the timestamp of accessed user which solves the non-repudiation and integrity verification issue. The author of [24] provides the secure group communication using identity-based encryption (IBE) and asymmetric group key management; this also provides secure data sharing among group members by assuring data confidentiality and privacy. If each group member uploads the files in larger size, the computational cost may be higher due to its forward and backward security service. The paper [5] uses the 128-AES encryption and SMS-based technique to increase the degree of confidentiality and privacy of user data in cloud computing. Author of this paper uses AES encryption which uses shared key between cloud user and CSP. So cloud user need to trust CSP.

From Table 2, we infer, the researchers have contributed to ensure that data are secure and safe in cloud computing during its transmission. But when we look at the threats (data location, data recycling, and data recovery) when data-at-rest in Table 2, the authors might feel difficult to counter these. Because firstly, when user data is outsourced to cloud, the CSP will not reveal the location of user data. The data will be kept on relocating to provide high-availability service. The problem arises when user and provider break SLA and would feel difficult in following acts and laws of data location of country. Secondly, the previous user removed the data which might be in bin directory. So the current user in that space may use recovery tool to recover the data of previous user from recyclebin. Hence, these are the open threats and challenges for the researchers to ensure the data security in cloud computing when data-at-rest.

Table 2 Summary of counter mechanisms with general observations

Paper	Topic discussed	Data-in-motion			Data-at-rest				Proc	Cons
		Spoofing	Eavesdropping	Man-in-middle	Data storage	Data recycling	Data relocation	Data recovery		
[11]	Data confidentiality, ABE, access control, bilinear pairing	×	√	√	√	×	×	×	Achieved data confidentiality	Computational cost can be due to bilinear pairing
[14]	Data security, VDCI indexing, confidentiality, integrity	×	√	√	√	×	×	×	Data privacy and confidentiality	Authorized user may not be able to access file due to VDCI threshold limit
[17]	Threshold cryptography, encryption	×	√	×	√	×	×	×	Part of key for each member of group, level of decryption to get full key	Part of keying for joining/Leaving member
[13]	Broad cast encryption, group signature	×	√	√	√	×	×	×	User authentication by OTP, group signature for each member of group	Regenerating group signature when a member leaves/join the group

(continued)

Table 2 (continued)

Paper	Topic discussed	Data-in-motion			Data-at-rest				Proc	Cons
		Spoofing	Eavesdropping	Man-in-middle	Data storage	Data recycling	Data relocation	Data recovery		
[6]	Data confidentiality, Meta data, multi cloud	×	√	√	√	×	√	×	Increased degree of data security	Computational cost may high due to splitting of encrypted file into equal size
[3]	Data confidentiality, Access control, Data privacy, Re-encryption	×	√	√	√	×	√	√	Data security without re-encryption, data confidentiality	Splitting of single key among group members
[1]	AES, RSA, Data storage security	×	√	√	√	×	×	×	Data privacy, Double encryption	Need to trust key management and storage service
[16]	HE, KDC, Paillier algorithm	×	√	√	√	×	×	×	Data privacy	Need to trust KDC, computational cost many high
[19]	Proxy based encryption, HE, data security	√	√	√	√	×	×	×	Data confidentiality, privacy and malware detection	Performance of system may come down due to malware detection technique

(continued)

Table 2 (continued)

Paper	Topic discussed	Data-in-motion			Data-at-rest				Proc	Cons
		Spoofing	Eavesdropping	Man-in-middle	Data storage	Data recycling	Data relocation	Data recovery		
[8]	Data encryption, data privacy, searching over encrypted data	×	√	√	√	×	×	×	Relevant results through ranking concept, data confidentiality, searching over encrypted data	Computational cost may high due to bilinear pairing
[12]	Data privacy, access control, attribute based encryption	×	√	√	√	×	×	×	User authentication, preserving user data privacy and confidentiality	Computation cost might be higher due to bilinear pairing
[2, 23]	Public key encryption, data confidentiality	×	√	√	√	×	×	×	Data confidentiality through certificate less PKE	Computation cost might be higher due to bilinear pairing

(continued)

Table 2 (continued)

Paper	Topic discussed	Data-in-motion			Data-at-rest				Proc	Cons
		Spoofing	Eavesdropping	Man-in-middle	Data storage	Data recycling	Data relocation	Data recovery		
[24]	Identity based encryption (IBE), key management, secure communication	×	√	√	√	×	×	×	Data confidentiality and privacy among group members using IBE	If each group member shares files in larger size, computational cost may be higher due to forward and backward security model
[5]	AES encryption, Data confidentiality	×	√	√	√	×	×	×	SMS based technique to increase the degree of data security	Secrete key is shared between user and provider

3 Open Challenges and Future Research Directions

The important and main challenges with respect to security of data-in-transit and data-at-rest in cloud are managing the number of keys used for encryption and decryption, and also reducing the computational cost of existing techniques due its bilinear pairing and re-encryption techniques. As we defined data recovery, data recycling, and data location threats of DCaP in Table 1, the researcher can take up these challenges and explore the counter techniques. Some of the challenges are describe below,

- The Computational cost, resources, and power usage are very expensive due to bilinear pairing and number theory concepts used for generating number of keys to encrypt and decrypt the data.
- Managing the number of keys of each user is one of the major challenges in the cloud computing system.
- Loss of control on data: Once user data are outsourced to cloud, users will loose control on their own data, because the CSP may perform any malicious activity on the data even though both parties are agreed on common SLA. Hence, loss of control on data leads to privacy of data which is one of major security challenges.
- Data recycling: CSP or any other user can recover the data of previous DO from recycle bin of storage server. This is very difficult task, because the cloud resources are shared and same storage space can also be shared among many users. So if one user leaves the cloud by removing his own data that data might be in recycle bin and can be recovered by the current user using some tools and software. Hence, it is a open challenge in cloud data security.
- Data location/relocation: Any user or organization, once they outsource their data to cloud, will not be knowing exact location of data. The location of data is hidden by CSP. The CSP may relocate the data across the part of different countries for data backup to give high availability of data to user. This issue arise when both user and providers break the SLA. So it essential for DO to know the location of their data.

4 Conclusion

Cloud computing has become a fastest growing and emerging technology for IT industry and individual users due to its unlimited storage service on-demand. Hence, CSP has to protect the cloud user data and credentials. This paper classified the possible cloud security issues based on cloud infrastructure entity and presented in taxonomy. This paper also identified the possible threats in data confidentiality and privacy and available mechanisms to counter the threats. The sub-classification of available counter mechanisms is also presented. We also tried post some open security challenges to enhance better data security in cloud computing so that any researcher can provide a better counter mechanism to ensure the data security in cloud during data-in-motion and data-at-rest. And finally, we summarised work in literature with general observation made on the proposed model in the literature to ensure the data confidentiality and privacy. From this, we infer that there is no much work has been done to counter the threats when data-at-rest.

The future work includes the identifying and describing the different possible threats on integrity, availability, duplication, and access control of user data along with available counter techniques.

References

1. Al-Jaberi, M.F., Zainal, A.: Data integrity and privacy model in cloud computing. In: 2014 International Symposium on, Biometrics and Security Technologies (ISBAST), pp. 280–284. IEEE (2014)
2. Al-Riyami, S.S., Paterson, K.G.: Certificateless public key cryptography. In: Asiacrypt, vol. 2894, pp. 452–473. Springer (2003)
3. Ali, M., Dhamotharan, R., Khan, E., Khan, S.U., Vasilakos, A.V., Li, K., Zomaya, A.Y.: Sedasc: secure data sharing in clouds. IEEE Syst. J. (2015)
4. Arockiam, L., Monikandan, S.: Efficient cloud storage confidentiality to ensure data security. In: 2014 International Conference on, Computer Communication and Informatics (ICCCI), pp. 1–5. IEEE (2014)
5. Babitha, M., Babu, K.R.: Secure cloud storage using aes encryption. In: International Conference on, Automatic Control and Dynamic Optimization Techniques (ICACDOT), pp. 859–864. IEEE (2016)
6. Balasaraswathi, V., Manikandan, S.: Enhanced security for multi-cloud storage using cryptographic data splitting with dynamic approach. In: 2014 International Conference on, Advanced Communication Control and Computing Technologies (ICACCCT), pp. 1190–1194. IEEE (2014)
7. Bensitel, Y., Romadi, R.: Secure data storage in the cloud with homomorphic encryption. In: 2016 2nd International Conference on Cloud Computing Technologies and Applications (CloudTech), pp. 1–6. IEEE (2016)
8. Cao, N., Wang, C., Li, M., Ren, K., Lou, W.: Privacy-preserving multi-keyword ranked search over encrypted cloud data. IEEE Trans. Parallel Distrib. Syst. **25**(1), 222–233 (2014)
9. Go, W., Kwak, J.: Dual server-based secure data-storage system for cloud storage. Int. J. Eng. Syst. Model. Simul. 1 **6**(1–2), 86–90 (2014)
10. Iliana, I., Daneva, M.: Cloud computing security requirements: a systematic review. In: 2012 Sixth International Conference on, Research Challenges in Information Science (RCIS), IEEE
11. Kumar, N.S., Lakshmi, G.R., Balamurugan, B.: Enhanced attribute based encryption for cloud computing. Procedia Comput. Sci. **46**, 689–696 (2015)
12. Liu, H., Ning, H., Xiong, Q., Yang, L.T.: Shared authority based privacy-preserving authentication protocol in cloud computing. IEEE Trans. Parallel Distrib. Syst. **26**(1), 241–251 (2015)
13. Marimuthu, K., Gopal, D.G., Kanth, K.S., Setty, S., Tainwala, K.: Scalable and secure data sharing for dynamic groups in cloud. In: 2014 International Conference on, Advanced Communication Control and Computing Technologies (ICACCCT), pp. 1697–1701. IEEE (2014)
14. Moghaddam, F.F., Yezdanpanah, M., Khodadadi, T., Ahmadi, M., Eslami, M.: VDCI: Variable data classification index to ensure data protection in cloud computing environments. In: 2014 IEEE Conference on Systems, Process and Control (ICSPC), pp. 53–57. IEEE (2014)
15. Nimje, A.R., Gaikwad, V., Datir, H.: Attribute-based encryption techniques in cloud computing security: an overview. Int. J. Comput. Trends Technol. (2013)
16. Prasad, K., Poonam, J., Gauri, K., Thoutam, N.: Data sharing security and privacy preservation in cloud computing. In: 2015 International Conference on Green Computing and Internet of Things (ICGCIoT), pp. 1070–1075. IEEE (2015)

17. Saroj, S.K., Chauhan, S.K., Sharma, A.K., Vats, S.: Threshold cryptography based data security in cloud computing. In: 2015 IEEE International Conference on Computational Intelligence and Communication Technology (CICT), pp. 202–207. IEEE (2015)
18. Seo, S.H., Nabeel, M., Ding, X., Bertino, E.: An efficient certificateless encryption for secure data sharing in public clouds. IEEE Trans. Knowl. Data Eng. **26**(9), 2107–2119 (2014)
19. Sirohi, P., Agarwal, A.: Cloud computing data storage security framework relating to data integrity, privacy and trust. In: 2015 1st International Conference on Next Generation Computing Technologies (NGCT), pp. 115–118. IEEE (2015)
20. Tseng, Y.M., Tsai, T.T., Huang, S.S., Huang, C.P.: Identity-based encryption with cloud revocation authority and its applications. IEEE Trans. Cloud Comput. (2016)
21. Wei, J., Liu, W., Hu, X.: Secure data sharing in cloud computing using revocable-storage identity-based encryption. IEEE Trans. Cloud Comput. (2016)
22. Wikipedia, f.e.: Homomorphic encryption. https://en.wikipedia.org/wiki/Homomorphic_encryption
23. Xu, L., Wu, X., Zhang, X.: CL-PRE: a certificateless proxy re-encryption scheme for secure data sharing with public cloud. In: Proceedings of the 7th ACM Symposium on Information, Computer and Communications Security, pp. 87–88. ACM (2012)
24. Zhou, S., Du, R., Chen, J., Deng, H., Shen, J., Zhang, H.: SSEM: Secure, scalable and efficient multi-owner data sharing in clouds. China Commun. **13**(8), 231–243 (2016)

Design and Development of a Knowledge-Based System for Diagnosing Diseases in Banana Plants

M. B. Santosh Kumar, V. G. Renumol and Kannan Balakrishnan

Abstract Farmers usually find it difficult to identify and treat various diseases in banana plants (BPs) because it demands a wide spectrum of tacit knowledge. This situation motivated the authors to design and develop a technology-assisted knowledge base (KB) system for farmers, in order to diagnose and treat diseases in BPs. As a preliminary step towards building the KB, a set of images of diseases in BPs were taken from the manual published by Vegetable and Fruit Promotion Council Keralam (VFPCK). These sets of images were used to collect data from the agricultural experts and experienced farmers about the symptoms and remedies of various diseases in BPs. The data was collected from the participants by conducting semi-structured interview and then analysed to design the KB system. Since the diagnosis of diseases was a subjective process, an inter-rater reliability check was done on the data, using Cohen's Kappa method. Then using this data, a KB system has been designed and developed as a mobile app named as 'Ban-Dis'. An initial usability study has been conducted on this prototype among a few farmers, and their feedbacks have been recorded. The study results are promising and warrant further enhancements to the system. The KB system would be more beneficial as indicated by the farmers if the interface was in vernacular language.

Keywords Banana plant (BP) diseases · Knowledge base (KB)
Mobile app · Cohen's Kappa · Reliable knowledge

M. B. Santosh Kumar (✉) · V. G. Renumol
Department of Information Technology, School of Engineering, Cochin University of Science and Technology, Cochin 682022, Kerala, India
e-mail: santo_mb@cusat.ac.in

V. G. Renumol
e-mail: renumolvg@gmail.com

K. Balakrishnan
Department of Computer Applications, Cochin University of Science and Technology, Cochin 682022, Kerala, India
e-mail: mullayilkannan@gmail.com

D. Reddy Edla et al. (eds.), *Advances in Machine Learning and Data Science*,
Advances in Intelligent Systems and Computing 705,
https://doi.org/10.1007/978-981-10-8569-7_25

1 Introduction

Banana plant (BP) was originated from South-East Asia and has been cultivated for more than ten thousand years worldwide [1]. The first traces were found in Papua New Guinea [2] which dates back to 7000 years. This belongs to monocotyledon class and Musaceae family. Natural banana crops have been produced with lot of nutritional values for human [3]. As time passed by, the varieties of banana had been brought to Indian subcontinent from South-East Asia and Papua New Guinea [4]. The production cycle for a normal banana plant is approximately 9–12 months depending on the climatic zones. The difference in size of the plants depends on the area of production and altitude of the plantation [5]. The genetic diversity of the banana plants is mainly due to the combination of natural reproduction and human selection.

In 2013, the world banana production was around 220 million tonnes, of which 40% were cooking banana varieties and 60% were dessert banana varieties. A survey on world banana production was conducted by Lescot and Charles [6], and their findings are shown in Fig. 1. The banana cultivators are nowadays present across the globe. The main production zones are located in Asia which represents approximately 45% of the world volume followed by Africa (about 25%) and the remaining in the South and Central America. From Table 1 given below, it is observed that India is the largest producer of banana [7].

Lot of diseases affect banana plants. A study conducted by Vegetable and Fruit Promotion Council Keralam (VFPCK) identified 19 common diseases in BPs [8]. The most harmful is the fungus-induced disease namely Sigatoka and Banana Wilt [9]. The diseases appear in the form of black streaks which destroy the leaves. Banana Wilt or Panama disease is caused mainly by soil and root fungus. This fungus normally remains in the soil for more than 30 years. Pests such as nematodes or black banana weevil disrupt the plant growth and topple it down. There are

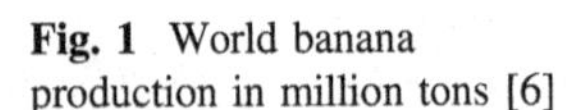

Fig. 1 World banana production in million tons [6]

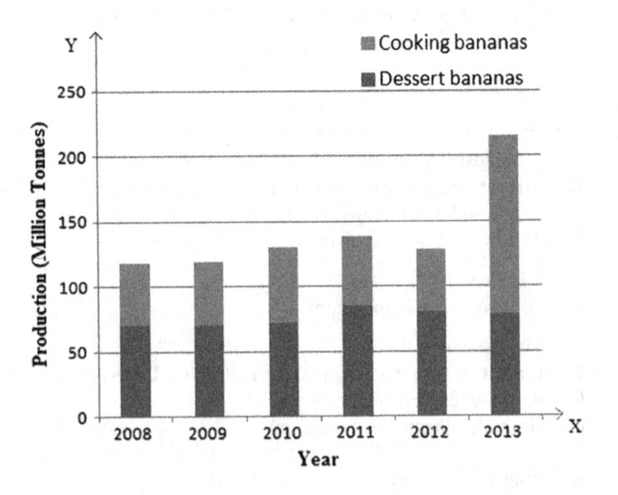

Table 1 Top ten countries in banana production (dessert and cooking) in tons as on 2013 [7]

Table	Dessert bananas (tons)	Cooking bananas (tons)	Total production (tons)
World total	78,860,773	54,831,192	133,691,995
India	17,075,000	10,500,000	27,575,000
China	11,506,238	569,000	12,075,238
Uganda	500,000	8,426,308	8,926,308
Philippines	5,790,091	2,855,658	8,645,749
Brazil	6,402,622	490,000	6,892,622
Ecuador	6,145,527	594,212	6,739,739
Colombia	2,587,625	2,817,740	5,405,365
Indonesia	3,289,115	2,070,000	5,359,115
Rwanda	250,000	3,013,462	3,263,462
Nigeria	315,000	2,907,000	3,222,000

bacterial diseases such as Moko disease that is spread through the soil, plantation tools or by other insects. There are viral diseases such as mosaic or bunchy top which cause losses to both commercial and village plantation. Thus, there is a need for identifying the various diseases affecting the banana plants and come up with a remedial measure depending on the symptoms of the diseases. However, farmers find it difficult to diagnose and take remedial measures to protect their BPs in time.

Banana ranks fourth as the world's most important starch crop after cassava and sweet potatoes since its yields of carbohydrates per unit area are very high. It is the fourth most widely grown food crop after rice, wheat and maize [10]. It is an important staple food in many Asian countries including India. Unfortunately, banana production has been on the decline nowadays. This can be attributed to a number of factors such as poor crop production and management practices, pests, diseases, declining soil fertility [11]. Overall, pests and diseases pose a serious threat to banana production. Some banana cultivations have been severely damaged by a wide range of pests and diseases, resulting in heavy yield losses to farmers. The important and widespread diseases include black and yellow Sigatoka leaf spots, cigar-end rot, banana bunchy top virus disease and post-harvest diseases. Banana thrips, nematodes and the banana weevil are among the most important pests of banana in most banana-growing regions. Yield losses in highlands are mainly due to weevil [12]. To a great extent, it is the ability of the farmers to identify pests and diseases affecting bananas. Normally, farmers are empowered to check the occurrence of the pests and diagnose the diseases at the initial stage itself so that the spread of the pests and disease to the crop can be avoided [13]. However, banana farmers usually find it difficult to identify and treat various diseases in BPs because it demands a wide spectrum of knowledge. Moreover, this knowledge is tacit and not readily available to farmers.

Further literature review was conducted to gather more information on similar systems and found that similar studies were conducted on various plants, for instance expert system in detecting coffee plant diseases [14], expert system for diagnosis of diseases of rice plants [15], expert system for diagnosing flower bulb diseases [16], expert system for plant disease diagnosis [17], expert system for the diagnosis of pests, diseases and disorders in Indian Mango [18]. But in all these expert systems the knowledge used was not tested for reliability. This situation motivated the authors to use reliable knowledge to design and develop a technology-assisted knowledge base (KB) system to diagnose diseases in BPs.

2 Design and Development of Knowledge Base System

There were different Small Agricultural Groups (SAGs) in every panchayat of every district of Kerala, India. These SAGs consist of 4–14 farmers, which are headed by an SAG leader. These SAG leaders get training on various precautionary steps to be taken when the agricultural products get infected with various diseases. The training to the SAG leaders is mainly given by VFPCK, Central Plantation Crops Research Institute (CPCRI) and agricultural experts from government and non-government organizations.

Kerala Agricultural University website [19] and various other portals have come up with different methods [20] to protect the banana plants. But the farmers located in remote areas do not get access to such training, which is normally conducted at the centres situated in the town limits. The knowledge imparted to these farmers by the agricultural experts may not be completely understood by the farmers. There also exists a lot of traditional knowledge [21] to identify and rectify various BP diseases. However, due to change in the climatic conditions there are chances for the traditional remedies not to work properly in banana plant cultivation.

Thus, the need for the development of some technology-assisted tool is very relevant and need of the hour. The easy availability and accessibility of the tool were considered for designing and developing the system. The farmers who have no access to the computers or iPads should also get access to this tool. Hence, it has been decided to develop an Android app which can be accessed through their smartphones. The architecture of the application is given in Fig. 2. There are two major components—the graphical user interface (GUI) and the knowledge base (KB).

The GUI was designed as a simple window, where farmers can select the symptoms as input to the system. Accordingly, the system will display the name of the disease. Then the farmer can click the 'remedy' button to get the solution steps. In order to develop the knowledge base part, a semi-structured interview was conducted among agricultural experts and experienced farmers. Their knowledge of BP diseases, related symptoms and remedies has been collected. It was decided to delegate one of the authors for conducting the interview. Authors identified a set of experienced farmers and agricultural experts as participants for the semi-structured interview. Before going to the field for data collection, the interviewer conducted a

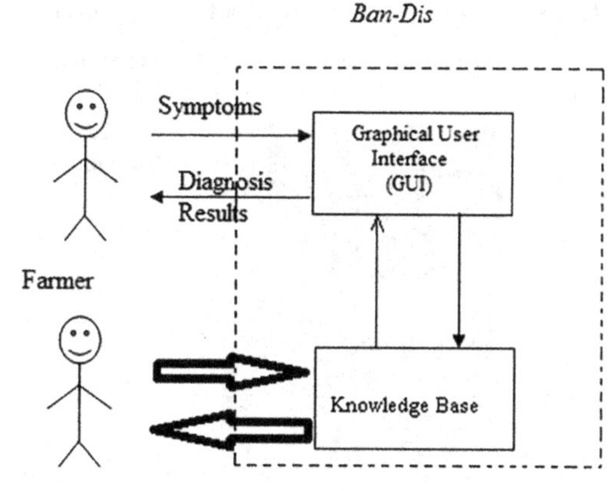

Fig. 2 Architecture of mobile app 'Ban-Dis'

study on various diseases in BP and their symptoms by referring the existing literature [22–25]. This awareness helped the interviewer to conduct the interview effectively. A set of interview questions were prepared to collect data on various diseases from the agricultural experts and experienced farmers. The quality and reliability of the data collected to a great extent depend on how effectively the data collecting procedures was conducted. Hence, the following measures were taken to increase the quality of the data collected:

- Properly explaining the main objective of the interview to the participants and the importance of giving correct answers.
- Casual way of collecting the data without revealing the details of the participants.
- Keeping the data collected confidentially and using the data for academic and research purpose only.
- Allowing the participants to express wilfully and sincerely without any compulsion.

The interviewer collected images of 19 diseases of banana plant from a manual published by VFPCK [8]. All the 19 diseases which indicated the most prominent diseases in BP were numbered and arranged sequentially. A data sheet with the 19 images of the diseases as shown in Table 2 was given to the participants.

During the data collection, the interviewer showed the data sheet to the participants and prompted them to identify the diseases. The interviewer also asked to suggest the common symptoms for each disease and the remedies for it. It was a semi-structured interview, so the interviewer interfered in between to collect more information while conducting the interview. The entire conversation was recorded as audio files for supporting the analysis process.

Table 2 Format for collecting data from participants

Sl. No.	Image of diseases	Name of diseases	Symptom of diseases	Remedial measures
1		<Filled during interview>	<Filled during interview>	<Filled during interview>
2		…	…	…
	…	…	…	…
19				

The data was collected from two districts of Kerala, namely Ernakulam and Trichur, since these two districts were easily accessible and closer to the location of the researchers. It took around 2 h for interviewing each participant. Feedbacks from a total of five participants were collected, of which two were agricultural experts (AE) from VFPCK and the other three were experienced farmers (EF). The details of the participants are shown in Table 3.

When the data has been analysed, it was noticed that the participants identified most of the diseases identically, but in few cases, the same image has been identified as different diseases by the participants. Moreover, the farmers' knowledge about the diseases was comparatively lesser than experts. They could not identify all the diseases, and they were ignorant of their symptoms and remedies. This indicates that the farmers are not much knowledgeable about the diseases in banana plants. Table 4 shows the difference in the names suggested by the agricultural experts and experienced farmers after observing the image.

Table 3 Details of participants

Participant	Age	Qualification	Experience in years
1 (AE1)	47	Post Graduate	10
2 (AE2)	38	Post Graduate	7
3 (EF1)	65	10th grade	22
4 (EF2)	62	10th grade	20
5 (EF3)	58	8th grade	18

Table 4 Name of the diseases suggested by the participants by observing the image

Sl. No.	Image	Name of the diseases identified by				
		AE1	AE2	EF1	EF2	EF3
1		Manavandu (cosmopolites sordidus)	Manavandu	Manavandu	Manavandu	Vella koomp
2		Thada thurappan (odoiporus longicollis)	Pindi puzhu	Thadathurappan	Pindi puzhu	Chellu
3		Vazha pen (pentalonia nigronervosa)	Vazha pen	Vazha pen	Vazha pen	<Could not Identify>
4		Ilatheeni puzhukkal	Puzhu	Pattala Puzhu	Puzhu	<Could not Identify>

(continued)

Table 4 (continued)

Sl. No.	Image	Name of the diseases identified by				
		AE1	AE2	EF1	EF2	EF3
…	…	…	…	…	…	…
19		Vellakoombu and vazhappani	Koombadachil and vazha pani	Koombu veluth thallal and vazhappani	<Could not Identify>	<Could not Identify>

The participants identified the diseases and symptoms and suggested remedial measures independently. Since this is a subjective process, the interviewer decided to calculate the inter-rater reliability using Cohen's Kappa method [26]. The answers from the participants were coded as shown in Table 5. The entry '1' indicates mutual agreement and '2' indicates disagreement.

The content of Table 5 was fed to the Statistical Package for Social Sciences (SPSS) tool, and Table 6 indicates the values of Cohen's Kappa for three categories namely:

- Expert 1 versus Expert 2
- Farmer 1 versus Farmer 2
- Expert 1 versus Farmer 1

The values of inter-rater reliability, Cohen's Kappa (K) between Expert1 and Expert2 with reference to the name of diseases, symptoms and remedial measures were found to be greater than 0.7, which indicate that the results are reliable. Also, the K values for Farmer1 versus Farmer2 were greater than 0.7, which again indicates that the values are reliable, whereas the Cohen's Kappa values for Expert1 versus Farmer1 were less (<0.7) when compared to the other two values which indicate that there is a mismatch in the knowledge level between farmers and experts in this area. Thus, this study reveals that the knowledge of the agricultural experts from VFPCK is more reliable than the experienced farmers. Thus, the knowledge base was built mainly using the data collected from agricultural experts. An Android-based mobile app named as 'Ban-Dis' has been developed using this knowledge base. Using this application, banana farmers can identify the diseases and come up with the required remedial measures based on the type of the symptoms exhibited. When farmers enter the symptoms of the diseases, the tool provides the accurate analysis of the diseases along with some images and pops up the desired remedies. The Android app can be downloaded from the Web link http://nss.cusat.ac.in/goorganic/home.php, and it is also available at http://indianrupeeservices.in/diseases/.

An initial usability study of the system was conducted among thirteen farmers, who were from Kerala and their mother tongue was Malayalam. Hence, they found it difficult to use the interface in English and suggested for making the interface in the regional language (Malayalam). Farmers also suggested providing facility to enhance the knowledge base with newly found diseases in BPs. The study results are promising and warrant further enhancements to the system. They also suggested incorporating different input modes such as audio, text, image. The KB system would be more beneficial as indicated by the farmers if the above said features are incorporated.

Table 5 Coded feedback from the participants which is given as input to SPSS tool

Sl. No.	Name of disease	Symptoms	Remedies											
	AE1	AE2	EF1	EF2	EF3	AE1	AE2	EF1	EF2	EF3	AE1	AE2	EF1	EF2	EF3
1	1	1	1	1	2	2	2	2	2	2	2	2	2	2	2
2	1	2	2	2	2	1	2	2	2	2	1	2	2	2	2
3	1	1	1	1	1	1	1	1	1	1	1	1	1	1	1
4	1	1	1	1	1	1	1	1	1	1	1	1	1	1	1
5	2	2	2	2	2	2	2	2	2	2	2	2	2	2	2
6	1	1	1	1	1	1	1	1	1	1	1	1	1	1	1
7	1	1	1	1	1	1	1	1	1	1	1	1	1	1	1
8	2	2	2	2	2	2	2	2	2	2	2	2	2	2	2
9	1	1	1	1	1	1	1	1	1	1	2	2	2	2	2
10	1	1	1	1	1	1	1	1	1	1	2	2	2	2	2
11	1	1	1	1	1	1	2	2	2	2	1	2	2	2	2
12	1	1	1	1	1	1	1	1	1	1	1	1	1	1	1
13	2	2	2	2	2	2	2	2	2	2	2	2	2	2	2
14	1	1	1	1	1	1	1	1	1	1	1	1	1	1	1
15	1	1	1	1	1	1	1	1	1	1	1	1	1	1	1
16	1	1	1	1	1	1	1	1	1	1	1	1	1	1	1
17	1	1	2	2	2	1	1	2	2	2	1	1	2	2	2
18	1	1	2	1	2	1	1	2	1	2	1	1	2	1	2
19	1	1	1	1	1	1	1	1	1	1	1	1	1	1	2

Table 6 Coefficient of inter-rater reliability (Cohen's Kappa) between the participants

Sl. No.	Category	K value for diseases	K value for symptoms	K value for remedies
1	Expert 1 versus Expert 2	0.826	0.732	0.776
2	Farmer 1 versus Farmer 2	0.872	0.890	0.895
3	Expert 1 versus Farmer 1	0.578	0.537	0.587

3 Conclusion and Future Work

Banana farmers usually face problems in diagnosing and treating diseases in BPs. Hence, the study focused to build a knowledge-based system to help farmers. The data, to build the required knowledge base for the system, was collected from agricultural experts and experienced farmers. The reliability of the collected data was found out by using Cohen's Kappa inter-rater reliability. An Android app named 'Ban-Dis' was designed and developed by using this reliable knowledge. With the help of this app, farmers can easily diagnose the disease and get remedial measures. An initial usability study has been conducted among a few farmers from Kerala, and their feedbacks warrant further enhancements to the system. Some of the future modifications to the system can be incorporating regional languages, user input in multimodality, etc. The same system can be used for some other crops by replacing the knowledge in the knowledge base accordingly.

Declaration Authors hereby declare that the study was not concerned with any experiment on human beings, animals, etc. Authors have also obtained consents from all individual participants included in the study.

References

1. Valmayor, R.V., Umali, B.E., Bejosano, C.P.: Summary of discussion and recommendation of the INIBAP Brisbane Conference. In: Banana Diseases in Asia and the Pacific, pp. 1–4 (1991)
2. Denham, T., Haberle, S.G., Lentfer, C., Fullagar, R., Field, J., Porch, N., Therin, M., Winsborough B., Golson, J.: Multi-disciplinary evidence for the origins of agriculture from 6950–6440 Cal BP at Kuk Swamp in the Highlands of New Guinea. Science (June Issue) (2003)
3. Amorim, Edson P.: Genetic diversity of carotenoid-rich bananas evaluated by Diversity Arrays Technology (DArT). Genet Mol Biol **32**(1), 96–103 (2009)
4. Price, N.S.: The origin and development of banana and plantain cultivation. In: Bananas and Plantains, pp. 1–13. Springer Netherlands (1995)
5. Robinson, J.C., Saúco, V.G.: Bananas and Plantains, vol. 19. Cabi (2010)
6. Lescot, T.: Genetic diversity of banana. Close-up. FRuiTReP231 (2015)

7. unctad.org. http://unctad.org/en/PublicationsLibrary/INFOCOMM_cp01_Banana_en.pdf (2017). Accessed 21 Sept 2017
8. VFPCK—Vegetable And Fruit Promotion Council Keralam. vfpck.org. N. p., 2017. Web, 21 Sept 2017
9. Álvarez, E., et al.: Current status of Moko disease and black sigatoka in Latin America and the Caribbean, and options for managing them (2015)
10. Bourke, R.M., et al.: Food production, consumption and imports. In: Food and Agriculture in Papua New Guinea, pp. 129–192 (2009)
11. Bogale, M., et al.: Survey of plant parasitic nematodes and banana weevil on Ensete ventricosum in Ethiopia. Nematologia Mediterranea **32**(2) (2004)
12. Rukazambuga, N.D.T.M., Gold, C.S., Gowen, S.R.: Yield loss in East African highland banana (Musa spp., AAA-EA group) caused by the banana weevil, Cosmopolites sordidus Germar. Crop Prot. **17**(7), 581–589 (1998)
13. Chowdhury, S.K.: Study on major insect pests and major diseases of banana of Malda, West Bengal, India. Indian J. Appl. Res. **5**(8) (2015)
14. Derwin Suhartono, W.A., Lestari, M., Yasin, M.: Expert system in detecting coffee plant diseases (2013)
15. Kalita, H., Sarma, S.K., Choudhury, R.D.: Expert system for diagnosis of diseases of rice plants: prototype design and implementation. In: International Conference on Automatic Control and Dynamic Optimization Techniques (ICACDOT). IEEE (2016)
16. Kramers, M.A., Conijn, C.G.M., Bastiaansen, C.: EXSYS, an expert system for diagnosing flowerbulb diseases, pests and non-parasitic disorders. Agric. Syst. **58**(1), 57–85 (1998)
17. Abu-Naser, S.S., Kashkash, K.A., Fayyad, M.: Developing an expert system for plant disease diagnosis. J. Artif. Intell. **3**(4), 269–276 (2010)
18. Prasad, R., Ranjan, K.R., Sinha, A.K.: AMRAPALIKA: an expert system for the diagnosis of pests, diseases, and disorders in Indian mango. Knowl.-Based Syst. **19**(1), 9–21 (2006)
19. Kerala Agricultural University. Towards Excellence in Agricultural Education. Kau.in. N. p., 2017. Web, 21 Sept 2017
20. Krishi Vikas Yojana. rkvy.nic.in. http://rkvy.nic.in/ (2017). Accessed 21 Sept 2017
21. Borborah, K.S., Borthakur, K., Tanti, B.: Musa Balbisiana Colla-Taxonomy, Traditional knowledge and economic potentialities of the plant in Assam, India. (2016)
22. O'Donnell, K., et al.: Multiple evolutionary origins of the fungus causing Panama disease of banana: concordant evidence from nuclear and mitochondrial gene genealogies. Proc. Natl. Acad. Sci. **95**(5), 2044–2049 (1998)
23. Su, H.J., Hwang, S.C., Ko, W.H.: Fusarial wilt of Cavendish bananas in Taiwan. Plant Dis. **70** (9), 814–818 (1986)
24. Deltour, P., et al.: Disease suppressiveness to Fusarium wilt of banana in an agroforestry system: influence of soil characteristics and plant community. Agric. Ecosyst. Environ. **239**, 173–181 (2017)
25. Banana Diseases and Their Control. Agropedia. Agropedia.iitk.ac.in. N. p., 2017. Web, 21 Sept 2017
26. McHugh, M.L.: Interrater reliability: the kappa statistic. Biochemia medica **22**(3), 276–282 (2012)

A Review on Methods Applied on P300-Based Lie Detectors

Annushree Bablani and Diwakar Tripathi

Abstract Deceit identification or detection has become a topic of study from past few decades. Detecting lie is not only a legal issue, but also moral and ethical issue. Various lie detectors like polygraph have been developed which check body temperature, heart rate, pulse rate, blood pressure, etc., to detect whether a person is telling truth or not. But these polygraph tests give an indirect and incomplete knowledge of deception, so directly measuring mental activity of subject using brain–computer interface (BCI) was adopted which identifies the mental state of subject and detects lie. These lie detectors use P300 component of event-related potential (ERP) generated during mental task and acquired using electroencephalography (EEG). In our paper we have presented a survey on state-of-the-art techniques applied for analyzing and classifying innocent and guilty subjects.

Keywords Brain–computer interface · P300 · Event-related potentials · Lie detectors

1 Introduction

Deception detection is trending and currently being used not only to identify if someone has committed a crime or not but also for patients having psychological disorders. Polygraph is traditional and physiological method to detect lies; these record respiration rate, blood pressure, heart rate, etc., for lie detection. These indirectly try to record brain activity of subject, so researchers tried to directly record the brain activity. There are various ways to record electrical activity undergoing into brain—one of them is EEG. EEG is noninvasive technique for acquiring electrical activity

A. Bablani (✉) · D. Tripathi
Department of Computer Science and Engineering,
National Institute of Technology Goa, Ponda 403401, Goa, India
e-mail: annubablani@nitgoa.ac.in

D. Tripathi
e-mail: diwakartripathi@nitgoa.ac.in

D. Reddy Edla et al. (eds.), *Advances in Machine Learning and Data Science*,
Advances in Intelligent Systems and Computing 705,
https://doi.org/10.1007/978-981-10-8569-7_26

generated by brain cells through scalp by placing metal electrodes. EEG is widely and most commonly used method especially for research purposes to record brain signals. EEG measures change in potential developed in neurons while some activity is performed or stimulus is received. One of such kind of potential is event-related potentials (ERPs).

ERPs are elicited during any cognitive, sensory, and motor events when detected on scalp and recorded using noninvasive technique. It identifies components of brain rhythms that are generated because of various information processing that takes place in brain like memorizing, decision making, analyzing. ERPs that are elicited because of sensory signals are described in terms of positive and negative peak values. When some physical stimulus is given, short-latency ERPs are generated (before 80 ms) which are called as exogenous or stimulus bound. During cognitive task, long-latency ERPs are generated which are known as endogenous.

P300 or P3 is one of the ERP components that occur when a cognitive task of determining any object is given to subject. P300 is a positive-going component which has latency of around 300 ms and is elicited when signals are correctly detected and its amplitude increases when detection is made with more accuracy [1]. This feature of P300 makes it useful in identifying targets without actually pointing them and which will be helpful for the persons with different abilities. Usually, P300 occur at C_z, P_z, and F_z electrode locations of standard EEG electrode placement map as shown in Fig. 1. A "Interrogative Polygraph Test" to reveal the concealed information was developed by Farwell et al. [2, 3], using P300 component of ERP in context of oddball paradigm. This test is also fondly known as *Guilty Knowledge Test* or *Concealed Information Test*. Oddball paradigm [4] is experimental design which is used in psychological research where successive stimulus is given to subject and activity generated in response to stimuli is recorded and observed. A series of events

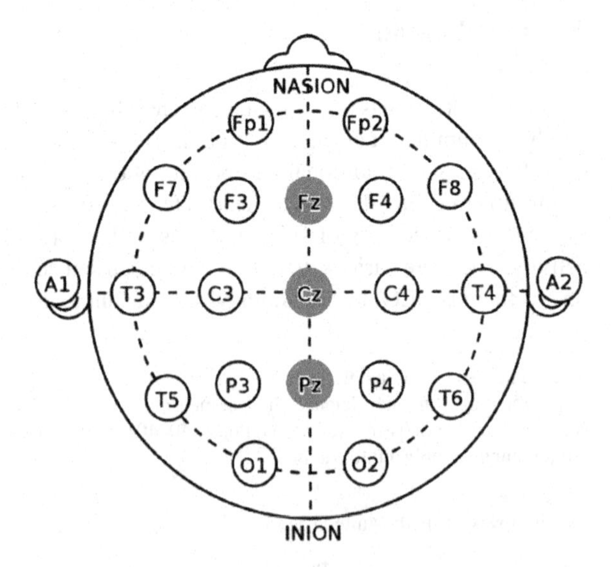

Fig. 1 Occurrence of P300 on C_z, P_z, and F_z locations in 10–20 electrode system

are presented in front of subject in their experiment, and classification techniques are used to place each event in one or two classes depending on the property of the events. Event classified into two or more categories are presented randomly in front of the subject, and one of the events is presented very rarely. This rarely presented stimuli will elicit the P300 component. Rare the occurrence of event and more its relevance to subjects, larger the P300 component is, or in other words, amplitude to P300 is inversely proportional to probability of occurrence to Stimulus and directly to relevance of stimulus. For lie detection, Farwell et al. [2] have presented a series of three stimuli; one stimulus occurs 17% of the time which was known to all subjects and labeled as *targets*, whereas other 83% is the stimulus is not known to subject or are *irrelevant*. Out of this 83% of irrelevant stimulus 17% is the stimulus is known only to guilty and the authorize personnel and are irrelevant to innocent. This rarely occurring third stimulus is called *probe* which generates P300 in brain and is used to identify guilty subject.

A variant of Concealed Information Test (CIT) was developed [5, 6], where not only EEG data, but conductance of skin and heart rate was simultaneously measured and used as features for deception detection. All these state-of-the-art methods record EEG signals during a given task performed by subject and then analyze these signals by applying various statistical analysis methods or machine learning techniques. Few of these analysis methods have been discussed in our paper.

2 Methods for Analyzing ERP Components

For discriminating between guilty and innocent, test similar to *Guilty Knowledge Test* [7] has been performed by researchers who worked on lie detectors. A task was created, which was given to all subjects and asked to focus on certain stimuli and perform a task; this stimuli is *target*, which will generate P300 in all subjects. The tasks were performed so that subject can pay attention on all the stimuli given. An oddball series is constructed, which has crime relevant and crime irrelevant items, and this will be helpful in distinguishing between guilty and innocent. For innocent subject, P300 is elicited only when *target* stimulus is given, and rest two stimuli (*irrelevant* and *probe*) will be unknown for innocent and will show same behavioral response, whereas for guilty subject *target* and *probe* stimuli will generate P300 response as probe stimulus is related to crime and known to guilty subject, and *irrelevant* will generate a different response, as this is unknown to guilty subject too. The aim is to develop a technique or method which will identify behavior of three responses and measure a difference between them, so that guilty and innocent are recognized. Some of the techniques used in the literature are discussed in our paper.

2.1 Bootstrapping

Bootstrapping is a technique which allows estimation of any distribution using some random sampling methods.

2.1.1 Correlation Analysis

Bootstrapping using correlation analysis [2, 8, 9] is used to estimate sampling distribution of two correlations. Before computing correlation between pairs of stimuli, an average overall stimuli (*target*, *probe*, and *irrelevant*) is computed, which is stored in a column vector, let us say *Ave*. A difference between *Ave* and *target*, *probe*, and *irrelevant* is calculated. Now, a cross-correlation coefficient between these stimuli is calculated as: (1) correlation between *probe* and *irrelevant*, denoted as C_1 and (2) correlation between *probe* and *target*, denoted as C_2. Next difference between these two correlation coefficients is computed $Diff = C_1 - C_2$. This same process is repeated for 100 iterations. A comparison between these two correlations is made:

- If $C_1 < C_2$, that means *probe* is more similar to *target* (P300 present), and *probe* is not similar to *irrelevant* (P300 absent); hence, the subject with this type of behavior is recognized as guilty.
- If $C_1 > C_2$, that means *probe* is more similar to *irrelevant* (P300 absent) and *probe* is not similar to *target* (P300 present); hence, the subject with this type of behavior is recognized as innocent.

2.1.2 Amplitude Difference

For distinguishing whether subject is guilty, we need to determine that P300 evoked by one stimuli is greater than other stimuli or not [2, 8–10]. This can be determined by identifying whether average of *probe* and average of *irrelevant* P300 is greater than zero. For each subject, only one average *probe* and one average *irrelevant* P300 are available. So, bootstrapping is used where distribution is estimated by random selection. A computer program is run, which randomly selects a set of waveform, let us say w_1, from *probe* set with replacement. It averages these waveforms and calculates P300 amplitude using maximum selection segment method [9, 11]. Later, from *irrelevant* set, a set of waveform was drawn randomly, let say w_2, with replacement; using this average P300 amplitude was calculated. The *irrelevant* average P300 is subtracted from *probe* average value and a difference value, let us say W_d is obtained which will contain as many values as number of iteration used in the process. Guilty decision is made if the value of W_d is greater than zero, otherwise innocent decision is made.

2.1.3 Reaction Time Analysis

This method is similar to amplitude difference method, difference is instead of amplitude or size of brainwaves, reaction times are considered [9, 12]. This method randomly samples average reaction time for *irrelevant* and for *probe* with replacement and calculates the difference between both reaction times. Guilty decision is made if the value of zero is more than 1.65 standard deviation (95% confidence), below mean of different distribution, otherwise innocent decision is made.

2.2 *Wavelet Transform*

Feature extraction using Fourier transform gives only frequency information as temporal information is lost during transformation. In 1989, Mallat [13] devised a *Multiresolution Theory*, in which he showed wavelet transform (WT), which not only gives frequency information but also provides temporal information of signal. WT provides precise measurements of when and to what degree transient events occur in a signal waveform and of when and how the frequency content of a waveform changes over time. For this, WT performs translation and scaling and generates functions called wavelets. Wavelet transformation can be performed in three ways: discrete wavelet transform (DWT), continuous wavelet transform (CWT), and multiresolution-based transform (MT). Abootalebi et al. [8] used discrete wavelet transform (DWT) to analyze P300 component to recognize guilty, where translation and scaling operations consist of discrete values. As mother wavelets, quadratic B-spline functions were used by authors because of their similarity with the evoked response. Extracted wavelet coefficients can be used as features of single-trial ERPs. These features have more information than simple features such as extracted by bootstrapping analysis of amplitude, latency, and correlation. For discriminating guilty subject from innocent one, classification was applied and authors have applied linear discriminant analysis (LDA) [14]. LDA handles data with unequal class frequency and evaluates their performance by randomly generating test data. It tries to provide separability by drawing decision region between classes.

2.3 *Independent Component Analysis*

It, also called *blind source separation* [15], decomposes data into various independent components according to their statistical interdependence, and then undesired components are removed. For analyzing the signal, independent component analysis (ICA) follows a linear transformation method. Let $X(t)$ represent signal at time t, $I(t)$ be the transformed independent component, and A be the mixing matrix, then

$$X(t) = I(t) \cdot A \tag{1}$$

Table 1 Detection rate of different methods

Detection rate (%)	Methods
74	Amplitude analysis
80	Correlation analysis
79	Wavelet transform
84.29	ICA with SVM

$$I(t) = A^{-1} \cdot X(t) \tag{2}$$

Gao et al. [16] in their work have applied ICA and decomposed EEG signals into two independent components (ICs); one IC is task related that is ERP generated, and other IC is task unrelated that is noise for lie detector. A template matching method based on Independent Component Analysis (ICA) has been applied for denoising ERP data and decomposing it into P300 component and non-P300 components. Finally after getting P300 ICs, various classifiers have been applied and their performance have been compared (Table 1).

3 Conclusion

P300 component of ERP helps in detecting the lie or mal-intent of the subject, as this component of ERP cannot be elicited or modified according to what subject wants brain to do, but instead P300 is elicited by itself when any related or relevant stimuli is presented to subject. Hence, P300 became very useful in determining who is guilty or who is innocent. In this paper, we have reviewed state-of-the-art methods to extract features of guilty and innocent subjects. A statistical analysis method, bootstrapping, has been widely used by many authors to determine presence and absence of P300 in guilty and innocent subjects. Other than statistical methods, some pattern recognition methods like ICA, wavelet transform have also been applied on ERP data to distinguish between various ERP features.

References

1. Hillyard, S.A., Kutas, M.: Electrophysiology of cognitive processing. Ann. Rev. Psychol. **34**(1), 33–61 (1983)
2. Farwell, L.A., Donchin, E.: The truth will out: interrogative polygraphy (lie detection) with event-related brain potentials. Psychophysiology **28**(5), 531–547 (1991)
3. Farwell, L.A., Richardson, D.C., Richardson, G.M., Furedy, J.J.: Brain fingerprinting classification concealed information test detects US Navy military medical information with P300. Front. Neurosci. **8** (2014)

4. Fabiani, M., Gratton, G., Karis, D., Donchin, E., et al.: Definition, identification, and reliability of measurement of the P300 component of the event-related brain potential. Adv. Psychophysiol. **2**(S1), 78 (1987)
5. Farahani, E.D., Moradi, M.H.: Multimodal detection of concealed information using Genetic-SVM classifier with strict validation structure. Inf. Med. Unlocked (2017)
6. Farahani, E.D., Moradi, M.H.: A concealed information test with combination of ERP recording and autonomic measurements. Neurophysiology **45**(3), 223–233 (2013)
7. Lykken, D.T.: The GSR in the detection of guilt. J. Appl. Psychol. **43**(6), 385 (1959)
8. Abootalebi, V., Moradi, M.H., Khalilzadeh, M.A.: A comparison of methods for ERP assessment in a P300-based GKT. Int. J. Psychophysiol. **62**(2), 309–320 (2006)
9. Rosenfeld, J.P., Soskins, M., Bosh, G., Ryan, A.: Simple, effective countermeasures to P300-based tests of detection of concealed information. Psychophysiology **41**(2), 205–219 (2004)
10. Rosenfeld, J.P., Labkovsky, E., Winograd, M., Lui, M.A., Vandenboom, Catherine, Chedid, Erica: The Complex Trial Protocol (CTP): a new, countermeasure-resistant, accurate, P300-based method for detection of concealed information. Psychophysiology **45**(6), 906–919 (2008)
11. Kubo, K., Nittono, H.: The role of intention to conceal in the P300-based concealed information test. Appl. Psychophysiol. Biofeedback **34**(3), 227–235 (2009)
12. Lukács, G., Gula, B., Szegedi-Hallgató, E., Csifcsák, G.: Association-based concealed information test: a novel reaction time-based deception detection method. J. Appl. Res. Mem. Cogn. (2017)
13. Mallat, S.G.: A theory for multiresolution signal decomposition: the wavelet representation. IEEE Trans. Pattern Anal. Mach. Intell. **11**(7), 674–693 (1989)
14. Fisher, R.A.: The use of multiple measurements in taxonomic problems. Ann. Eugen. **7**(2), 179–188 (1936)
15. Herault, J., Jutten, C., Denker, J.S.: Space or time adaptive signal processing by neural network models. In: AIP Conference Proceedings, vol. 151, pp. 206–211. AIP (1986)
16. Gao, J., Liang, L., Yang, Y., Gang, Y., Na, L., Rao, N.N.: A novel concealed information test method based on independent component analysis and support vector machine. Clin. EEG Neurosci. **43**(1), 54–63 (2012)

Implementation of Spectral Subtraction Using Sub-band Filtering in DSP C6748 Processor for Enhancing Speech Signal

U. Purushotham and K. Suresh

Abstract Implementation of novel algorithms on Digital Signal Processing [DSP] processor to extract the speech signal from a noisy signal is always of immense interest. Speech signals are usually complex that requires processing of signal in short frames, and thus DSP processors are widely used to process the speech signals in mobile phones. The performance of these devices is comparatively very well in noisy conditions, as compared with traditional processors. The speech signal is degraded by either echo or background noise and as a result, there exists a requirement of digital voice processor for human–machine interfaces. The chief objective of speech enhancement algorithms is to improve the performance of voice communication devices by boosting the speech quality and increasing the intelligibility of voice signal. Popular speech enhancement algorithms rely on frequency domain approaches to estimate the spectral density of noise. This paper proposes a method of assessment, wherein frequency components of noisy speech signal are filtered out using multiband filters. The multiband filters are developed in C6748 DSP processor. Experimental results demonstrate an improved Signal to Noise Ratio [SNR] with fewer computations.

Keywords Intelligibility · Spectral density · Multiband · DSP processor

1 Introduction

Information in the form of speech always goes along with some quantity of noise. The interfering unwanted signal usually corrupts the quality and intelligibility of speech. During speech communication, background noise surrounding the source of

U. Purushotham (✉)
Department of Electronics and Communication, PES University,
Bengaluru 560085, India
e-mail: purushothamu@pes.edu

K. Suresh
SDM Institute of Technology, Ujire, Dakshina Kannada 574240, India
e-mail: ksece1@gmail.com

D. Reddy Edla et al. (eds.), *Advances in Machine Learning and Data Science*,
Advances in Intelligent Systems and Computing 705,
https://doi.org/10.1007/978-981-10-8569-7_27

speech is the key element for signal degradation. Speech enhancement process not only makes the human listening easy but also solves other problems related to efficient transmission of signal. The spectral subtraction (SS) proposed by Boll [1] is a straightforward evaluation of short-term spectral magnitude. The principle of SS is to take away the magnitude spectrum of noise from that of the degraded speech assuming noise to be uncorrelated and additive to the speech signal. The typical SS technique [2–6] significantly reduces the intensity of noise but leads the way for distortion in speech signal which is commonly known as musical noise. The incorrectness in estimation of short-time noise spectrum leads to musical noise. Unlike white Gaussian noise, real-world noise such as vehicle noise does not have plane spectrum, i.e., man-made noise has non-uniform distribution [6]. Hence, an educated guess is necessary to find suitable factors that will lessen the vital magnitude of noise spectrum from every frequency bin and to shun critical subtraction of speech to get rid of a large amount of the residual noise.

Extensive emphasis has been given in development of algorithms for different DSP processors, which are used for variety of applications. Digital signal processors handle complex digital processing algorithms to get better quality of signals. Development of new DSP techniques and processors has enabled the efficient implementation of speech signal enhancement algorithms.

In this paper, we propose a sub-band filtering-based spectral subtraction that can be easily implemented in DSP processor. These algorithms trim down the aforementioned distortions and maintain good speech quality. Section 2 discusses some of the standard approaches for speech enhancement in DSP processor; Sect. 3 presents the proposed technique, describing the accomplishment of the proposed method; Sect. 4 gives the results; and conclusions and comments are given in Sect. 5.

2 Approaches for Speech Enhancement Using DSP Processor

There are plentiful approaches to retrieve the enhanced speeches. Among them, frequency domain approaches are popular than temporal domain. Over the years, spectral subtraction-based frequency domain approaches are widely used for real-time applications. There exist diverse versions of SS type algorithms to extract clean speech. During the process of subtracting average noise, the phase of the noisy speech is kept unchanged, and it is assumed that the phase distortion is not perceived by human ear [5]. Sub-band filtering [7] is fundamentally used to improve convergence rate in distinction to full band result. Ephraim and Malah [8, 9] have developed an algorithm that utilizes the MMSE criteria that use models for the distribution of the spectral components of speech and noise signals.

DSP processors are integrated with ADC, DAC along with digital interpolation filters. These processors have the capability to transfer data word lengths ranging from 16 to 32 with sampling rate varied from 8 to 96 kHz. The processing action is carried out at low cost and low power consumption.

Cai et al. [10] have proposed a TMS320VC5509, a hardware platform to develop algorithm for noise reduction using optimized SS method. The central processing unit of this processor handles 400 million multiplication and accumulation in 1s with a maximum clock rate of 200 MHz. Tiwari et al. [11] have proposed complex speech enhancement algorithm that combines spectral subtraction and multiband frequency compression in order to decrease the effects of increased intraspeech masking. This algorithm is implemented in 16-bit fixed point DSP processor TMS320C5515. Yermeche et al. [12] have proposed sub-band beam-forming process for multichannel filtering operations on array of input signals considering sub-band frame independently. It is implemented in floating point DSP processor ADSP21262.

3 Proposed Methodology

In mobile communication, real-time disturbance such as car, train, or airport noise has dissimilar fading effects over the entire spectrum. Therefore, it is necessary to understand the characteristics of this colored noise which will indicate the scaling aspects for each frequency bin [13]. In this paper, speech signal is segmented to avoid destructive subtraction. During this procedure, it is found that some segments of speech enclose an unvoiced speech segment which matches residual noise. Hence, distinctiveness of unvoiced speech is acquired from analysis to eliminate residual noise. The algorithm for the proposed method is given here;

Algorithm:
Step 1: The noisy signal is passed through multirate rate system.
Step 2: At the second level, spectrum of resultant signal is obtained.
Step 3: Estimation of noise is carried out using a distinct window.
Step 4: Calculate δ and α, utilizing noise estimated in previous step.
Step 5: Spectral subtraction is carried out using results of step 3 and 4.
Step 6: Combine the signals of all the four bands using appropriate weights for each band, to get noise-free signal.

If $H(e^{jw}) = H(z)$ represents the frequency response of the filter, with M as the highest power then it is given by

$$H(z) = az^{M} + bz^{M-1} + cz^{M-2} + dz^{M-3} + ez^{M-4} \tag{1}$$

which can be decomposed into even and odd powers of z as shown in Fig. 1.

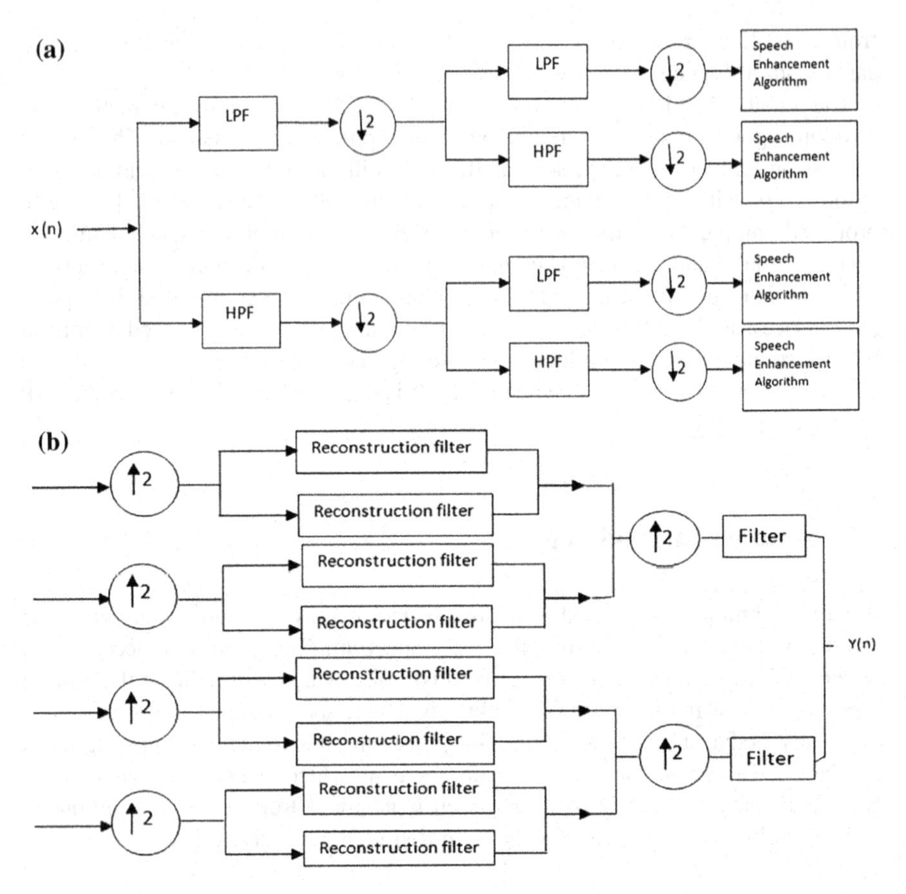

Fig. 1 **a** Speech enhancement using analysis filter bank. **b** Speech reconstruction using synthesis filter bank

$$H_1(z) = az^M + cz^{M-2} + ez^{M-4} \tag{2}$$

$$H_2(z) = bz^{M-1} + dz^{M-3} \tag{3}$$

Similarly, further decomposition of H_1 and H_2 leads to H_{11}, H_{12} and H_{21}, H_{22}, respectively. These filters are used to divide the spectrum into L number of bands. The modified approach of spectral subtraction divides the speech spectrum into L non-overlapping bands. To estimate the clean speech, spectral subtraction is achieved separately in each frequency band as shown by following equation:

$$|S_i(k)|^2 = |Y_i(k)|^2 - \alpha_i \delta_i |D_i(k)|^2 \tag{4}$$

Table 1 Variations in tuning factor

Sl. No	Type of noise	Frequency f_i	δ_i
1.	Car	$f_i < 1$ kHz	1.12
2.	Airport		0.89
3.	Train		1.08

k represents the frequency bin.
α_i is the over-subtraction factor, and
δ_i is a tuning factor.

To customize the noise removal properties, tuning factors can be varied for different types of noise as shown in Table 1.

Similarly for other frequency ranges, tuning factors can be adjusted instead of constant 2.5 and 1.5 [13]. Whereas α_i is a band-specific over-subtraction that depends on segmental SNR of the *i*th frequency band which is intended as:

$$SNR_i(db) = 10\,log\frac{\sum |Y_i(k)|^2}{\sum |D_i(k)|^2} \tag{5}$$

The proposed sub-band filtering is implemented in TMS320C6748 DSP processor. Effective processing of audio signals is done using C6748. A general function McASP: audio serial port is optimized for multichannel audio applications. The audio serial port is used for various kinds of data streaming such as TDM, I2S protocols, and DIT data steaming. The audio serial port has two different sections for transmission and reception. These sections are provided with master clocks to operate synchronously and independently [14].

4 Experimental Results

The modified algorithm is demonstrated using speech utterance taken from IEEE Noizeus database [15]. Speech signal of 8 kHz sampling frequency with SNR of 5 and 10 dB with background car noise is considered for illustration. Two sets of results are demonstrated in the paper, first with non-real time and second with DSP processor using TI platforms with code composer.

In Table 2, SNR values are tabulated for two sentences and SNR is calculated for MMSE and proposed methodology, respectively. The performance of sub-band filtering SS approach is better than MMSE approach. Figure 2a demonstrates time domain plot of the noisy signal and its sub-bands, Fig. 2b illustrates transfer characteristics of the LPF and HPF and also its response, and Fig. 3a, b illustrates the output of analysis and synthesis filter bank, respectively.

For comparative purposes, we also have plotted the frequency and time domain of noisy and synthesized signal in Fig. 4a, b. Spectrogram of clean and enhanced signal is illustrated in Fig. 5a, b. Informal listening tests show improvement in

Table 2 Comparison of SNR

Input SNR	MMSE		Proposed method	
Speaker	Male	Female	Male	Female
5 db	10.6665	11.3821	11.1081	11.4354
10 db	14.7504	14.3721	14.7690	14.4514

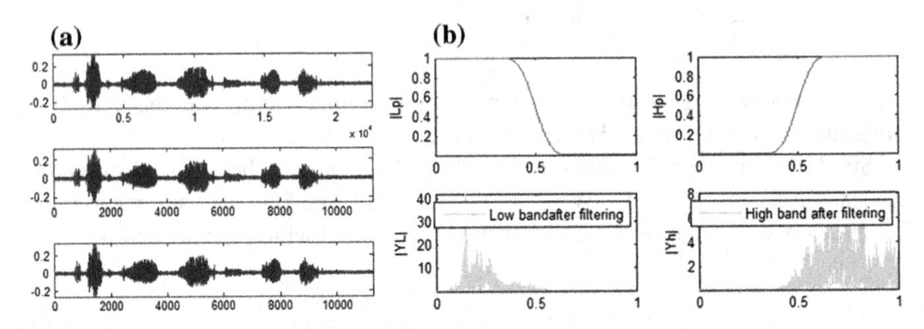

Fig. 2 **a** Speech utterance with background car noise (first) and its sub-bands (second and third). **b** Filter response (first) and spectrum of sub-bands (second)

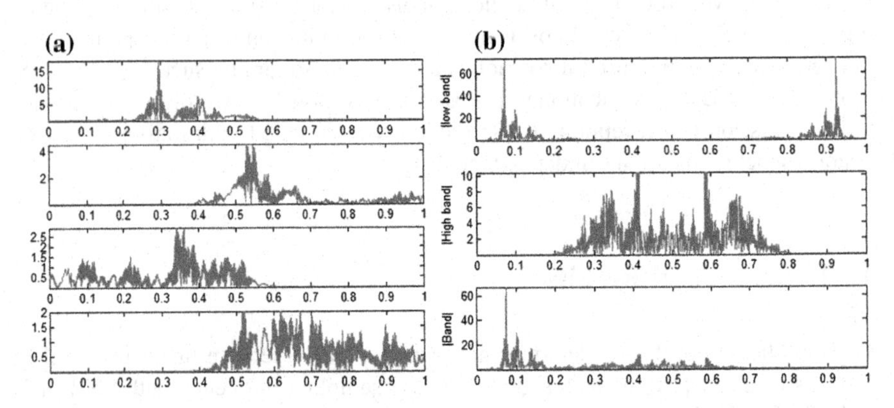

Fig. 3 **a** Filter response of four sub-bands after the second stage of filtering. **b** Final two bands after synthesis

speech quality indicating modified SS approach yielding very good speech quality with very little trace of musical noise. Implementation of sub-band filters is shown in Figs. 6, 7, and 8. Figure 6 illustrates the clean, noise component, and noisy speech signal. Figure 7 demonstrates the noisy signal and enhanced signal. i.e., output of the proposed method. Figure 8 illustrates the memory allocation.

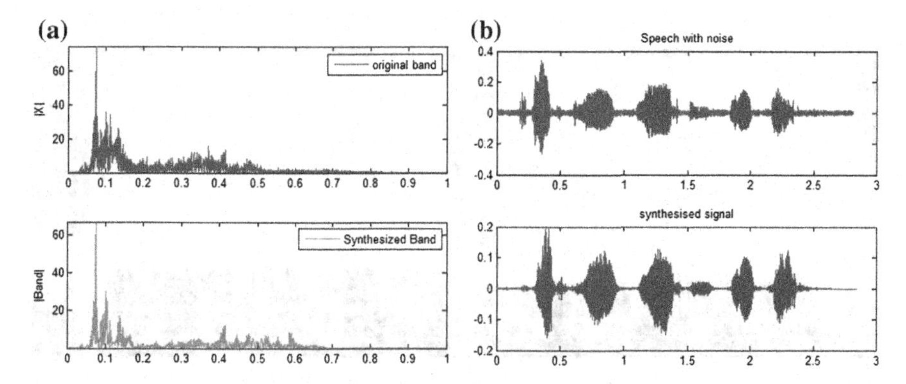

Fig. 4 **a** Comparison of original and synthesized band in frequency domain. **b** Comparison of speech and synthesized signal

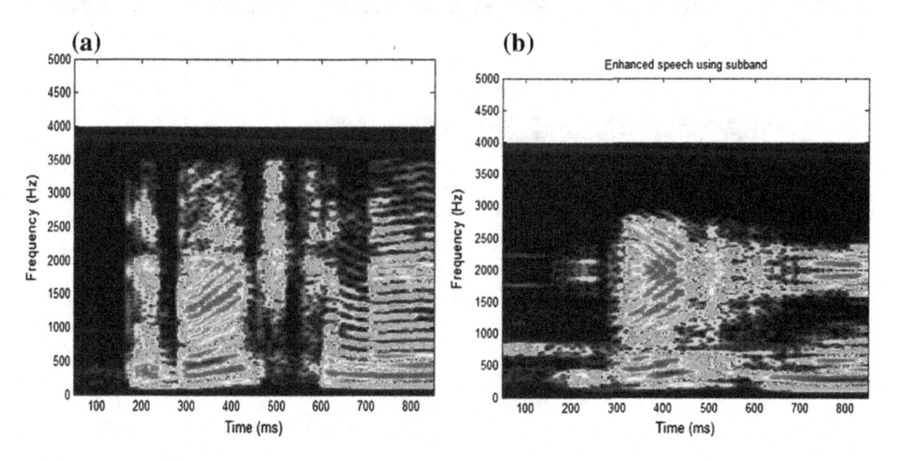

Fig. 5 **a** Spectrogram of clean speech. **b** Spectrogram of signal obtained using modified spectral subtraction using sub-bands

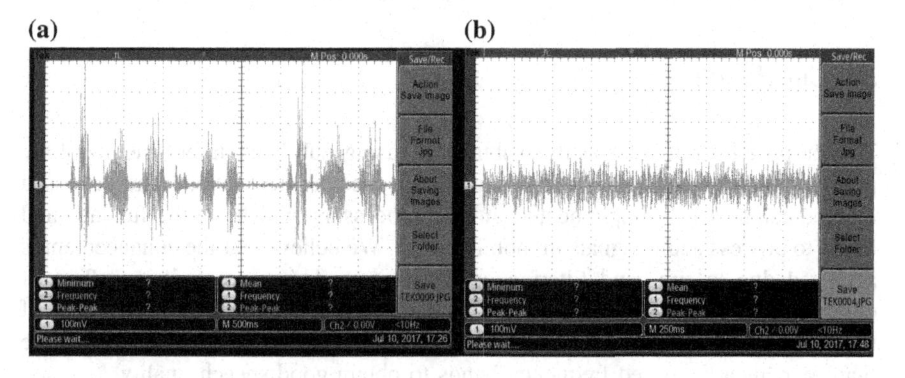

Fig. 6 **a** Speech utterance of clean signal. **b** Noise component in the speech

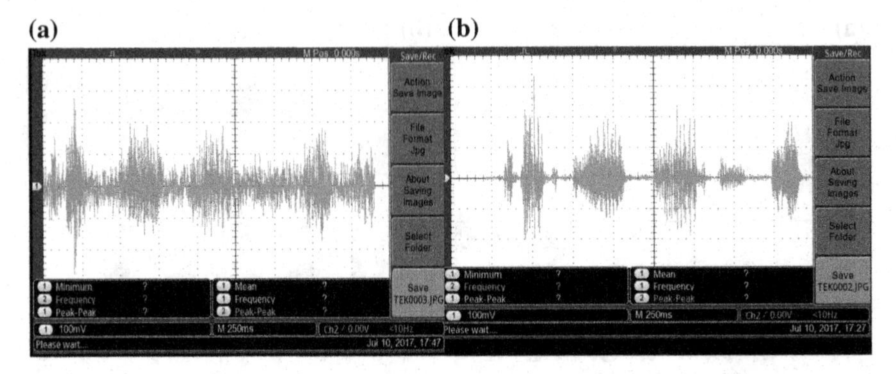

Fig. 7 **a** Speech utterance with background car noise. **b** Speech utterance of enhanced signal

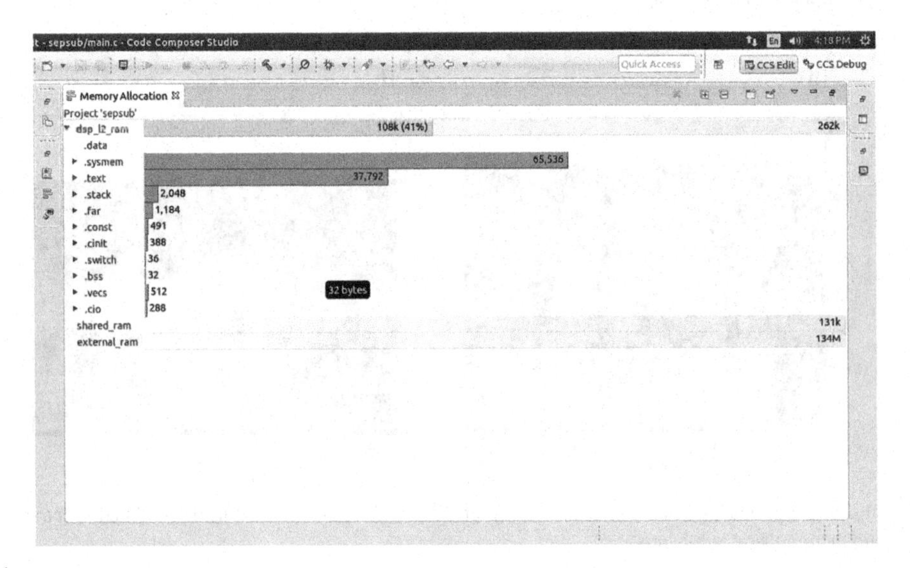

Fig. 8 Memory allocation for the modified SS using sub-band filtering

5 Conclusion

The proposed modified spectral subtraction using sub-band filters methodology provides better results than traditional power spectral subtraction. The proposed method can be easily implemented in DSP processors, which are the fundamental entity to process audio signals in mobile phone. We believe that the enhancement is achieved due to sub-band filtering approach, considering non-uniform effect of colored noise on the spectrum of speech. The added computational complexity of the algorithm is minimal. There is always a scope of improvement considering more than four linearly spaced frequency bands to obtain good speech quality.

References

1. Boll, S.F.: Suppression of acoustic noise in speech using spectral subtraction. IEEE Trans. Acoust. Speech Signal Process. **27**, 113–120 (1979)
2. Lockwood, P., Boudy, J.: Experiments with a nonlinear spectral sub tractor (NSS), hidden markov models and the projection, for robust speech recognition in cars. Speech Commun. **11** (2–3), 215–228 (1992)
3. Soon, I., Koh, S., Yeo, C.: Selective magnitude subtraction for speech enhancement. In: Fourth International Conference High Performance Computing in the Asia-Pacific Region, vol. 2, pp. 692–695 (2000)
4. He, C., Zweig, G.: Adaptive two-band spectral subtraction with multi-window spectral estimation. ICASSP **2**, 793–796 (1999)
5. Virag, N.: Single channel speech enhancement based on masking properties of the human auditory system. IEEE Trans. Speech Audio Process. **7**, 126–137 (1999)
6. Krishnamoorthy, P., Mahadeva Prasanna, S.R.: Enhancement of noisy speech by spectral subtraction and residual modification. IEEE Trans. Signal Process. 1–5 (2006)
7. Shin, J.W., Yoo, J.W., Park, P.G.: Variable step-size sign subband adaptive filter. IEEE Signal Process. Lett. **20**(2) (2013)
8. Epharin, Y., Malah, D.: Speech enhancement a minimum mean square error log–spectral amplitude estimator. IEEE Trans. Acoust. Speech Signal Process. **ASSP-32**, 1109–1121 (1984)
9. Epharin, Y., Mallah, D.: Speech enhancement using a minimum-mean square error log-spectral amplitude estimator. IEEE Trans. Acoust. Speech Signal Process. **ASSP-333**, 443–445 (1985)
10. Cai, Y., Yuan, J., Ma, X., Hou, C.: Low power embedded speech enhancement system based on a fixed-point DSP. In: 2009 Eighth IEEE International Conference on Dependable, Autonomic and Secure Computing, Chengdu, pp. 132–136 (2009)
11. Tiwari, N., Waddi, S.K., Pandey, P.C.: Speech enhancement and multi-band frequency compression for suppression of noise and intra speech spectral masking in hearing aids. In: 2013 Annual IEEE India Conference (INDICON), Mumbai, pp. 1–6 (2013)
12. Yermeche, Z., Sallberg, B., Grbic, N., Claesson, I.: Real-time DSP implementation of a sub band beam forming algorithm for dual microphone speech enhancement. In: 2007 IEEE International Symposium on Circuits and Systems, New Orleans, LA, pp. 353–356 (2007)
13. Purushotham, U., Suresh, K.: Feature extraction in enhancing speech signal for mobile communication. In: 2016 IICIP, Delhi, India, pp. 1–5 (2016)
14. http://www.ti.com/lit/ug/spruh79c/spruh79c.pdf
15. IEEE Subcommittee: IEEE recommended practice for speech quality measurements. IEEE Trans. Audio Electroacoust. **AU-17**(3), 225–246 (1969)

RETRACTED CHAPTER: In-silico Analysis of LncRNA-mRNA Target Prediction

Deepanjali Sharma and Gaurav Meena

Abstract Long noncoding RNAs (lncRNAs) constitutes a class of noncoding RNAs which are versatile molecules and perform various regulatory functions. Hence, identifying its target mRNAs is an important step in predicting the functions of these molecules. Current lncRNA target prediction tools are not efficient enough to identify lncRNA-mRNA interactions accurately. The reliability of these methods is an issue, as interaction site detections are inaccurate quite often. In this paper our aim is to predict the lncRNA-mRNA interactions efficiently, incorporating the sequence, structure, and energy-based features of the lncRNAs and mRNAs. A brief study on the existing tools for RNA-RNA interaction helped us to understand the different binding sites, and after compiling the tools, we have modified the algorithms to detect the accessible sites and their energies for each interacting RNA sequence. Further RNAstructure tool is used to get the hybrid interaction structure for the accessible lncRNA and mRNA sites. It is found that our target prediction tool gives a better accuracy over the existing tools, after encompassing the sequence, structure, and energy features.

Keywords Long non-coding RNA · Accessible sites · RNA structure · Target prediction · Machine learning

1 Introduction

Ribonucleic acid (RNA) molecules are single-stranded nucleic acids composed of nucleotides adenine (A), guanine (G), cytosine (C), and uracil (U). RNA plays a

---

The original version of this chapter was revised: The chapter has been retracted. The retraction note to this chapter is available at https://doi.org/10.1007/978-981-10-8569-7_39

D. Sharma (✉) · G. Meena
Central University of Rajasthan, Ajmer, India
e-mail: deepanjalisharma2013_imscs@curaj.ac.in

G. Meena
e-mail: gaurav.meena@curaj.ac.in

D. Reddy Edla et al. (eds.), *Advances in Machine Learning and Data Science*, Advances in Intelligent Systems and Computing 705,
https://doi.org/10.1007/978-981-10-8569-7_28

major role in protein synthesis as it is involved in the transcription, and translation of the genetic code to produce proteins. Messenger RNAs (mRNAs) in cells code for proteins. Apart from this, major part of the RNA world is composed of the noncoding RNAs (ncRNAs) that are transcribed from DNA but not translated into proteins. ncRNAs include long (>200 nucleotides long) as well as small ncRNAs (~30 nucleotides long). Small ncRNAs consist of microRNAs (miRNAs), small interfering RNAs (siRNAs), PIWI-interacting RNAs (piRNAs), etc. Long ncRNAs are the long noncoding RNAs (lncRNAs). In general, ncRNAs regulate gene expression at the transcriptional and post-transcriptional level. In this work, We have focused on the lncRNAs. LncRNAs are versatile molecules which have different regulatory roles in a biological system. They also have crucial roles in disease biology. Recent studies [1] have shown that lncRNAs are differentially expressed in various types of cancers including leukemia, breast cancer, hepatocellular carcinoma, colon cancer, and prostate cancer. In other diseases, such as cardiovascular diseases, neurological disorders, and immune-mediated diseases, lncRNA appears to be deregulated. LncRNAs involved in nearly all aspects of gene regulation, including X chromosomal inactivation [2] gene imprinting, epigenetic regulation, nuclear and cytoplasmic trafficking, transcription, mRNA splicing and translation. It should be mentioned that lncRNAs can serve as decoys, preventing the access of transcription factors [3]. Hence it is essential to understand the various modes of interaction of lncRNA with other RNA molecules to unveil its function. Unlike mRNAs which are localized in the cytoplasm, lncRNAs tend to be localized in the nucleus. This tendency suggest that lncRNAs exert their functions by interacting with precursor mRNAs (pre-mRNAs). Thus, it is important to understand the sequence and structural features of the mRNAs and lncRNAs as well as binding parameters between them to predict the interaction between lncRNAs and pre-mRNAs mentioned in Fig. 1.

Although many bioinformatic tools have been developed for RNARNA interactions, such as IntaRNA [5], GUUGle [6], RactIP [7], LncTar [8], RIblast [9],

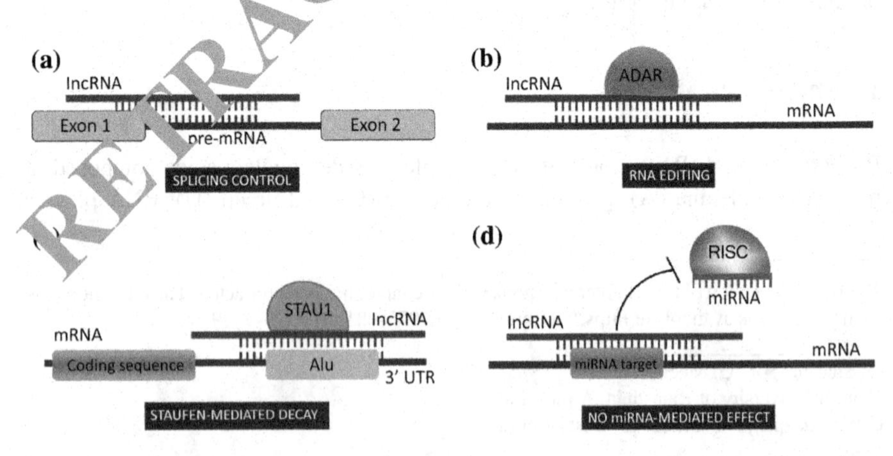

Fig. 1 Possible roles of lncRNA-mediated interactions in transcript processing, stability control, and expression regulation [4]

RIsearch [10], and RNAup [11], most of them are unable to meet the requirement of large-scale prediction of the RNA targets for lncRNAs. Most of these tools mentioned above have size issues and do not highlight any quantitative standard to determine whether two RNA molecules are accessible to each other or not. Hence, it is essential to identify all the accessible regions of query and target RNA sequences for the ease of target prediction. The large number of user-defined parameters required for target prediction makes the user very cautious while doing this analysis. The results have not been fully verified yet for most of the predicted target pairs. The absence of hybrid structure information in the above-mentioned tools for target prediction boost us to go ahead for designing our own target prediction tool.

In this work, We have devised an algorithm which will give approximate accessible energies for long RNA sequences and also energy- and structure-based features for the hybridization energy calculation.

2 Materials and Methods

2.1 *Input Dataset*

Human coding (mRNA) and ncRNA transcript information have been downloaded from Refseq database [12]. We have used experimentally validated lncRNA-mRNA information [8] for target prediction analysis.

2.2 *Computational Resources*

Since analyzing and storing of the long RNA sequence needs extensive computational resources, we have used a DELL precision server (10 GB RAM, Octa core processor) for this analysis. We have opted UNIX environment for our work. Perl and Python scripts have been used to write the code required for our analysis. We took help of awk and bash scripts for creating and modifying the database format.

2.3 *Discriminating Between the Nucleotide Composition of the Target and the Query Sequences*

In this problem, lncRNA (which is a ncRNA) is the query sequence and mRNA (which is the coding sequence) is the target sequence. It is very important to understand the nucleotide composition of both the target and the query sequences in order to elucidate effective binding region between the two. Knowing that the initial biologically important functional information is present at the sequence level, we have developed a Perl script *evaluating the composition-based features like mono, di, tri, and tetra* for both ncRNA and mRNA datasets.

We have used machine learning algorithms like Support Vector Machine (SVM), decision tree for discriminating the coding, and ncRNAs on the basis of compositional features. Coding and ncRNAs obtained from NCBI were classified into positive (+) and negative (−) datasets, respectively. Further 75% of the dataset was used as training dataset, and 25% was used as testing data. Feature extraction method was performed as follows.

We have considered 4 composition-based features such as mono, di, tri, and tetra (as mentioned above). There are 4 combinations for mononucleotide feature, 16 combinations for dinucleotide feature, 64 combinations for trinucleotide feature, and 256 combinations for tetra nucleotide feature. Thereafter we proceeded in the following way:

- Calculating the average of each of the composition-based features (as mentioned above) for both the positive and negative datasets.
- Calculating the difference of the average of each of the composition-based features (as calculated in previous step) between coding and noncoding sequences.
- Calculating the mean of these differences (as calculated in previous step).
- Selecting the features whose mean of the difference values (calculated in step 3) > than the difference value (calculated in step 2).

SVM models were trained using different kernels (poly, linear, gamma) and the model was evaluated using the evaluation matrix based on the Eqs. 1–3.

$$Sensitivity = \frac{TP}{TP + FN} * 100 \tag{1}$$

$$Specificity = \frac{TN}{TN + FP} * 100 \tag{2}$$

$$Accuracy = \frac{TN + TP}{TN + FP + TP + FN} * 100 \tag{3}$$

TP = True Positive, TN = True Negative, FP = False Positive, FN = False Negative

2.[illegible] Identifying Accessible Nucleotides Using Stem Probability

Inaccessible and accessible regions of an RNA sequence are discriminated with the help of stem probability (σ). Stem probability for each nucleotide is calculated using ParaSor algorithm [13] . This algorithm can exhaustively compute structural features such as structural profiles [14] and globally consistent γ-centroid structures [15], as well as conventional base pairing probabilities, stem probabilities, and accessibilities. It divides dynamic programming (DP) matrices into smaller pieces, such that each piece can be computed by a separate computer node without losing the connectivity information between the pieces.

The nucleotides having stem probability greater than the threshold value are grouped as accessible nucleotides, and the nucleotides whose stem probability is less than the threshold value are grouped as paired ones.

2.5 Calculating the Accessible Energy Needed for Efficient LncRNA-Target Binding

Accessible energy is the energy required to prevent the RNA sequences from forming intramolecular base pairs. It can be calculated by utilizing a partition function algorithm [13]. A segment with high accessible energy has a higher probability of intramolecular base pairs and hence less chance to form intermolecular base pairs. Hence accessible energy at the site of interaction of the query and target RNA should be less than that of the inaccessible region so as to facilitate intermolecular base pair formation between the two RNAs. In simple words, accessible energy can be understood as the measure of intramolecular bond strength of an RNA sequence. After evaluating the accessible nucleotides (as mentioned above), seed length (a region where query and target binding initiates) has to be specified by the user for selecting the intervals. Thereafter, the sequence can be divided into such continuous intervals of a specific window size, and then the accessible energy using the traditional methods [16] can be calculated for each window; the sum of the accessible energies of each window length can be defined as the total accessible energy of the sequence. The above procedure will be repeated for calculation of the accessible energy for the target sequence. We have used ParaSoR [13], which is an efficient tool for calculating accessible energies for long RNA sequences. We took up the dataset corresponding to experimentally validated lncRNA–mRNA interactions reported in LncTar [8] and calculated the accessible energies for lncRNAs and target mRNAs pairs. The algorithm to calculate the accessible energy for interactive and noninteractive region is described in Algorithm 1.

2.6 Calculating the Hybridization Energy Needed for Efficient LncRNA-Target Binding

Hybridization energy is the free energy derived from intermolecular base pairs between two segments. This can be calculated as the sum of stacking energies and loop energies in the base-paired structure based on a nearest-neighbor energy model. While calculating hybridization energies, intramolecular base pairs were not taken into consideration. The hybridization energy is evaluated considering all accessible regions of the query and target. We have used Bifold [17] (*which incorporates the secondary and tertiary hybrid structures*) to calculate the hybridization energies for two RNA sequences.

2.7 Calculating Interaction Energy Between LncRNA and mRNA

Interaction energy (I) can be defined as the sum of the accessible energies (Acc) of query and target RNA sequences and the hybridization energy (H) between them. Interaction energy (I) between two segments xs and ys is the sum of the accessible energy of xs, accessible energy of ys and hybridization energy between xs and ys. The workflow for calculating the interaction energy between the lncRNA and target mRNA is shown in Fig. 2.

Algorithm 1 Calculation of the accessible energies for interacting and non-interacting sites of the query and the target

1. `Query_Start Query_end Target_start Target_end` ← LncTar(query,target)
2. `Range,Acc` ← ParaSoR(query))
3. `st` ← value of `Query_start`
4. `end` ← value of `Query_end`
5. `Rng` ← value of `Range` where `Range` is obtained from ParaSoR
6. `Acc` ← value of `Acc` where `Acc` is accessible energies obtained from ParaSoR for corresponding window
7. `mair` ← where `mair` is minimum of accessible energies between [st,end] obtained
8. `mawr` ← where `mawr` is minimum of accessible energies of whole region
9. **function** VALIDATION(`st end Rng Acc`)
10. mawr=min(Acc)
11. mair=min(Acc[st:end])
12. **if** *mair* > *mawr* **then**
    Return Output=Invalid;
13. **else**
    Return Output=Valid;
14. Repeat steps 2 to 6 again for Target
    Call Validation(st end Rng Acc)

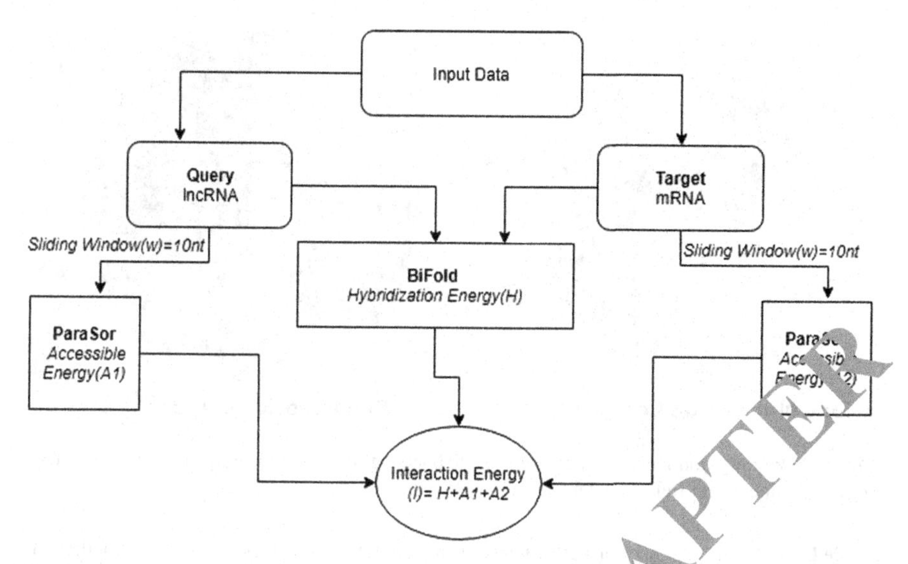

Fig. 2 Workflow for calculating the interaction energy between lncRNA and its target mRNA

3 Results

1. **Discriminating between the nucleotide composition of the target and the query sequences**: The in-house Perl script calculates the length of the sequence, AT count, GC count, and sequence features of the corresponding transcript. The mononucleotide and dinucleotide frequency composition of ncRNAs has been shown in Fig. 3a, b respectively. Similarly the mononucleotide and dinucleotide

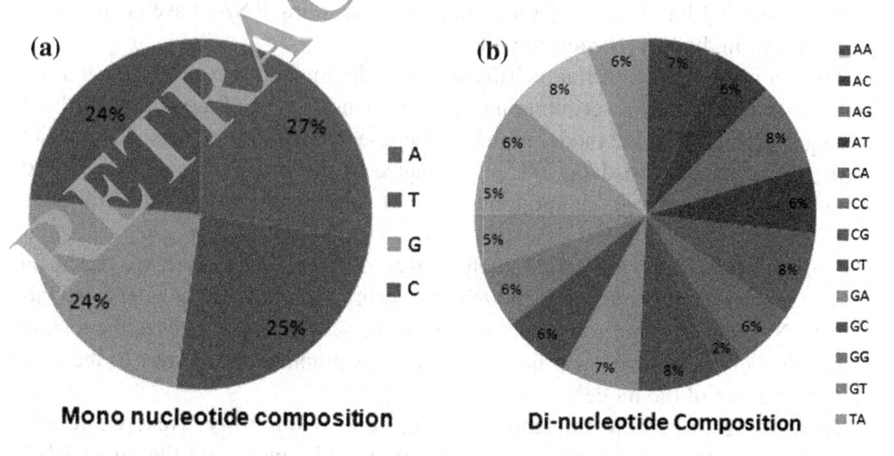

Fig. 3 **a** Average mononucleotide frequency and **b** average dinucleotide composition frequency of mRNA transcripts obtained from RefSeq database

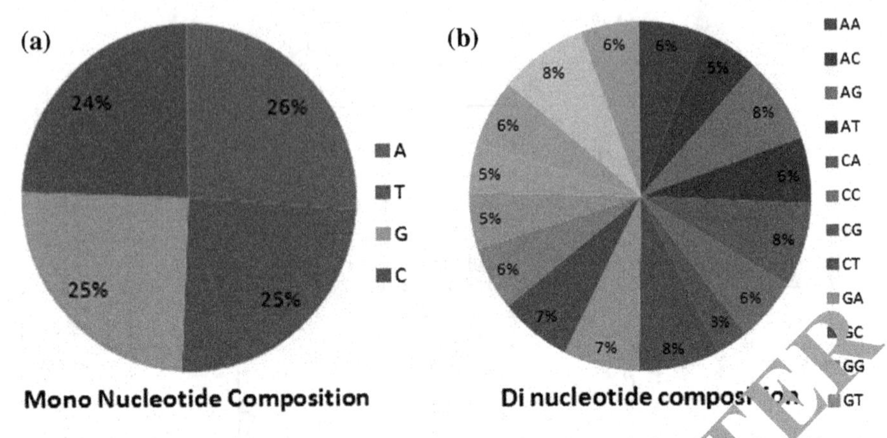

Fig. 4 **a** Average mononucleotide frequency and **b** average dinucleotide frequency of ncRNA transcripts obtained from RefSeq database

Table 1 SVM-based prediction performances using different nucleotide compositions for discrimination between target and query RNAs. Mono (MNC), DI (DNC), TRI (TNC), TETRA (TTNC) are corresponding nucleotide compositions

Sequence features	Accuracy (%)	Sensitivity (%)	Specificity (%)
MNC	58.98	36.07	70.14
DNC	61.83	41.03	71.97
TNC	65.64	45.[illegible]1	75.59
TTNC	67.94	[illegible]38	77.47

frequency composition of mRNA has been shown in Fig. 4a, b respectively. The compositional-based features for coding and noncoding RNAs have been briefly cataloged in the **Supplementary file**.

The feature selection technique (discussed in the method section) is carried out to build a model for discriminating the query (lncRNA) and the target (mRNA) sequences. The trained model based on the selected feature set is used to predict the result for the test data. The performance of the model is calculated based on the Eqs. 1–3. The evaluation matrix for the predicted model is mentioned in Table 1.

It is observed from the above analysis that this selected feature set based on nucleotide composition is not enough to discriminate between the query and target RNAs due to lower sensitivity which can be seen in Fig. 5. Hence, the feature set must consider the repeat families as well as graphical properties to increase the accuracy of the model.

2. **Identifying accessible nucleotides using stem probability**: Next we moved on to identify the accessible nucleotides within the query and the target RNA sequences so as to initiate binding between them. Nucleotides are categorized as accessible and inaccessible based upon the threshold value. As stem probability

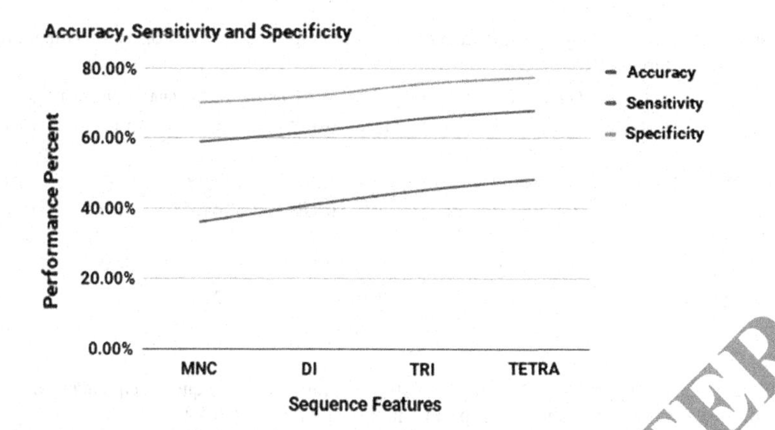

Fig. 5 Graph representing the performance of SVM model for discriminating query and target RNAs on the basis of sequence features

decreases, the overlapping of the accessible nucleotides of query and target RNAs within the experimentally validated interaction region between lncRNA and mRNA (as taken from LncTar) also decreases (this has been checked for HIF1A-AS1 and HIF1A1). At $\sigma = 0.6$, we observed a maximum overlap for both query and target sequences. The above analysis has been performed for a single validated pair (mentioned in LncTar). We need to check the same for other validated pairs mentioned in LncTar for precise selection of the cutoff for stem probability.

3. **Calculating the accessible energy needed for efficient lncRNA-target binding**: Thereafter we have calculated the accessible energies of the query and the target RNAs. Using ParaSor, the accessible energies of the interacting and noninteracting sites of query (lncRNA) and its target (mRNA) are calculated. Table 2 reports the minimum accessible energies for the interacting sites Q_{int_min} and T_{int_min} of the query and target, respectively. It is found that for all the experimentally validated lncRNA-mRNA pairs where the exact interaction site is reported, the $Q_{int\_min} < Q_{nint\_min}$ for the query and $T_{int\_min} < T_{nint\_min}$ for the target. *This proves the fact that accessible energy at the site of interaction of the query and target RNAs is less than that of the inaccessible region which facilitates intermolecular base pair formation between the two RNAs.*
For 5 lncRNA–mRNA interactions in LncTar, the exact interaction site is not reported. Hence for these, we have calculated the accessible energies for the query and the target RNAs for each segments of both (of window length 10) and have reported the regions with minimum accessible energy, i.e., Q_{int_min} and T_{int_min} (for both in the query and the target) as shown in Table 3. These regions can be designated as the sites of interaction of the lncRNAs with their targets for these cases (Fig. 6).

Table 2 Accessible energies of the interacting and noninteracting sites of 5 experimentally validated lncRNA-target mRNA pairs

Query	Target	Qstart	Qend	Tstart	Tend	Q_int_min	Q_nint_min	T_int_min	T_nint_min
HIF1A-AS2	HIF1A	808	2051	1	1244	0.0407	0.12797	0.404	0.0419
EMX2OS	EMX2OS1	1	145	2764	2908	2.076	0.3958	0.06	0.0419035
HIF1A-AS1	HIF1A	1	423	3533	3955	0.39586	1.0644	0.6867	0.04253
BDNF-AS	BDNF	1	935	9042	3976	0.35	0.29774	0.348	0.028794
NPPA-AS1	NPPA	972	1662	1	691	0.182	0.321184	0.13	0.028794

Table 3 Designating the interaction sites within the target and the query sequences based on the minimum accessible energies for 5 experimentally validated lncRNA-target mRNA pairs

Query	Target	Qstart	Qend	Tstart	Tend	Q_int_min	T_int_min
BC200	ARC	1	200	59	258	[illegible]	0.9288
BACE1-AS	BACE1	1	840	4096	4935	[illegible]	0.22637
PEG3-AS1	PEG3	1	1314	2410	3723	[illegible]	0.1026
BC200	MAP1B	1	200	2058	[illegible]	0.0227	0.18864
BC200	CAMKA21 1	200	4669	4868	[illegible]	0.1766	

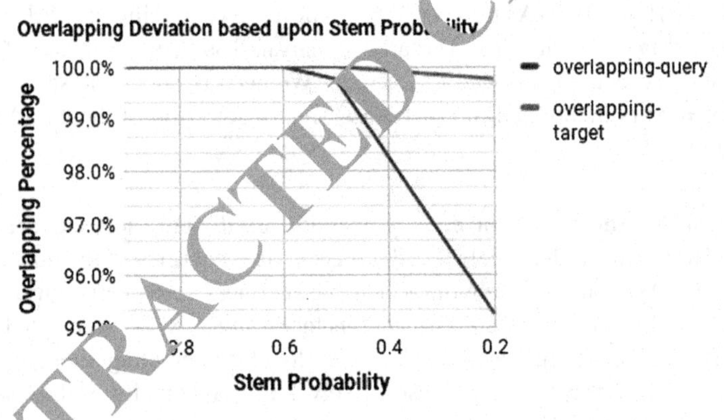

Fig. 6 Thresho[illegible] for stem probability for discriminating accessible and inaccessible regions for query [illegible]) and target (mRNA) using the lncTar output

4. **Calculating hybridization energy between lncRNA and mRNA**: Bifold [17] calculates the hybridization energies for two RNA sequences (query and target), and it also incorporates the secondary and tertiary hybrid structures. The efficiency of the analysis is par with other existing lncRNA-target prediction methods. The hybridization energy calculated for the validated lncRNA-mRNA targets is provided in Table 4.

Table 4 Hybridization energy between lncRNA-mRNA

Query	Target	Hybridization energy
HIF1A-AS2	HIF1A	−6158.526557
EMX2OS	EMX2OS1	−2509.043746
HIF1A-AS1	HIF1A	−679.7820264
BDNF-AS	BDNF	−2620.165492
NPPA-AS1	NPPA	−296.2256511
BC200	ARC	−84.8714098
BACE1-AS	BACE1	−662.324779
PEG3-AS1	PEG3	−6720.87946
BC200	MAP1B	−251.353148
BC200	CAMKA21	−352.8093336

Table 5 Interaction energy between lncRNA-mRNA

Query	Target	Interaction energy
HIF1A-AS2	HIF1A	−903.317
EMX2OS	EMX2OS1	−[illegible]4.3084
HIF1A-AS1	HIF1A	−107.476
BDNF-AS	BDNF	−187.887
NPPA-AS1	NPPA	−49.7978
BC200	ARC	−10.6164
BACE1-AS	BACE1	−86.068
PEG3-AS1	PEG3	−911.62
BC200	MAP1B	−11.8859
BC200	CAMKA21	−18.1752

5. **Calculating interaction energy between lncRNA and mRNA**: Summing up both the accessible energies of the query and the target RNAs and hybridization energy using Bifold, the interaction energy is calculated. The interaction energy for the experimentally validated interactions of lncRNA-target mRNA pairs is shown in Table 5.

4 Discussion and Conclusion

We have studied the nucleotide compositions of the target (mRNA) and the query (lncRNA) sequences in order to find out the sequence-based features within them which will have an effect on target binding. We have also tried to discriminate between them based upon the features using machine learning model [18]. From this analysis, it is concluded that nucleotide composition-based feature is not enough for characterizing the query and the target sequences. It is important to check the accessibility parameters in order to zoom into the target binding story of lncRNAs. Hence we took up the experimentally validated lncRNA-target mRNA information from the lncTar and checked the accessible nucleotides and the accessible energy of the target and query sequences in order to discriminate between the accessible and inaccessible regions in the query and target sequences. We also calculated the hybridization energy between the two and thereafter calculated the interaction energy for lncRNA-target mRNA pairs constituting the validated dataset. This work will serve as the building block to calculate and select the features that can be used for designing lncRNA-target prediction protocol. As a future scope of this work, our aim is to build up the training dataset using greater number of experimentally validated dataset and eventually build up the target prediction model using machine learning algorithms.

Acknowledgements We would like to thank Dr. Zhumur Ghosh (Assistant Professor, Bose Institute) and Sibun Parida (Research Associate, Bioinformatics Center) for their valuable support.

References

1. Available at https://www.lncrnablog.com/what-are-lncrnas/
2. Lee, J.T., Bartolomei, M.S.: c, imprinting, and long noncoding RNAs in health and disease. Cell **152**(6), 1308–1323 (2013)
3. Rinn, J.L., Chang, H.Y.: Genome regulation by long noncoding RNAs. Annual Rev. Biochem. **81**, 145–166 (2012)
4. Szczeniak, M.W., Makaowska, I.: lncRNA-RNA interactions across the human transcriptome. PloS One **11**(3), e0150353 (2016)
5. Busch, A., Richter, A.S., Backofen, R.: IntaRNA: efficient prediction of bacterial sRNA targets incorporating target site accessibility and seed regions. Bioinformatics **24**(24), 2849–2856 (2008)
6. Gerlach, W., Giegerich, R.: GUUGle: a utility for fast exact matching under RNA complementary rules including GU base pairing. Bioinformatics **22**(6), 762–764 (2006)
7. Kato, Y., et al.: RactIP: fast and accurate prediction of RNA-RNA interaction using integer programming. Bioinformatics **26**(18), i460–i466 (2010)
8. Li, J., et al.: LncTar: a tool for predicting the RNA targets of long noncoding RNAs. Brief. Bioinfo. **15**(5), 806–812 (2014)
9. Fukunaga, T., Hamada, M.: RIblast: an ultrafast RNA RNA interaction prediction system based on seed and-extension approach. Bioinformatics (2017)
10. Wenzel, A., Akbali, E., Gorodkin, J.: RIsearch: fast RNA RNA interaction search using a simplified nearest-neighbor energy model. Bioinformatics **28**(21), 2738–2746 (2012)
11. Hofacker, I.L.: RNA secondary structure analysis using the Vienna RNA package. In: Current Protocols in Bioinformatics, pp. 12–22 (2009)
12. Available at https://www.ncbi.nlm.nih.gov/projects/genome
13. Kawaguchi, R., Kiryu, H.: Parallel computation of genome-scale RNA secondary structure to detect structural constraints on human genome. BMC Bioinfo. **17**(1), 203 (2016)
14. Fukunaga, T., Ozaki, H., Terai, G., Asai, K., Iwasaki, W., Kiryu, H.: CAPR: revealing structural specificities of RNA-binding protein target recognition using CLIP-seq data. Genome Biol. **15**(1), 16 (2014)
15. Hamada, M., Kiryu, H., Sato, K., Mituyama, T., Asai, K.: Prediction of RNA secondary structure using generalized centroid estimators. Bioinformatics **25**(4), 46573 (2009)

16. Kiryu, H., et al.: A detailed investigation of accessibilities around target sites of siRNAs and miRNAs. Bioinformatics **27**(13), 1788–1797 (2011)
17. Available at http://rna.urmc.rochester.edu/RNAstructureWeb
18. Panwar, B., Amit, A., Gajendra, P.S.R.: Prediction and classification of ncRNAs using structural information. BMC Genom. **15**(1), 127 (2014)

Energy Aware GSA-Based Load Balancing Method in Cloud Computing Environment

Vijayakumar Polepally and K. Shahu Chtrapati

Abstract Cloud computing environment is based on a pay-as-you-use model, and it enables hosting of prevalent applications from scientific, consumer, and enterprise domains. Cloud data centers consume huge amounts of electrical energy, contributing to high operational costs to the organization. Therefore, we need energy efficient cloud computing solutions that cannot only minimize operational costs but also ensures the performance of the system. In this paper, we define a framework for energy efficient cloud computing based on nature inspired meta-heuristics, namely gravitational search algorithm. Gravitational search algorithm is an optimization technique based on Newton's gravitational theory. The proposed energy-aware virtual machine consolidation provision data center resources to client applications improve energy efficiency of the data center with negotiated quality of service. We have validated our approach by conducting a performance evaluation study, and the results of the proposed method have shown improvement in terms of load balancing, energy consumption under dynamic workload environment.

Keywords Load balancing · Cloud computing · Virtual machine migration Gravitational search algorithm

1 Introduction

1.1 Cloud Computing Environment

At present, cloud computing is an attracting research topic, which stands superior over all other growing technical ideas. The main reason behind this growth is

V. Polepally (✉)
Kakatiya Institute of Technology and Science, Warangal, Telangana, India
e-mail: vijaykumarpolepally@gmail.com

K. Shahu Chtrapati
JNTUH College of Engineering, Manthani, Peddapalli, Telangana, India

D. Reddy Edla et al. (eds.), *Advances in Machine Learning and Data Science*,
Advances in Intelligent Systems and Computing 705,
https://doi.org/10.1007/978-981-10-8569-7_29

because of the resources and services provided by cloud computing to the clients [1]. Cloud computing is a term, which provides distributed computing, storing, sharing, and accessing of data through the Internet. Cloud computing is an environment that enables user-friendly services like on-demand access with a plenty of resources available in the data center. At present, cloud computing is attracting most of the organizations and clients as it provides low-cost and high-quality services on-demand basis [2]. The services to the clients are from various domains, and the cloud servers have the facility to control and monitor all the processes involved in the cloud environment, like client request, data outsourcing, and so on [3]. Thus, this cloud computing promotes mass consumption and delivery of the requested services and protocols. Upon client requests, data center provides hardware, software, and data services. These collaborative services performed based upon the demand faces a lot of challenges. The most significant challenges include the poor QoS, interoperability problems, and scalability issues. These challenges that affect the cloud servers will result in reliability problems, problems in efficient resource utilization, load balancing, and so on. Among all, load balancing is a major issue in the cloud computing environment [4].

1.2 Literature Survey

Load balancing is of great importance in the cloud environment as it reduces the interaction between the various virtual machines, which in turn reduces the energy consumption. The virtual machines are controlled using the large servers present in the data centers, and they share a large amount of resources provided by the servers. These virtual machines may get loaded with the resource consumption in certain cases, and a perfect model is required, which computes the exact physical host for load balance [5]. Load balancing with the virtual machine migration (VMM) strategy is a solution to achieve the perfect load balance in the cloud network. VMM strategy is simply the migration of the virtual machine from the over loaded physical machine to the other under loaded physical machine. In addition, it is essential to determine the physical host to undergo migration. Various technologies have been employed to determine the exact physical host to migrate the loaded virtual machine. After determining the physical host, load balancing is done [6, 7]. However, this physical position does not hold good for the varying load. Therefore, in the dynamic cloud environment it is clear that for varying load, the physical host varies, which poses the need to determine the optimal physical host for the migration[8, 9].

The organization of the paper is as follows: Introduction and literature survey were presented in Sect. 1. Section 2 is demonstrating about the cloud system model with its load, energy, and quality of service models. The proposed algorithm is presented in Sect. 3 followed by conclusion in Sect. 4.

2 System Model

Cloud data center is an integration of several physical machines, in turn, each physical machine is partitioned into several virtual machines based on the requirement of the resources like memory, CPU time, energy consumption, network band width. Cloud environment performance is a multi-objective function in which the most important resources are to be optimally utilized. In our proposed method, load balancing is achieved by migrating the VM of the overloaded PM to another under loaded PM by considering the performance metrics quality of service (QoS) and energy consumption.

2.1 Energy Model

Energy utilization is the major concern concentrated in this section. The objective of energy model is to maintain the efficient energy utilizing capability. Energy modeling is more significant as it poses the necessity to utilize minimum energy while attaining maximum utilization of resources.

Figure 1 representing a model for the effective utilization of typical cloud data center resources like CPU, memory, and energy consumption. Initially, PM1 with VM1 and VM2, PM2 with VM3 and VM4, PM3 with VM4 and VM6 are using some amount of resources, respectively. This leads to more energy consumption. Therefore, migration of virtual machines optimally on to other physical machines will significantly reduce the energy consumption which will improve the overall performance of the cloud computing environment in terms of energy utilization.

Energy model of the system is shown below

$$e = \frac{1}{|PM| * |VM|} \left[\sum_{i=1}^{|PM|} \sum_{j=1}^{|VM|} N_{ij} e_{\max} + (1 - N_{ij}) R_{ij}^{U} e_{\max} \right] \quad (1)$$

where PM is the total number of physical machines in the data center, VM is the total number of virtual machines in a particular machine. R_{ij}^{U} is the resource utilized by the *i*th physical machine of the *j*th virtual machine, N_{ij} is the fraction of energy consumed by the *i*th physical machine of the *j*th virtual machine, $e_{\max}$ is the maximum energy consumed, *e* is the total energy consumed during migration.

The resources utilized by the *i*th physical machine of the *j*th virtual machine is shown below

$$R_{ij}^{U} = \frac{1}{2} \left[\frac{C_{ij}^{U}}{C_{ij}} + \frac{M_{ij}^{U}}{M_{ij}} \right] \quad (2)$$

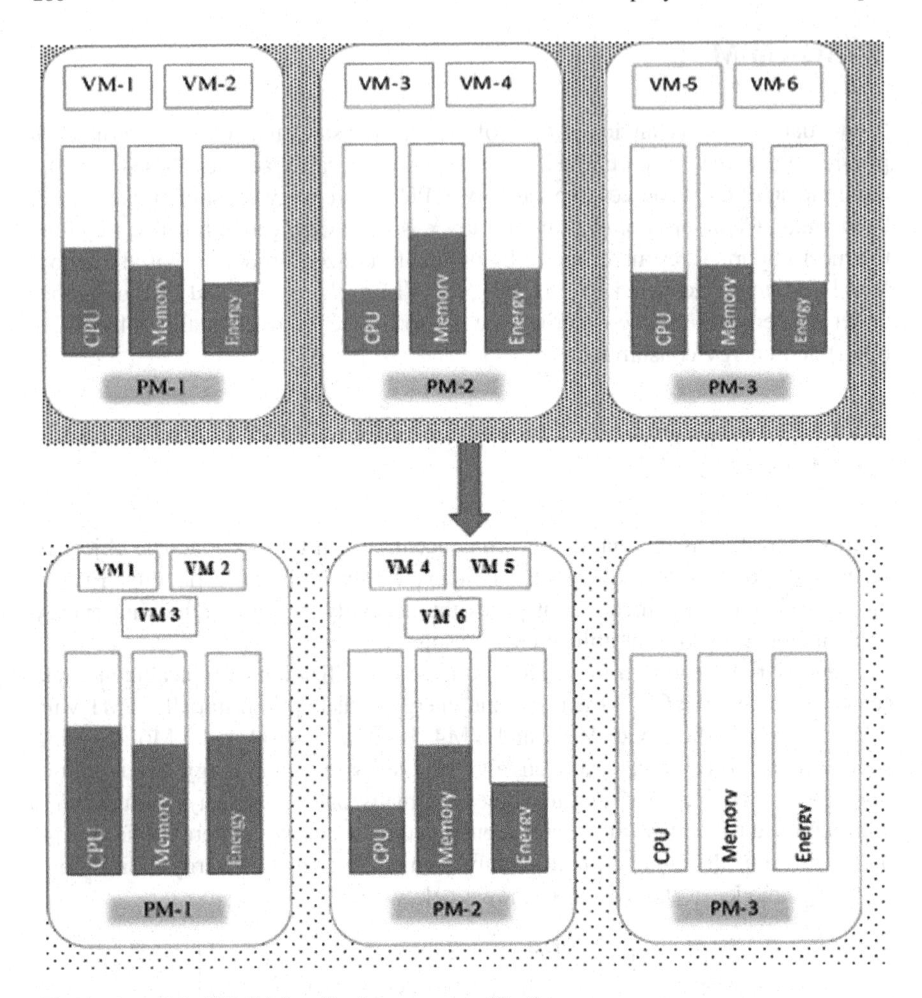

Fig. 1 A model of VMM for effective resource utilization

where C_{ij}^{U} is the number of CPU utilized in the *i*th physical machine of the *j*th virtual machine, C_{ij} is the total number of CPU in the *i*th physical machine of the *j*th virtual machine, M_{ij}^{U} is the memory used by the *i*th physical machine of the *j*th virtual machine, and M_{ij} is the total memory available in the *i*th physical machine of the *j*th virtual machine.

2.2 QoS Model

QoS modeling is significant in cloud computing environment. QoS determines the overall performance of the network. Hence, it is necessary to perform all the functions involved in the network with high quality. QoS of the cloud network depends upon the QoS of the memory in the ith physical machine and the QoS of the CPU in the ith physical machine. The QoS of the cloud network is formulated from the following equation as

$$QoS^{t} = \frac{1}{2}[QoS^{c} + QoS^{m}] \tag{3}$$

where QoS^{t} is the total quality of service of the cloud network, QoS^{c} is the quality of service of the CPU in the ith physical machine, and QoS^{m} is the quality of service of the memory in the ith physical machine.

The computation of the QoS for the CPU and memory of the ith physical machine is given using the expressions below

$$QoS^{C} = \frac{1}{|P|}\sum_{i=1}^{|P|}\frac{C_i^d}{C_i} \tag{4}$$

where C_i^d corresponds to the unutilized CPU in the ith physical machine and C_i corresponds to the total CPU in the ith physical machine.

$$QoS^{M} = \frac{1}{|P|}\sum_{i=1}^{P}\frac{M_i^d}{M_i} \tag{5}$$

where M_i^d is the unutilized memory present in the ith physical machine and M_i is the total memory present in the ith physical machine.

2.3 Load Model

Load modeling enables to schedule the user requested resource without any traffic. This section carries the analysis of the varying loads. The condition of the load in the cloud network is the key factor behind load modeling. The cloud data center provides all the relevant data requested by the user based upon the priorities set to them during the time of arrival. During this process, load calculation of the overall cloud network is essential and the calculation of load as in Eq. (6). Initially, a load threshold is fixed for detecting the loaded condition. Let Lth be the load threshold and L_D be the calculated load of the network. When the calculated load is found less than the threshold load, this states that the network is not under overload, but when the calculated load is found greater than the threshold load, the condition states that the system is under overload. A load of each physical machine varies from time to time.

$$L_M = \frac{1}{|P| * |V|} \left[\sum_{i=1}^{|P|} \sum_{j=1}^{|V|} \left(\frac{C_{ij}^U}{C_{ij}} + \frac{M_{ij}^U}{M_{ij}} \right) \right] * \frac{1}{2} \tag{6}$$

where C_{ij}^U is the CPU utilized by the jth virtual machine in the ith physical machine, C_{ij} is the total CPU utilized in ith physical machine of the jth virtual machine, M_{ij}^U is the memory used by the ith physical machine of the jth virtual machine, M_{ij} is the total memory available in the ith physical machine of the jth virtual machine, $|P|$ is the total number of physical machines present in the network, and $|V|$ is the total number of virtual machines present in the network.

3 Proposed Load Balancing Method Based on GSA

Figure 2 shows the block diagram of the proposed method. Basically, in the cloud computing environment, there is a lot of user requests seeking the relevant resource from the data center. The cloud data center holds all the relevant data to satisfy the

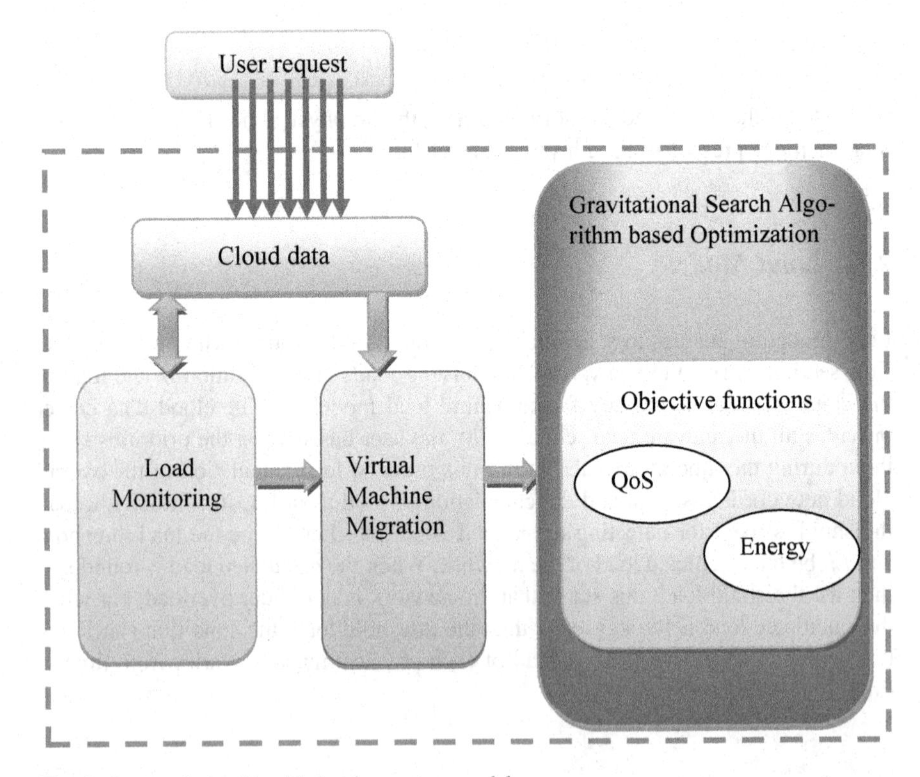

Fig. 2 Proposed cloud load balancing system model

requested users. The exchange of the resource to the users normally follows this simple procedure. However, when overload is detected, virtual machines cannot handle the request, which enables the need for the load monitor. Therefore, the load monitor monitors the load conditions using a load balancing mechanism. The load balancing mechanism employs a simple calculation to compute the load. If the computed load is greater than the load threshold, then the network undergoes load balancing using the VMM strategy. The proposed algorithm is an optimization algorithm that performs continuous iteration to undergo the optimized migration of virtual machines keeping in mind the considered two performance metrics, namely quality of service and energy onsumption.

Proposed Load Balancing Algorithm.

1. Initialization of the cloud model
2. Determine the load on the PMs
3. If (Load > Threshold)
4. Perform Virtual Machine Migration.//call GSA
5. End
6. Update the variables
7. End

3.1 Fitness Calculation

Fitness calculation is an important parameter in the VMM strategy, which identifies the best migrating location. Fitness is the key factor for performing the effective migration of the virtual machines without affecting the system performance. Equation (7) shows the formula for performing the fitness calculation.

$$fit = \alpha * QoS + \beta(1 - e) \tag{7}$$

QoS is the quality of service.

α and β are the constants.

The fitness calculation uses the maximization concept, which means the value obtained from the computation of the fitness function should be maximum. The fitness function depends on the QoS and the energy of consumption. Energy consumption is a significant factor that contributes to a better fitness value. The energy consumed while utilizing the resources should be minimum. However, the resource utilization should remain maximum.

Fitness calculation is possible only when the resources present in the loaded physical machine are greater than the old resources present in the machine else the fitness function is zero.

$$Fitness = \begin{cases} fit, \text{Re}\,source\ of\ L_M > Old\,\text{Re}\,sources \\ 0, Otherwise \end{cases}$$

Equation (7) determines the best fit values, and the other factors like the migration cost and the quality of service are determined for the best fit value.

3.2 GSA Algorithm

According to the gravitational theory, all the objects are determined using their masses and the communication between them is due to the gravity [10]. This theory introduces the parameters namely, mass, acceleration, velocity, and energy. Thus, in this algorithm, the cloud computing environment is considered, which possess some

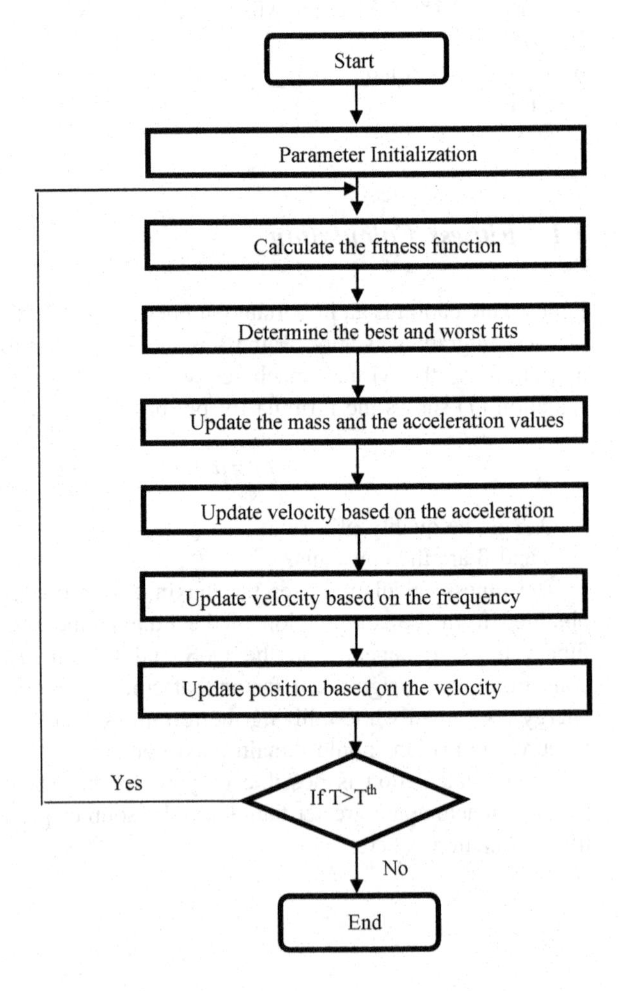

Fig. 3 Flowchart for gravitational search algorithm

physical machines and the virtual machines. When huge requests reach the data center, the physical machines become overloaded. Hence, the need for the migration of virtual machine arises. Moreover, the migration location is very significant as it influences the QoS, migration time, and energy.

The best migration position depends on the quality of service and energy. It is necessary to attain the maximum resource utilization capacity with the consumption of very less amount of energy. Initially, parameter initialization and randomization are done. The fitness function determines the best and worst fit values, and they are prioritized based on the fitness values. Among these prioritized values, the best value is selected and subjected to mass and acceleration computations. The velocity update is possible using the calculated acceleration value, and the frequency-based velocity value is determined. At the same time, update the best position to perform virtual machine migration with the frequency-based velocity computation. On the other hand, determine the energy consumption and quality of service for the selected locations. Finally, the location to migrate the virtual machine is determined, which is found to utilize the minimum energy. Thus, a highly energy efficient and high-quality migration are carried out using this proposed algorithm to determine the optimized location for VMM. Figure 3 shows the flowchart that represents the steps involved in the migration algorithm.

4 Conclusion

Cloud computing resources are in huge demand due to the convenience of using the resources based on the need. This high demand for resources from the variety of cloud users is leading to the unforeseen load on the cloud data center which may cause the cloud system failure. Hence, the load balancing methods have prominent role in smooth functioning of cloud environment. There are numerous load balancing techniques that are proposed by several computing scientists among which nature-inspired optimization algorithm-based load balancing is significant. Gravitational search algorithm is a population-based optimization techniques used for multi-objective function load balancing problem. The GSA is used to identify the VM for migration by satisfying multiple objectives like CPU time, memory, and energy consumption. The performance of the proposed GSA-based VMM load balancing algorithm is significantly better than some of the existing algorithms. In future, we are planning to consider other performance metrics in load balancing.

References

1. Zhan, Z.-H., Liu, X.-F., Gong, Y.-J., Zhang, J., Chung, H.S.-H., Li, Y: Cloud computing resource scheduling and a survey of its evolutionary approaches. ACM Comput. Surv. (CSUR) **47**(4), 63 (2015)

2. Dhinesh Babu, L.D., Venkata Krishna, P.: Honey bee behaviour inspired load balancing of tasks in cloud computing environments. Appl. Soft Comput. **13**(5), 2292–2303 (2013)
3. Tang, M., Pan, S.: A hybrid genetic algorithm for the energy-efficient virtual machine placement problem in data centers. Neural Process. Lett. **41**(2), 211–221 (2015)
4. Beloglazov, A., Abawajy, J., Buyya, R.: Energy-aware resource allocation heuristics for efficientmanagement of data centers for cloud computing. Future Gener. Comput. Syst. **28**(5), 755–768 (2012)
5. Merkle, D., Middendorf, M., Schmeck, H.: Ant colony optimization for resource-constrained project scheduling. IEEE Trans. Evol. Comput. **6**(2), 333–346 (2002)
6. Domanal, S.G., Reddy, G.R.M.: Optimal load balancing in cloud computing by efficient utilization of virtual machines. In: Proceedings of the Sixth International Conference on Communication Systems and Networks (COMSNETS), pp. 1–4, 10 Feb 2014
7. Ahmad, R.W., Gani, A., Hamid, S.H.A., Shiraz, M., Yousafzai, A., Xia, F.: A survey on virtual machine migration and server consolidation frameworks for cloud data centers. J. Netw. Comput. Appl. **52**, 11–25 (2015)
8. Ghomi, E.J., Rahmani, A.M., Qader, N.N.L.: Load-balancing algorithms in cloud computing: a survey. J. Netw. Comput. Appl. (2017)
9. Jiang, S., Ji, Z., Shen, Y.: A novel hybrid particle swarm optimization and gravitational search algorithm for solving economic emission load dispatch problems with various practical constraints. Int. J. Electr. Power Energy Syst. **55**, 628–644 (2014)
10. Rashedi, E., Nezamabadi-Pour, H., Saryazdi, Saeid: GSA: a gravitational search algorithm. Inf. Sci. **179**(13), 2232–2248 (2009)

Relative Performance Evaluation of Ensemble Classification with Feature Reduction in Credit Scoring Datasets

Diwakar Tripathi, Ramalingaswamy Cheruku and Annushree Bablani

Abstract Extensive research has been done on feature selection and data classification. But it is not clear which feature selection approach may result in better classification performance on which dataset. So, the comparative performance analysis is required to test the classification performance on the dataset along with feature selection approach. Main aim of this work is to use various feature selection approaches and classifiers for the evaluation of performances of respective classifier along with feature selection approach. Obtained results are compared in terms of accuracy and G-measure. As in many studies, it is shown that ensemble classifier has better performance as compared to individual base classifiers. Further, five heterogeneous classifiers are aggregated with the four ensemble frameworks as majority voting and weighted voting in single and multiple layers as well and results are compared in terms of accuracy, sensitivity, specificity, and G-measure on Australian credit scoring and German loan approval datasets obtained from UCI repository.

Keywords Classification · Credit scoring · Ensemble framework · Feature selection

1 Introduction

In the previous decade, the credit business has a quick improvement and it is turning into a vital issue in financing, debt and credit card, and so forth. For instance, the credit card has turned into an extremely important financial product. Credit

D. Tripathi (✉) · R. Cheruku · A. Bablani
Department of Computer Sciene and Engineering, National Institute of Technology Goa, Ponda 403401, Goa, India
e-mail: diwakarnitgoa@gmail.com

R. Cheruku
e-mail: rmlswamigaud@nitgoa.ac.in

A. Bablani
e-mail: annubablani@nitgoa.ac.in

D. Reddy Edla et al. (eds.), *Advances in Machine Learning and Data Science*, Advances in Intelligent Systems and Computing 705,
https://doi.org/10.1007/978-981-10-8569-7_30

scoring is a procedure of calculating the risk of credit products [6] using previous data and statistical techniques. The fundamental concentration of credit scoring model to decide that credit applicant has a place with either legitimate or suspicious customer group.

Credit scoring is not a single step process, and it is a periodical process which is conducted by the banking and financial institutions into various steps such as application scoring, behavioral scoring, collection scoring, and fraud detection [7]. Application and behavioral scoring are similar, but difference is that application scoring is accomplished for the assessment of the authenticity or suspiciousness of new applicant's, and behavioral scoring is same for existing customers. Collection scoring is about to separate the existing customers into different groups (early, middle, late recovery), to put more, moderate or less level of attention on these groups. Fraud detection models rank the customer as indicated by the relative likelihood that a candidate might be unscrupulous. In this paper, our main focus is on the application scoring problem.

A wide range of classification techniques have already been proposed such as quadratic discriminant analysis (QDA), decision tree (DT), naive bayes (NB), multilayer feedforward neural network (MLFFNN), probabilistic neural network (PNN), distributed time delay neural network (DTNN), time delay neural network (TDNN). A wide range of feature selection techniques have already been proposed such as stepwise regression (STEP), classification and regression tree (CART), correlations based (CORR), multivariate adaptive regression splines (MARS), t-test, and neighborhood rough set (NRS). But it is currently unclear from the literature which classification technique is the most appropriate for which dataset, and a combination of classification algorithm with feature selection can produce the most representative results on a specific dataset.

Conventional credit scoring models are based on individual classifier or a simple combination of abovementioned classifiers which tend to show moderate performance. Thus, ensemble classification is a strong approach to produce a near to optimal classifier for any problem [8]. Generally, the result of combination of diverse classifiers has better classification performance [2–5]. Most popular ways to combine the base classifiers are majority voting and weighted voting [2, 3]. A multilayer classifiers ensemble framework [4, 5] is used based on the optimal combination of heterogeneous classifiers. The multilayer model overcomes the limitations of conventional performance bottlenecks by utilizing an ensemble of five heterogeneous classifiers.

Hence, the aim of this paper is to conduct a comparative analysis on various classification techniques and ensemble classifier frameworks with various feature selection approaches based on two real-world credit scoring datasets.

Rest of the paper is organized as follows: Sect. 2 describes the architecture for multilayer ensemble classifier. Section 3 presents experimental results obtained from the various feature selection algorithms and relative performance of different feature selection techniques followed by classification and ensemble framework performance. Further, it is followed by concluding remarks and references.

2 Ensemble Framework

As in many studies [2–5, 8] show that ensemble classifier has better performance as compared to base classifiers. In case of the ensemble classifier, outputs predicted by base classifiers are aggregated for final output against a particular sample. Most of the cases ensemble classifiers are aggregated by Majority Voting (MV) and Weighted Voting (WV). Authors in articles [4, 5] have proposed multilayer ensemble framework which has multiple layers with different classifiers at different layers. And each layer's classifiers are aggregated individually; further, the output predicted by lower layers is forwarded to next higher layer for the aggregation and at last, layer predicted output is a final output.

In this paper, we have used a multilayer ensemble classifier and single-layer ensemble layer ensemble classifier with five classifiers. Multilayer ensemble classifier is shown Fig. 1. MLFFNN, NB, QDA, DTNN, and TDNN classifiers are combined in single and multilayer, and in both cases, these classifiers are aggregated by majority voting and weighted voting. In case of majority voting approach, same weights are assigned to each classifier and output is the class which has the highest votes. But, in case of weighted voting approach the highest weight is assigned to classifier with the highest accuracy and vice versa, and final output is weighted sum of the outputs predicted by the base classifiers.

Multilayer ensemble framework is shown in Fig. 1, MLFFNN, NB, and QDA which are used in first layer; further, the predicted outputs of these classifiers are aggregated by the Combiner-1. Output predicted by Combiner-1, DTNN and DTNN are combined by the Combiner-2. Output calculated by Combiner-2 is the final output. In this work majority voting and weight voting are used as Combiner-1 and

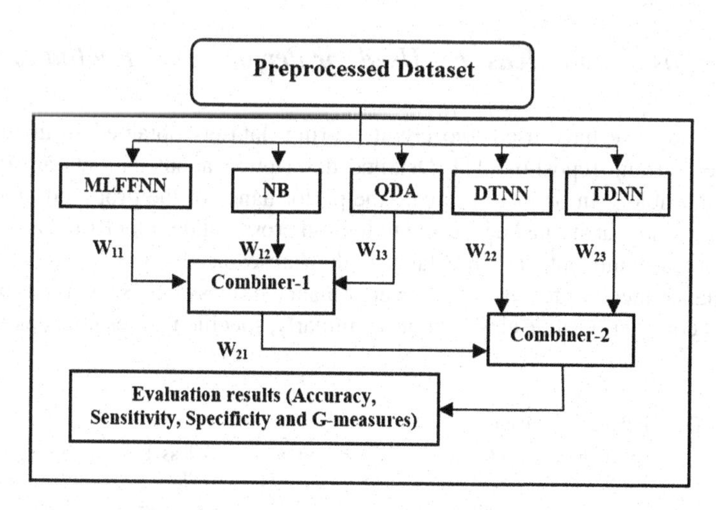

Fig. 1 Multilayer ensemble framework

Combiner-2. In case of weighted voting, Combiner-1 and Combiner-2 aggregate the output predicted by associated classifiers using Eq. (1) [10].

$$O = W_1 * X_1 + W_2 * X_2 + \cdots + W_n * X_n \quad (1)$$

where W_i and X_i are the weight and predicted output of the ith classifier, respectively.

In this work, as we have used weighted voting approach with heterogeneous multilayer ensemble framework. In case of weighted voting, a weight is assigned to each base classifier. For weight assignment, classification accuracy is used as parameter to calculate the weights for the classifiers. A classifier with the highest accuracy is assigned the highest weight and the lowest weight is assigned to a classifier with the lowest accuracy. These weights are calculated by Eq. (2). Initially, equal weights are assigned to each base classifier, and then dataset is applied for classification to calculate the accuracy. Further, the weights are updated (Eq. (2)) and this procedure is repeated for n iteration and mean of the updated weights of n iterations is assigned to the respective classifier.

$$W_U^{i,j} = W_o^{i,j} + \frac{1}{2} log(\frac{Acc_j}{1 - Acc_j}) \quad (2)$$

where $W_U^{i,j}$ and $W_o^{i,j}$ represent the weight updated and old weight at ith iteration for jth classifier, and Acc_j represents the accuracy obtained by jth classifier.

3 Simulation Results

3.1 Datasets and Measures Used for Performance Evaluation

In this study, we have used two credit scoring datasets obtained from the UCI machine learning repository [1]. Detailed description about datasets are demonstrated in Table 1. In order to evaluate the performance of the proposed model, all the datasets are partitioned according to tenfold cross-validation (10-FCV).

If there is significant class imbalance in the dataset, accuracy is not sufficient as a performance measure to evaluate the performance of classifier. Sensitivity measure does not consider positive class samples; similarly, specificity does not consider any

Table 1 Descriptions about datasets used

Dataset	Samples	Classes	Features	Class-1 /Class-2	Categorical /Numerical
Australian	690	2	14	383/317	6/8
German	1000	2	20	700/300	13/7

negative class samples. Sensitivity and specificity consider only single class data so these are also not effective for the proposed model evaluation. G-measure is a metric that considers both the positive and negative class accuracies to compute the score. It can be interpreted as geometric mean of sensitivity and specificity. All aforementioned measures are given in Eqs. 3–6.

$$Accuracy = \frac{TP + TN}{TP + TN + FP + FN} \tag{3}$$

$$Sensitivity(recall) = \frac{TP}{TP + FN} \tag{4}$$

$$Specificity = \frac{TN}{TN + FP} \tag{5}$$

$$G\text{-}measure = \sqrt{sensitivity * specificity} \tag{6}$$

where TP represents the actual positive cases which are observed as positive, TN represents the actual negative cases which are observed as negative, FP represents the actual positive cases which are observed as negative, and FN represents the actual negative cases which are observed as positive by the classifier.

3.2 Preprocessing and Feature Selection Results

Before feature selection, dataset is preprocessed, and the steps are as follows:

1. Data cleaning: In this step, whole dataset is considered and checked for the missing values. If some missing values are in a sample, samples with missing values are eliminated.
2. Data transformation: As in credit scoring dataset have combination of numerical and categorical attributes, to convert the categorical attribute's values into numerical values data transformation is required. Because some classification algorithms do not work on heterogeneous attributes. A unique integer number replaces the unique categorical value in each feature set.
3. Data normalization: As in the previous step, categorical data values are replaced by numerical values. Categorical attributes which are transformed as numerical have small range of numerical values. But in case of numerical attribute (except than transformed attributes) have a long range of values. So to make them in balanced range, normalization is done.

Further, the normalized dataset is applied for the feature selection. In this work, we have used six feature selections as STEP, CART, CORR, MARS, T-test, and NRS approaches. Results obtained by these feature selections are listed in Tables 2 and 3 with Australian and German dataset, respectively [9, 11].

Table 2 Features selected on Australian dataset

Method	Features	No of features
NRS	8, 9, 14	3
STEP	7, 8, 9, 12, 14	5
CART	5, 7, 8, 9, 10, 14	6
MARS	3, 5, 7, 8, 10, 14	6
CORR	2, 3, 4, 7, 8, 9, 10, 14	8
T-Test	2, 3, 4, 5, 6, 7, 8, 9, 10, 12, 13, 14	12

Table 3 Selected Features in German dataset

Method	Features	No of features
STEP	1, 2, 3, 7, 8, 15, 20	7
CART	1, 2, 3, 4, 5, 9, 10, 18	8
CORR	1, 2, 3, 5, 6, 7, 12, 13, 14, 15, 20	11
MARS	1, 2, 3, 4, 5, 6, 9, 15, 16, 17, 20	11
T-Test	1, 2, 3, 4, 5, 6, 7, 8, 9, 10, 12, 13, 14, 15, 20	15
NRS	1, 2, 3, 4, 6, 7, 8, 9, 11, 12, 13, 14	12

3.3 Comparative Analysis on Classification Results

Extensive research has been done on feature selection method and on data classification. But it is not clear which feature selection approach may perform better classification performance on which dataset. So the comparative performance analysis is required to test the classification performance on the dataset along with feature selection approach.

For the comparative analysis of classifiers and features selection approach, we have used seven classifiers as QDA, MLFFNN, DT, TDNN, DTNN, PNN, and NB. Mean of the tenfold cross-validation results are used to compare in terms of accuracy and G-measure. Figure 2a, b depicts the comparative graphs for aforementioned feature selection results with classifiers on Australian dataset in terms of accuracy and *G*-measure, respectively. Figure 3a, b shows the comparative graph for aforementioned feature selection results with classifiers on German dataset in terms of accuracy and *G*-measure, respectively.

As the results are depicted in Fig. 2, it is clear that without applying features selection DTNN has the highest and TDNN has second highest accuracy and *G*-measure in Australian dataset. Similarly and in case of German dataset, conclusion is similar as Australian dataset. And PNN has the lowest accuracy and *G*-measure with both the datasets.

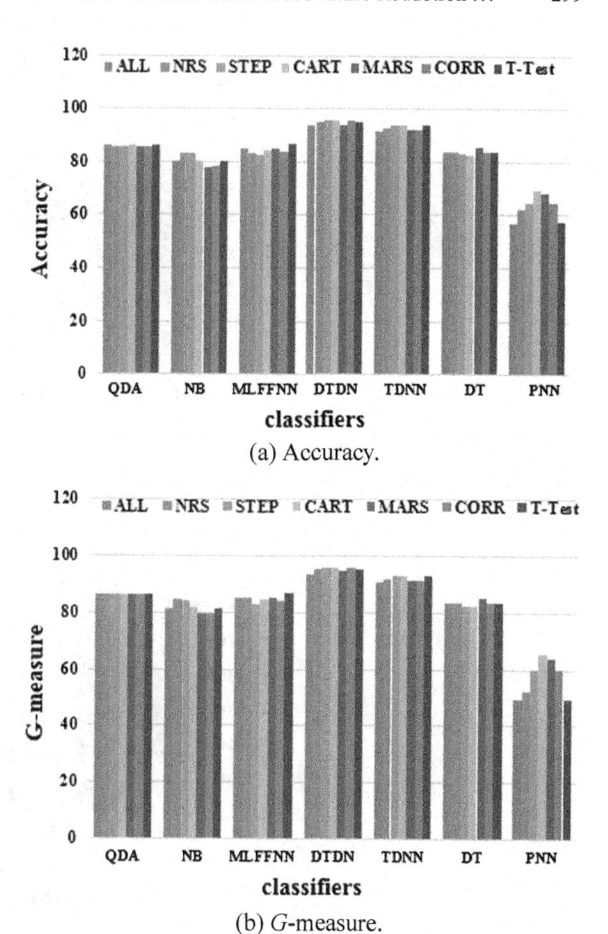

(a) Accuracy.

(b) *G*-measure.

Fig. 2 Comparison graph on Australian dataset

T-test-based feature selection performs best when used with QDA and MLFFNN. NRS and STEP-based feature selections have same accuracy but in case of *G*-measure, NRS performs best with NB and DT. With DTNN and PNN, CART shows the best performance. CART and STEP-based feature selection have similar performances for Australian dataset, and they have best values as compared to other feature selection algorithms. In case of DT, MARS based feature selection has best performance, whereas for PNN, CART-based feature selection has best performances. For Australian dataset, we conclude that CART with DTNN performs best in terms of accuracy and *G*-measure.

For German dataset, T-test-based feature selection performs best with QDA, MLFFNN, and DTNN, CART-based feature selection with NB, STEP-based feature selection with TDNN and PNN, and CORR-based feature selection with DT. Hence for German dataset, we come to a conclusion that STEP with TDNN has best performances in terms of accuracy and *G*-measure.

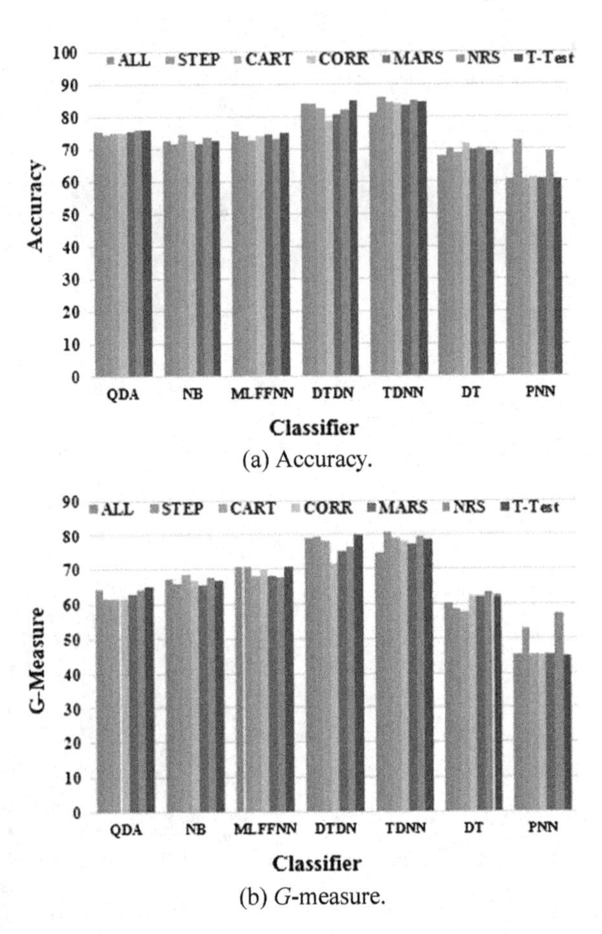

Fig. 3 Comparison graph on German dataset

(a) Accuracy.

(b) *G*-measure.

3.4 *Comparative Analysis on Ensemble Framework Results*

This section presents the results obtained by various ensemble classification frameworks along with the features selected by various approaches. For the comparative analysis, here we have used single-layer ensemble framework with majority voting and weighted voting as an aggregation methods. As in the previous section, multilayer ensemble framework is explained with two layers. In layer-1, three classifiers MLFFNN, QDA, NB are used, and in layer-2, DTNN and TDNN are used. Further, these classifiers are aggregated by majority voting approach and weighted voting approaches. All aforementioned approaches are applied with feature selection approaches and results obtained with respective ensemble framework along with respective features selection approach are illustrated in Table 4 with respective dataset.

From the results, it is clear that multilayered ensemble framework with weighted voting as aggregation function has the best performances as compared to other

Table 4 Performance comparison of ensemble frameworks with feature selection approaches

Method (FS)	Ensemble approach	Australian dataset				German dataset			
		Accuracy	Sensitivity	Specificity	*G*-measure	Accuracy	Sensitivity	Specificity	*G*-measure
All	MV	90.35	94.32	86.89	90.52	80.74	92.96	54.66	70.90
	WV	91.26	95.36	86.91	90.92	80.96	93.65	55.46	72.06
	LMV	91.12	94.79	87.18	90.80	81.17	81.09	81.33	81.03
	LWV	93.15	99.37	86.58	92.69	83.82	95.46	59.00	74.59
Step	MV	91.84	93.96	91.42	92.68	79.68	92.03	53.33	69.83
	WV	92.04	93.83	90.12	91.90	80.88	93.26	54.53	71.31
	LMV	94.42	98.43	90.18	94.21	83.82	84.53	82.33	83.23
	LWV	94.58	99.06	89.85	94.34	86.27	96.25	65.00	78.61
CART	MV	91.35	95.33	87.12	91.04	77.23	96.09	37.00	58.54
	WV	91.35	95.33	87.12	91.04	77.93	96.59	37.23	59.96
	LMV	92.20	94.42	89.85	92.01	83.82	86.56	78.00	81.85
	LWV	93.31	97.83	88.52	93.00	82.97	96.25	54.66	71.29
CORR	MV	92.00	95.33	88.45	91.76	78.51	91.71	50.33	67.44
	WV	92.26	95.78	88.85	92.24	78.51	91.71	50.33	67.44
	LMV	93.17	95.70	90.45	92.91	78.61	79.06	77.66	78.11
	LWV	93.77	98.77	88.45	93.38	82.55	93.75	58.66	73.00

(continued)

Table 4 (continued)

Method (FS)	Ensemble approach	Australian dataset				German dataset			
		Accuracy	Sensitivity	Specificity	*G*-measure	Accuracy	Sensitivity	Specificity	*G*-measure
MARS	MV	91.68	96.27	86.79	91.36	78.61	93.75	46.33	65.50
	WV	92.45	96.62	86.88	91.62	78.93	93.99	46.63	68.29
	LMV	92.99	96.27	89.52	92.73	80.31	80.62	79.66	79.99
	LWV	93.45	99.06	87.52	93.07	83.19	95.00	58.00	73.69
T-Test	MV	91.82	95.12	88.00	91.49	79.89	92.43	54.33	70.86
	WV	92.02	95.67	88.12	91.72	80.31	93.43	52.33	68.78
	LMV	92.04	94.45	89.45	91.80	82.76	83.28	81.66	82.19
	LWV	94.07	99.06	88.79	93.74	86.27	96.56	64.33	78.14
NRS	MV	92.20	94.76	89.45	92.01	77.65	91.25	48.66	66.38
	WV	92.85	94.96	89.79	92.33	77.98	92.02	49.52	67.50
	LMV	94.44	98.17	90.45	94.17	81.27	80.31	83.33	81.69
	LWV	94.76	99.40	89.79	94.47	86.27	96.78	66.00	79.92

ensemble frameworks among both datasets. In case of Australian dataset, NRS-based feature section has the best performance and STEP-based feature selection has next best performance in terms of accuracy, sensitivity, and *G*-measure. For German dataset, STEP, T-test, and NRS-based feature selection have the best performance in terms of accuracy and NRS has the best sensitivity.

4 Conclusion

In this work, we have done comprehensive comparative analysis on various feature selection approaches along with various classification algorithms. Results obtained are compared in terms of classification accuracy and *G*-measure. Further, the ensemble framework with single and multilayer with majority voting and weighted is also analyzed for comparative study in terms of classification accuracy, sensitivity, specificity, and *G*-measure. Aforementioned approaches are applied on Australian and German credit scoring datasets.

CART and STEP with DTNN have the highest accuracy and G-measures, respectively, on Australian dataset. Also, DTNN shows highest accuracy with all feature selection approaches as compared to other six classifiers in Australian dataset. STEP with TDNN has the highest accuracy and *G*-measure in German dataset. Overall we can conclude that DTNN followed by TDNN and have the best accuracy among all classifiers in case of Australian dataset, whereas for German dataset, we observed that TDNN followed by DTNN and have the best accuracy.

From the results obtained by ensemble framework, it is clear that NRS-based feature selection approach and layered weighted voting approach have the highest accuracy, sensitivity, and *G*-measure in Australian dataset. While in case of German dataset STEP, T-Test, and NRS-based feature selection approaches with layered weighted voting have same accuracy, NRS with same shows the highest *G*-measure. Layered weighted voting approach has the best performance as compared to other three ensemble approaches.

References

1. Asuncion, A.: D.N.: UCI Machine Learning Repository (2007). http://www.ics.uci.edu/sim/mlearn/MLRepository.html
2. Ala'raj, M., Abbod, M.F.: Classifiers consensus system approach for credit scoring. Knowl.-Based Syst. **104**, 89–105 (2016)
3. Ala'raj, M., Abbod, M.F.: A new hybrid ensemble credit scoring model based on classifiers consensus system approach. Expert Syst. Appl. **64**, 36–55 (2016)
4. Bashir, S., Qamar, U., Khan, F.H.: Intellihealth: a medical decision support application using a novel weighted multi-layer classifier ensemble framework. J. Biomed. Inf. **59**, 185–200 (2016)
5. Bashir, S., Qamar, U., Khan, F.H., Naseem, L.: Hmv: a medical decision support framework using multi-layer classifiers for disease prediction. J. Comput. Sci. **13**, 10–25 (2016)
6. Mester, L.J., et al.: Whats the point of credit scoring? Bus. Rev. **3**(Sep/Oct), 3–16 (1997)

7. Paleologo, G., Elisseeff, A., Antonini, G.: Subagging for credit scoring models. Eur. J. Oper. Res. **201**(2), 490–499 (2010)
8. Parvin, H., MirnabiBaboli, M., Alinejad-Rokny, H.: Proposing a classifier ensemble framework based on classifier selection and decision tree. Eng. Appl. Artif. Intell. **37**, 34–42 (2015)
9. Ping, Y., Yongheng, L.: Neighborhood rough set and svm based hybrid credit scoring classifier. Expert Syst. Appl. **38**(9), 11300–11304 (2011)
10. Triantaphyllou, E.: Multi-criteria decision making methods. In: Multi-criteria Decision Making Methods: a Comparative Study, pp. 5–21. Springer (2000)
11. Yao, P.: Hybrid classifier using neighborhood rough set and SVM for credit scoring. In: International Conference on Business Intelligence and Financial Engineering, 2009. BIFE'09, pp. 138–142. IEEE (2009)

Family-Based Algorithm for Recovering from Node Failure in WSN

Rajkumar Krishnan and Ganeshkumar Perumal

Abstract Wireless sensor network (WSN) is an emerging technology. A sensor node runs on battery power. If power drains, then there is a possibility of node failure. Node failure in wireless sensor networks is considered as a significant phenomenon. It will affect the performance of the entire network. Recovering the network from node failure is a challenging mission. Existing papers proposed number of techniques for detecting node failure and recovering network from node failure. But the proposed work consists of an innovative family-based solution for node failure in WSN. In a family, there are n numbers of persons. If anyone is in sick, then another person/s of the same family will take over the responsibility until he/she recovered from illness. In the same way, the proposed family tree based algorithm is written in three level of hierarchy. It deals with who is taking care of the task until the low-power node is recovered. This work is simulated using Network Simulator 2 (NS2) and is compared with two existing algorithms, namely LeDiR algorithm and MLeDir algorithm. The simulation results show that the proposed method gives better performance than the existing methods in terms of delivery ratio, end-to-end delay, and dropping ratio.

Keywords Wireless sensor network · Battery power · Node failure detection Family-based algorithm · Node failure recovery

1 Introduction

A WSN comprises of an arrangement of scaled-down minimal effort sensors [1]. The sensors fill in as remote information obtaining gadgets for all the effective performing artist hubs that procedure the sensor readings and set forward a proper

R. Krishnan (✉) · G. Perumal
PSNA College of Engineering and Technology, Dindigul, Tamil Nadu, India
e-mail: rajkumardgl@psnacet.edu.in

G. Perumal
e-mail: drpganeshkumar@gmail.com

D. Reddy Edla et al. (eds.), *Advances in Machine Learning and Data Science*, Advances in Intelligent Systems and Computing 705,
https://doi.org/10.1007/978-981-10-8569-7_31

action [2]. All sensor systems are portrayed by the necessity for vitality productivity, adaptability, and adaptation to non-critical failure [3]. The fundamental idea inside topology administration is the capacity to keep up a completely associated network constantly in the middle of the lifetime of WSN [4].

One approach to preserve power in a node is to enable it to rest while not being used. Topology administration plans limit the system vitality utilization to protect power [5, 6]. The choice of which node to be permitted to rest is basic issues. So time-delicate information is not lost or postponed [7]. A distributed fault location systems have been proposed in [8]. In the proposed system, family-based tree topology is used. Tree structure adapts the best when the system is separated into a lot of branches. The signals being passed by the root node are gotten by every node in the meantime [7]. Node disappointments and ecological dangers cause regular change of topology, communication disappointment, and partition of network. Such bothers are significantly more successive than those found in conventional remote systems. In genuine frameworks, over 80% of the flaws are discontinuous shortcomings [9, 10].

A versatile online disseminated answer for fault finding in sensor networks is proposed [11]. Based on the time of arrival of packet in the queue, the fault node is identified [12]. Powerful checking of network execution is basic for administrators [13]. The sensors it selves can stop working because of battery consumption. This is the motivation of this research.

The objectives of this work are to identify the level of the low-power node and call the corresponding algorithm to manage network due to low power.

This paper is structured as follows: Sect. 2 deals with the related works Sect. 3 presents the proposed system. Section 4 shows the experimental results. Section 5 discusses the conclusion.

2 Related Work

A more noteworthy battery limit is frequently profitable, especially in applications, for example electric vehicles, where it will moderate buyer worries about driving extent [14]. Y. V. Mikhaylik et al. made study on batteries. Li-S batteries still experience the ill effects of quick corruption and high self-release [15].

Ostensible limit Qmax (kWh) that shows the appraised limit of the battery and roundtrip effectiveness E (%) which demonstrates the level of the vitality going into the battery that can be stepped retreat [16].

Coulomb counting:

Examining the charge which is coming inside and going out of the battery is the widely recognized method for SOC estimation from the below expression.

$$SoC = SoC_0 + \frac{1}{C_N} + \int_{t0}^{t} (I_{batt} - I_{loss})dr \quad (1)$$

C_N is the related capacity, I_{batt} is the current of battery, and I_{loss} is the consumption of current by loss [17].

The battery characteristics force that the battery substance is not required in the expression

$$Q_j^t \geq S_j \forall t \quad (2)$$

S_j is the minimum condition of charge determined as input, and Q_j^t is energy content of battery at time t [18]. Abbasi et al. proposed the LeDiR method for tolerating fault in a node [1]. C. A. Subasini et al. implemented the MLeDir method [7].

3 Proposed Work

Figure 1 shows three-level tree formations. In the proposed system, root is called as grandfather (A). The level 1 node is called as son or daughter node (B, C, D, E, and F). The leaf nodes are called as grandchildren of root node or son/daughter of level 1 nodes (G, H, I, J, K, L, M, N, O, P, and Q).

Terminologies:
LPN: low-power node, BP: battery power, LP: low power, L2N: level 2 node, L1N: level 1 node, L0N: level 0 node, msg: message, n: node, T: threshold, HR: help request, RH: ready to help, CF: cousin family, CSN: cousin sibling node, M: member, SHICD: stop help I can do, Max: maximum, S/D: son/daughter, SN: sibling node, DP: deputation, SLP: sibling of low-power node, GS/GD: grandson/ granddaughter.

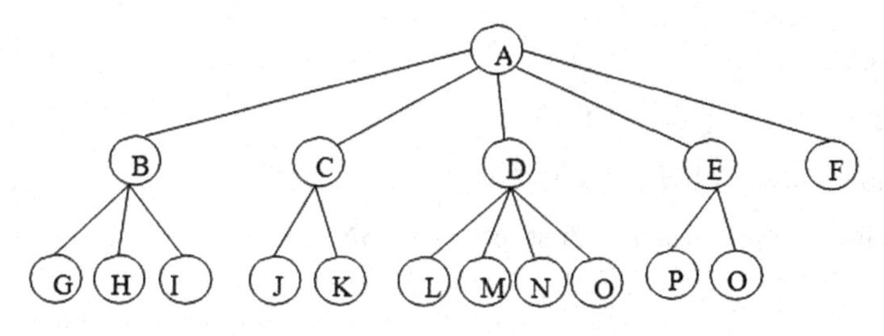

Fig. 1 Tree structure of nodes

LPN Algorithm:

if (BP of n<T)

n send LP_{msg} to Admin

if (LP node is at level 2)

Call the L2N algorithm

else if (LP node is at level 1)

Call the L1N algorithm

else

Call the L0N algorithm

If the power of any node is less than the value of 20 dBm, then the node is called as LPN. Based on the level, the corresponding algorithm is called.

L2N-Algorithm:

do{

LP node sends HR_{msg} to nearest siblings

If (LP node receives RH_{msg} from siblings)

LPN Hand over processing to first received sibling

else

LP node Send HR_{msg} to its immediate ancestor

If (LP node receives RH_{msg})

Hand over all the processing to ancestor

else{

i=1;// 0th family already checked

while (i<CFmax){

j=0;

while (j<Mmax){

LP node Send HR_{msg} to i^{th} CF

If (LP node receives RH_{msg} from i^{th} CF)

LPN Hand over processing to first received CSN

j++;

}i++;

} }}while (LP node gets rectified by admin)

For example, if the node G is the LPN, then it sends HR message to nodes H and I. For node G, it sends the HR message to B and tries to receive positive reply from parent. For example, node G first sends help request message to the nearest CF nodes C, J, and K, then next CF nodes D, L, M, N, and O, and so on up to the last CF node F. At last, it will try with grandfather node A.

L1N-Algorithm:

do{

LP node send HR_{msg} to immediate S/D node

If (LP node receives RH_{msg} from S/D node)

LPN Hand over processing to first received S/D

else{LP node send HR_{msg} to its siblings

If (LP node receives RH_{msg} from siblings)

LPN Hand over processing to first received sibling

for (i=0;$i<SN_{Max}$;i++){LP node sends DP_{msg} to i^{th} sibling

for (j=0;$j<S/D_{Max}$;j++){

SLP node sends HR_{msg} to its S/D nodes // send with ID of LP node}

If (SLP receives RH_{msg} from S/D node){

SLP selects S/D node which sends the RH_{msg} first

SLP forward S/D RH_{msg} to LP node // send with ID of S/D node

Hand over processing to that selected S/D node

set flag=1;}

If (flag==1)

Exit (0); // exit from outer for loop

}if (flag==0)

LPN sends HR_{msg} to ancestor & Ancestor take care of processing.

}while (LP node gets rectified by admin)

Node B is the LPN, then it sends HR message to nodes G, H, and I. For node B, it sends the HR message to nodes C, D, E, and F and tries to receive reply from sibling. For example, for node B, the first cousin is C. So it communicates with C, and C communicates with its child J and K. Next, it will try with cousin D with its child L, M, N, and O, etc. If there is no RH message from any of its cousin children through cousin nodes, then it sends HR to ancestor node. Ancestor node will take care of all processing for LP node.

L0N-Algorithm:

do{

LP node send HR_{msg} to immediate S/D node

If (LP node receives RH_{msg} from S/D node)

LPN Hand over processing to first received S/D

else{S/Ds of level 0 node flood HR_{msg} to its child

For all S/D nodes of level 0 node{

if (S/Ds of L0N receives first RH_{msg} from any child)

It conveys to L0N and inform to other childs

}if (L0N receives RH_{msg} from GS/GD via S/D)

LP Hand over processing to first received GS/GD

}while (LP node gets rectified by admin);

LPN sends HR message to immediate son or daughter nodes. For example, the node A sends help request message to nodes B, C, D, E, and F. For node A, its S/D nodes of level-0 are B, C, D, E, and F, and they flood the HR message to its child. For example, node B floods HR message to its child G, H, and I. For example, node C floods HR message to its child J and K. Node D floods HR message to its child nodes L, M, N, and O. Node E floods to P and Q. If the LPN receives RH message from its grandchildren through level 1 nodes, then based on the first come first serve method it hands over all the processing. During the above processes, LPN gets rectified by administrator, then LPN sends SHICD message to node which is already taken in-charge for this node. Collect all processing information from the in-charge node and continue to process itself.

Input for LPN algorithm is set of sensor nodes, and output is network efficiency in terms of delivery ratio, end-to-end delay, dropping ratio, and network lifetime. L2N, L1N, and L0N algorithms are sub-algorithms which are invoked by LPN algorithm, and input for these algorithms are level 2 nodes, level 1 nodes, and level 0 nodes, respectively.

4 Experimental Results

The proposed tree-based algorithm is compared with LeDiR algorithm and MLeDir algorithm. This paper analyzes the delivery ratio with number of nodes, number of nodes with end-to-end delay, number of nodes with drop ratio, number of nodes with overheads, and number of nodes alive with simulation time.

The Fig. 2 shows that the comparison of three algorithms with the parameters delivery ratio and number of nodes. It is proved that the delivery ratio has been increasing in the proposed scheme than the existing two algorithms.

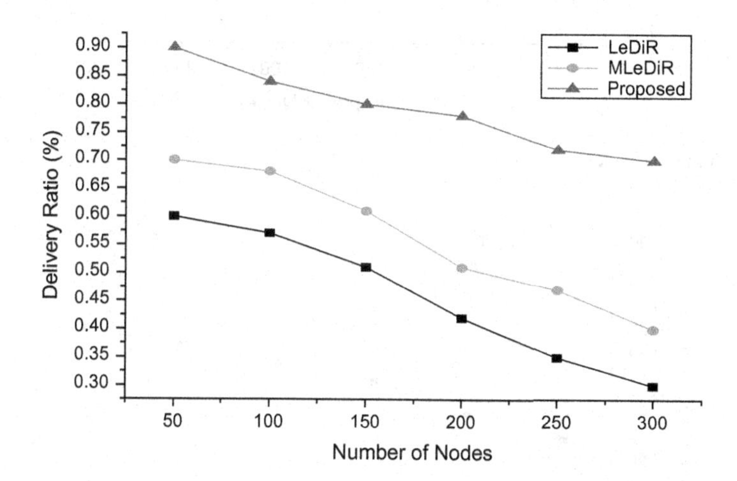

Fig. 2 Delivery ratio versus number of nodes

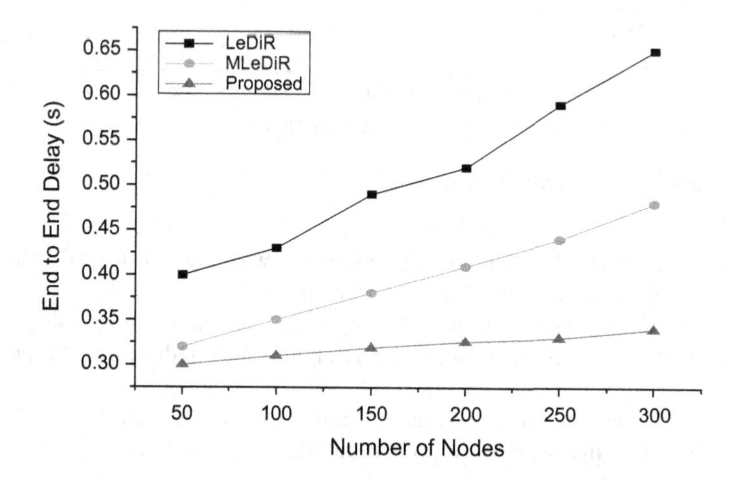

Fig. 3 Number of nodes versus end-to-end delay

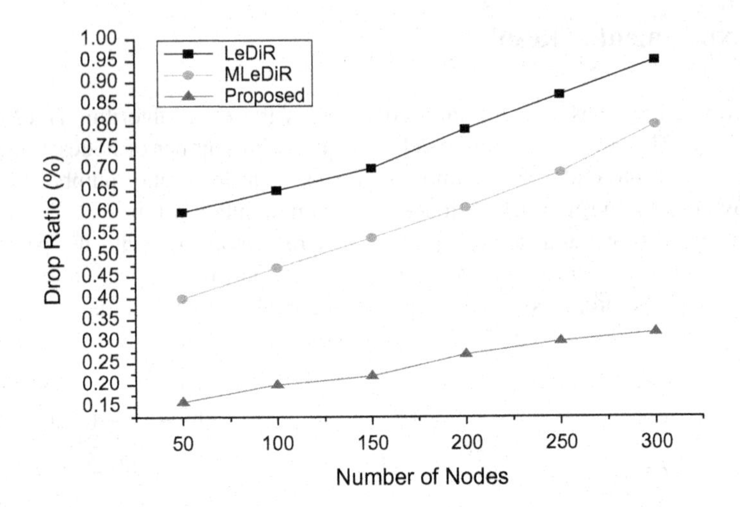

Fig. 4 Number of nodes versus drop ratio

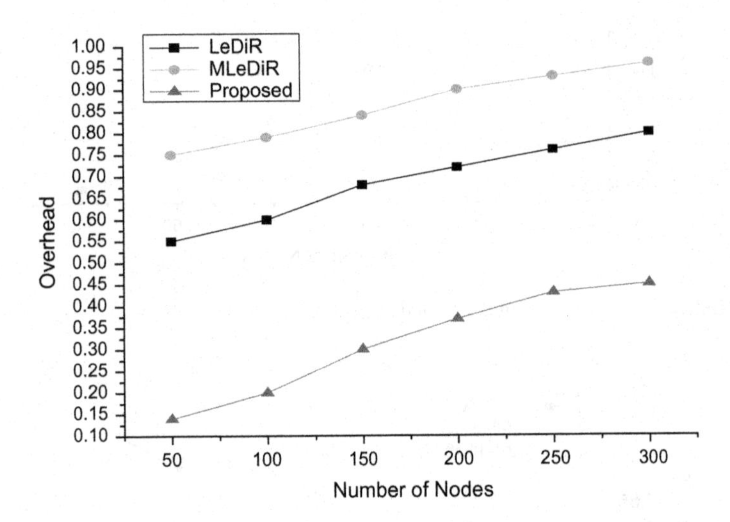

Fig. 5 Number of nodes versus overhead

Figure 3 gives the result of number of nodes with the end-to-end delay. The delay has been reduced in the proposed system.

Figure 4 presents the conclusion of comparing three protocols by using number of nodes with drop ratio. Dropping rate is significantly reduced in the proposed algorithm.

The Fig. 5 gives the analysis chart of number of nodes with overhead. In the proposed scheme, the overhead ratio is reduced.

5 Conclusion

Node failure in wireless sensor networks, due to battery discharge, is addressed in the proposed work. Family-based tree structure algorithm is implemented. This novel algorithm is simulated by using NS2 simulator and is compared with two existing algorithms, namely LeDiR algorithm and MLeDiR algorithm. The simulation results show that the family tree-based method gives better performance in terms of delivery ratio, end-to-end delay, and dropping ratio.

References

1. Abbasi, A.A., Younis, M.F., Baroudi, U.A.: Recovering from a node failure in wireless sensor actor networks with minimal topology changes. IEEE Trans. Veh. Technol. **62**(1) (2013)
2. Akyildiz, I.F., Kasimoglu, I.H.: Wireless sensor and actor networks: research challenges. Ad Hoc Netw. J. **2**(4), 351–367 (2004)
3. Arunanshu, M., Pabitra, M.: Research article: detection of node failure in wireless image sensor networks. ISRN Sens. Netw. **1**, 1–8 (2012)
4. Stankovic, J.A., Abdelzaher, T.F., Lu, C., Lui, S., Hou, J.C.: Real-time communication and coordination in embedded sensor networks. In: Proceedings of the IEEE, vol. 01, no. 7, July (2003)
5. Schurgers, C., Tsiatsis, V., Ganeriwal, S., Srivastava, M.: Optimizing sensor networks in the energy-latency-density design space. IEEE Trans. Mob. Comput. **1**(1) (2002)
6. Schurgers, C., Tsiatsis, V., Ganeriwal, S., Srivastava, M.: Topology management for sensor networks: exploiting latency and density. In: Proceedings of the MOBIHOC'02, Lausanne, Switzerland. ACM, June 9–11 (2002)
7. Subasini, C.A., Chandra Sekar, A.: Automatic recovering node failure in wireless sensor actor networks. Int. J. Eng. Technol. (IJET) **7**(1), 212–221 (2015). ISSN: 0975-4024
8. Chen, J., Kher, K., Somani, A.: Distributed fault detection of wireless sensor networks. In: Proceedings of the Workshop on Dependability Issues in Wireless Ad Hoc Networks and Sensor Networks (DIWANS'06), pp. 65–72. ACM, USA (2006)
9. Horst, R., Jewett, D., Leno ski, D.: The risk of data corruption in microprocessor-based systems. In: The Twenty-Third International Symposium on Fault-Tolerant Computing, Toulouse, France, pp. 576–585, 22–24 June (1993)
10. Siewiorek, D.P., Swmlz, R.S.: The Theory and Practice of Reliable System Design. Digital Equipment Corporation (1982)
11. Arunanshu, M., Pabitra, M.: Online fault diagnosis of wireless sensor networks. Cent. Eur. J. Comp. Sci. **4**(1), 30–44 (2014)
12. Manisha, W., Kanak, S.: How to detect failure node in a selected network. Int. J. Sci. Res. Publ. **4**(1) (2014)
13. Ma, L., Ting, H., Ananthram, S., Don, T., Kin, K.L.: On optimal monitor placement for localizing node failures via network tomography. Perform. Eval. **91**, 16–37 (2015)
14. Propp, K., Daniel, J.A., Abbas, F., Stefano, L., Vaclav, K.: Kalman-variant estimators for state of charge in lithium-sulfur batteries. J. Power Sources **343**, 254–267 (2017)
15. Mikhaylik, Y.V., Kovalev, I., Schock, R., Kumaresan, K., Xu, J., Affinito, J.: High energy rechargeable Li-S cells for EV application: status, remaining problems and solutions. ECS Trans. **25**(35), 23–34 (2010)

16. Manwell, J.F., McGowan, J.G.: Lead acid battery storage model for hybrid energy systems. Sol. Energy **50** (1993)
17. Piller, S., Perrin, M., Jossen, A.: Methods for state-of-charge determination and their applications. J. Power Sources **96**, 113–120 (2001)
18. Bordin, C., Harold, O.A., Andrew, C., Isabel, L.G., Chris, J.D., Daniele, V.: A linear programming approach for battery degradation analysis and optimization in off grid power systems with solar energy integration. Renew. Energy **101**, 417–430 (2017)

Classification-Based Clustering Approach with Localized Sensor Nodes in Heterogeneous WSN (CCL)

Ditipriya Sinha, Ayan Kumar Das, Rina Kumari and Suraj Kumar

Abstract In wireless sensor network, random and dense deployment of sensor nodes results in difficulties for sink node to detect the location of them without GPS. The inclusion of GPS for all sensor nodes increases the deployment cost. Energy is another constraint in wireless sensor network during data forwarding. In this paper, the proposed protocol CCL has applied the modified version of DV-hop technique to detect the location of sensor nodes without using GPS. Here, event-based clustering is designed to save the energy of nodes, which is classified using support vector machine. Packet is forwarded to the sink node by greedy forwarding technique. Packet loss is also removed by involving an antivoid approach called twin rolling ball technique. Simulation results show that the performance of CCL is enhanced with compared to LEACH, HEED, EEHC, DV-hop, and advanced DV-hop.

Keywords DV-hop · Twin rolling ball · Support vector machine
Localization

D. Sinha (✉) · R. Kumari
National Institute of Technology Patna, Patna, India
e-mail: ditipriya.cse@nitp.ac.in

R. Kumari
e-mail: rinakri08@gmail.com

A. K. Das
Birla Institute of Technology Mesra, Patna Campus, Patna, India
e-mail: das.ayan777@gmail.com

S. Kumar
National Institute of Technology Meghalaya, Shillong, India
e-mail: surajgate2015@gmail.com

D. Reddy Edla et al. (eds.), *Advances in Machine Learning and Data Science*,
Advances in Intelligent Systems and Computing 705,
https://doi.org/10.1007/978-981-10-8569-7_32

1 Introduction

In wireless sensor network (WSN), the energy-constrained sensor nodes are deployed randomly in remote areas to serve the applications such as disaster management, environmental monitoring, medical application, forest fire management, and home automation systems. The random and dense deployment of sensor nodes makes it difficult for sink node to find the location of sensor nods without Global Positioning System (GPS) [1]. The use of GPS increases the deployment cost. Thus, localization of sensor nodes is one of the objectives of the authors. In most of the applications, the sensor nodes are worked for the sensing of events and send the data regarding that event to the well-localized sink node. Packets may be dropped at the time of data forwarding due to void problem [2]. The solution of this problem is another objective of this paper. In remote areas, regular recharge of sensor nodes is almost impossible. Thus, the authors are motivated to implement energy-efficient routing protocol to forward the data from sensor nodes to sink node. In this paper, a novel protocol, called CCL, is designed to meet the above-stated demands.

2 Related Work

It is observed from the state-of-the-art study that many researchers working in the field of WSN have focused on localization of sensor nodes without GPS. The authors of [3] have used DV-hop technique to find their location. In DV-hop algorithm, GPS is assigned to some sensor nodes, called anchor nodes, and rest of the nodes without GPS are known as unknown nodes. Anchor nodes are aware of their location, and unknown node finds their location by applying DV-hop with the help of anchor node.

Though location of unknown nodes can be calculated by using traditional DV-hop algorithm, it may cause error in estimated location coordinate. Improved DV-hop [4] has been proposed to correct the mentioned error. In [4], the first step is same as of traditional DV-hop and it mainly focuses on second and third steps. In the second step instead of simple hop size, it calculates average hop size, which is used to calculate the distance. In the third step, 2D hyperbolic location algorithm is used to estimate the location of unknown nodes. Improved DV-hop increases the location accuracy with the increase in communication cost. An advanced DV-hop [1] technique is proposed to reduce the computational cost. The location accuracy is improved by the implementation of improved DV-hop localization algorithm [5].

In WSN, energy efficiency is another concern of the researchers. The dense deployment of sensor nodes causes data redundancy which drains the energy very quickly and increases routing overhead. Many cluster-based approaches are designed to control redundancy. LEACH [6] and HEED [7] are examples of equal clustering algorithm, whereas ULEACH [8], UHEED [9], EEUC [10], ULCA [11], EDUC [12], and EAUFC [13] are examples of unequal clustering algorithm. The examples of hybrid clustering algorithm are EEHC [14], HADCC [15], and HK-mean [16].

Many of the researchers have used greedy forwarding technique [17] to forward the packet from source to destination. GPSR [2] is introduced to solve the void problem. The drawbacks found in GPSR have been reduced in BOUNDHOLE [18]. The advantage of BOUNDHOLE algorithm lies in its independence. GAR [17] is designed to give guarantee for delivery of packet to the destination successfully. BPR [19] is proposed to deal with the problem of GAR.

3 Proposed Work

The following subsections describe the proposed protocol CCL.

3.1 Definitions

3.1.1 Support Vector Machine

Support vector machine is a supervised learning machine based on statistical learning theory. It is like linear learning machine that maps their input vectors to a feature space by the use of kernels and computes the optimal hyperplane as described in Fig. 1.

Training data are $(x_1, y_1), \ldots, (x_m, y_m) \in R_n \times \{+1, -1\}$, where n-dimensional vectors are xi and their labels are yi. Labels divide the data into two classes +1 and −1. A function, $F(x) = y: R_n \rightarrow \{+1, -1\}$, separates each vector in either class +1 or −1.

3.1.2 Energy Model

The total energy consumption E_{Ri} for *i*th node can be calculated as the summation of energy required to receive data packets, process that data, and transmit the processed data to the next hop node. This can be described as:

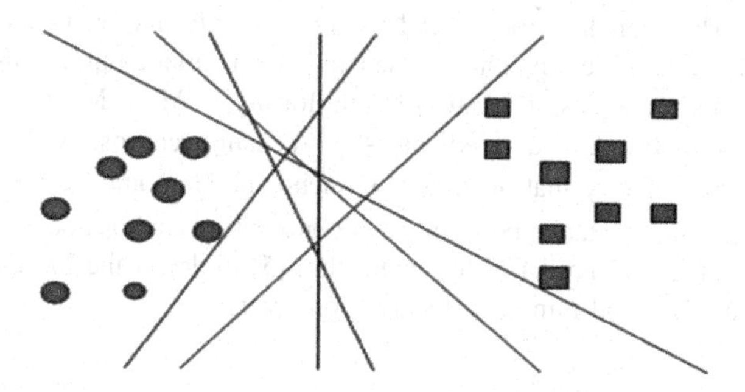

Fig. 1 Optimal hyperplane

$$E_{Ri} = E_{TX}(k, d) + E_{RX}(k) + E_{PX}(k) \quad (1)$$

where $E_{TX}(k, d)$ is the energy used to send k bits of data to the destination and can be computed as:

$$E_{TX}(k, d) = \begin{cases} KE_{elec} + K\varepsilon_{fs}d^2, & d \leq d_0 \\ KE_{elec} + K\varepsilon_{amp}d^2, & d > d_0 \end{cases} \quad (2)$$

where d is the distance between source and destination nodes, E_{elec} is the energy used for transmitting/receiving a single bit data, f_s and ε_{amp} are the amplifier characteristic constants in the free-space propagation model and the two-ray ground reflection model, respectively, and $d_0 = \sqrt{\frac{\varepsilon_{fs}}{\varepsilon_{amp}}}$ is the distance used to divide two kinds of path loss model. The energy consumed $E_{RX}(k)$ to receive a data packet for k bits of data can be calculated as:

$$E_{RX}(k) = k \times E_{elec} \quad (3)$$

Processing energy $E_{PX}(k)$ is constant. The total energy consumption for kth route containing n number of nodes can be computed as:

$$E_k = \sum_{i=1}^{n} E_{r_i} \quad (4)$$

3.2 Description

The proposed protocol is divided into the following four phases.

3.2.1 Anchor Node Selection

GPS is provided to few of the sensor nodes, called anchor nodes, at the time of deployment. These anchor nodes send its location information to the sink node. In CCL, at the time of deployment, the sensing area is divided into uniform sensing regions. Suppose N nodes are distributed uniformly in M × M region randomly, then the whole network is divided into $\frac{M \times M}{N}$ sensing regions. Anchor nodes are deployed in such a way that at least one anchor node should be present in each region. The average distance between each consecutive anchor nodes is $\sqrt{M \times M}$. CCL modifies the advanced DV-hop algorithm [5] to detect the location of sensor nodes without GPS and minimize localization error.

3.2.2 Location Discovery of Individual Nodes

Each anchor node broadcasts beacon packets which consist of id of anchor node, hop count initialized with zero, and its location coordinate. Each unknown node creates a table containing the field $(x_i,\ y_i,\ hop_i)$ for each anchor node who is deployed at the location $(x_i,\ y_i)$. The minimum number of hops between that unknown nodes to ith anchor node is hop_i. When a node receives multiple packets from the anchor node through different paths, the packet that contains with least hop count will be chosen as hop count value of the table. Average size of one hop is computed from one anchor node to another anchor node as in Eq. 5.

$$E_{Hopsize_i} = \left(\sum_{i \neq j} \sqrt{(X_i - X_j)^2 + (Y_i - Y_j)^2} \right) \Big/ \sum_{i \neq j} h_{min_{i,j}} \tag{5}$$

where (x_i, y_i) and (x_j, y_j) are the coordinates of anchor nodes i and j, respectively, and $h_{min_{i,j}}$ is the minimum hop count between anchor nodes i and j. After computing the hop size, each anchor node broadcasts its hop size with its id in the network. Any particular node receives the first arrived message (hop size), transmits it to its neighbor, and increases the hop count value by 1. Thus, all the nodes receive least hop count of their nearest anchor node. Each unknown node calculates distance to each individual anchor nodes with the help of Eq. 6:

$$dist_{ua} = E_{Hopsize_i} \times hop_{p_i} \tag{6}$$

where hop_{p_i} is the number of hops between unknown node p and the anchor node i. $Hopsize_i$ is the hop size of an anchor node. Location estimation of unknown node (x, y) is measured by n anchor nodes using the following n equations:

$$\sqrt{(x - x_1)^2 + (y - y_1)^2} = d_1 \tag{7}$$

$$\sqrt{(x - x_n)^2 + (y - y_n)^2} = d_n \tag{8}$$

The last equation is subtracted from first (n-1) equations in order to reduce error, and the above system of equations can be represented in the form of matrix like $AX = B$, where

$$A = \begin{bmatrix} -2(x_1 + x_n) & -2(y_1 + y_n) & 2 \\ -2(x_2 + x_n) & -2(y_2 + y_n) & 2 \\ -2(x_{n-1} + x_n) & -2(y_{n-1} + y_n) & 2 \end{bmatrix}, B = \begin{bmatrix} d_1^2 + d_n^2 - (Q_1 + Q_n) \\ d_2^2 + d_n^2 - (Q_2 + Q_n) \\ d_{n-1}^2 + d_n^2 - (Q_{n-1} + Q_n) \end{bmatrix}$$

$$\text{and } X = \begin{bmatrix} x \\ y \\ p \end{bmatrix}$$

Now, X can be computed by applying Eq. (9):

$$X = (A'A)^{-1}A'B \tag{9}$$

where A' represents the transpose of matrix A.

The estimated coordinate is updated to increase location accuracy. Assume that new location is (x', y') and estimated coordinate of unknown node is (x'', y''). x' and y' can be written in the form of x''&y'', as $x' = t \times x''$ and $y' = t \times y''$. The value of t can be estimated by Eq. (10)

$$P = (x')^2 + (y')^2 \tag{10}$$

The updated coordinate (X, Y) of unknown node can be calculated as $X = \frac{x' + x''}{2}$ and $Y = \frac{y' + y''}{2}$.

3.2.3 Cluster Formation and Cluster Head Selection

CCL designs event-based clustering applying SVM. Sensor nodes are divided into two types: (i) active sensor nodes and (ii) inactive sensor nodes. Gaussian kernel is used in the process of classification of active and inactive sensor nodes in WSN. The input space is $X \in Rn$ for a number of node distributions. Set of nodes for classification is $xi \subseteq X$. Radius and center of spherical search space are R0 and a0 respectively. The problem can be designed as $\{|\varphi(xj) - a|\} \cdot \{|\varphi(xj) - a|\} \le R$, $\forall j = 1, \ldots, n$ using simple dot product $< \varphi(xj) - a, \varphi(xj) - a > \le R$, $\forall j = 1, \ldots, n$, where φ (n) is a mapped function which maps active node to the feature space. This represents the optimization problem for clustering. After analyzing the problem, the following cases can be declared —$\{|\varphi(xj) - a|\} \cdot \{|\varphi(xj) - a|\} > R$; i.e., vector xi lies outside the sphere. It means that xi is the inactive nodes. $\{|\varphi(xj) - a|\} \cdot \{|\varphi(xj) - a|\} \le R$; i.e., vector xi lies on the sphere and they will be classified when we remap them into original space. It means that xi is the active nodes.

After assigning the cluster, we need to decide which node is in which class. We check for each pair of node; when the pairs of node lie in the different classes, the line that unites them is transposed outside the sphere in the higher space. The connected component of the graph of adjacency matrix A represents a cluster of same class. Now, each node belongs to the cluster and calculates its competition bid value (CV_i) for itself as:

$$CV_i = E_{R_i} + \frac{1}{d_i} \tag{11}$$

where, for any node i, E_{Ri} is its remaining energy which is computed by applying Definition 3.1.2, and d_i is the distance of node i from base station B.

All the nodes exchange their CV_i value among themselves. The highest CV_i value of node in each round is elected as cluster head. Each node sends sensed data to cluster head, and CH aggregates the collected data and sends it to base station applying greedy forwarding technique.

3.2.4 Greedy Forwarding of Aggregated Data to the Sink Node

In CCL, data are forwarded by applying greedy forwarding algorithm. It selects the next hop based on two conditions: (i) the next hop node should within the communication range of the node that currently holds the packet and (ii) the selected next hop node must be at nearest geographical location from the destination node. In Fig. 2, let us consider that any source node wants to send the packet to destination node D and, at the time of forwarding, the packet reaches to node A.

Though destination D is nearest to node A among its neighbors, D is in out-of-communication range of A. Hence, all the packets are dropped at node A. This problem is known as void problem. To reduce the void problem, the twin rolling ball technique has been applied, which is described in Fig. 3. N_s forwards the packet to node N_1 in order to reach the destination N_D by applying greedy forwarding technique. At node N_1, void problem occurs and thus N_1 is not able to forward the packet further. A ball is hinged at the stuck node N_1, and a cycle is formed with R/2 radius by rotating that ball in both clockwise and anticlockwise directions. No intersecting node is detected in the anticlockwise rotation, whereas in clockwise rotation N_4 is encountered and is selected as the next hop after N_1. Again at node N_4, the ball is rotated in both the directions. The process continues until the destination is reached.

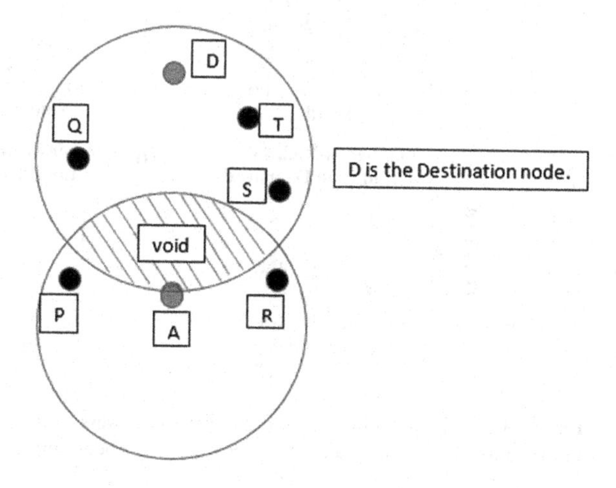

Fig. 2 Void problem

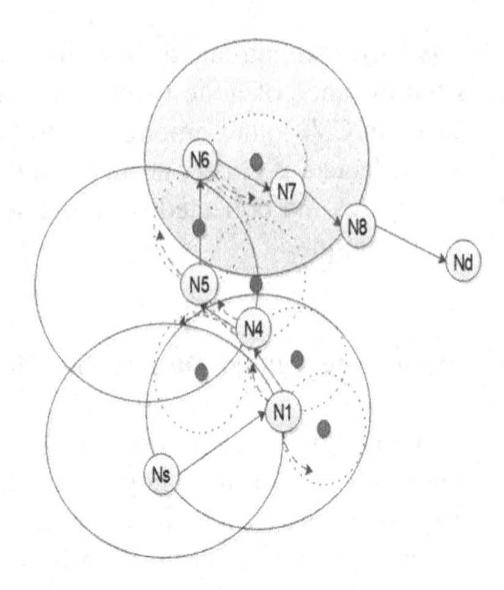

Fig. 3 Twin rolling ball technique

4 Simulation Results

The performance of proposed protocol has been simulated by MATLAB (R2013a) tool. The performance of proposed algorithm has been compared with cluster-based protocols such as LEACH [6], HEED [8], and EEHC [14]. The localization performance for detecting the position of unknown nodes is compared with DV-hop [3] and advanced DV-hop [1].

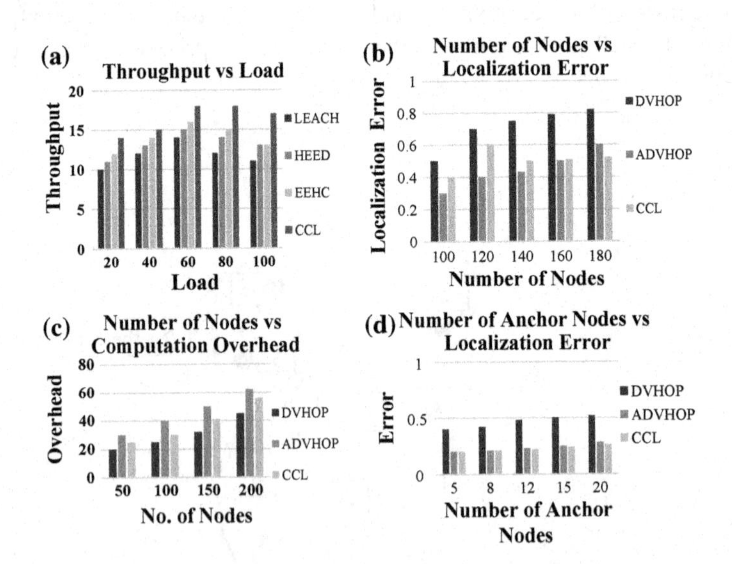

Fig. 4 **a** Throughput versus load. **b** Number nodes versus localization error. **c** Number of nodes versus computation overhead. **d**Number of anchor nodes versus localization error

Figure 4a, c shows that the performance of CCL is better than other cluster-based protocols with increased load of the network. On the other hand, Fig. 4b, d shows that localization error of CCL is decreasing with increased number of nodes compared to other localization-based protocols.

5 Conclusion

CCL is playing an inevitable role in WSN for cluster formation and location detection. A brief review of the well-known methodology in this area shows that very little attention has been provided by the researchers toward localization of sensor nodes and event-based cluster formation for reducing implementation cost and saving energy in the network. CCL has modified DV-hop technique to localize the sensor nodes without GPS. SVM classification technique is applied to save the energy of sensor nodes in the network. Data transmission is done through greedy forwarding technique, and void problem is minimized by applying twin rolling ball technique. Thus, CCL designs a novel approach to reduce the deployment cost as well as saves energy to increase the network lifetime.

References

1. Kumar, S., Lobiyal, D.K.: An advanced DV-Hop localization algorithm for wireless sensor networks. Wirel. Pers. Commun. 1–21 (2013)
2. Karp, B., Kung, H.T.: GPSR: greedy perimeter stateless routing for wireless networks. In: Proceedings of the 6th Annual International Conference on Mobile Computing and Networking. ACM (2000)
3. Du, X.: DV-Hop localization algorithms in wireless sensor networks (2016)
4. Zhang, X., Xie, H., Zhao, X.: Improved DV-Hop localization algorithm for wireless sensor networks. J. Comput. Appl. **27**(11), 2672–2674 (2007)
5. Zhang, D., et al.: Research on an improved DV-Hop localization algorithm based on PSODE in WSN. J. Commun. **10**(9) (2015)
6. Handy, M.J., Haase, M., Timmermann, D.: Low energy adaptive clustering hierarchy with deterministic cluster-head selection. In: 4th International Workshop Mobile and Wireless Communications Network. IEEE (2002)
7. Kour, H., Sharma, A.K.: Hybrid energy efficient distributed protocol for heterogeneous wireless sensor network. Int. J. Comput. Appl. **4**(6), 1–5 (2010)
8. Majadi, N.: U-LEACH: a routing protocol for prolonging lifetime of wireless sensor networks. Int. J. Eng. Res. Appl. **2**(4), 1649–1652 (2012)
9. Ever, E., et al.: UHEED-an unequal clustering algorithm for wireless sensor networks (2012)
10. Chen, G., et al.: An unequal cluster-based routing protocol in wireless sensor networks. Wirel. Netw. **15**(2), 193–207 (2009)
11. Sohrabi, K., et al.: Protocols for self-organization of a wireless sensor network. IEEE Pers. Commun. **7**(5), 16–27 (2000)
12. Yu, J., Qi, Y., Wang, G.: An energy-driven unequal clustering protocol for heterogeneous wireless sensor networks. J. Control Theor. Appl. **9**(1), 133–139 (2011)

13. Singh, J., Mishra, A.K.: Clustering algorithms for wireless sensor networks: a review. In: 2nd International Conference on Computing for Sustainable Global Development (INDIACom). IEEE (2015)
14. Kumar, D., Aseri, T.C., Patel, R.B.: EEHC: energy efficient heterogeneous clustered scheme for wireless sensor networks. Comput. Commun. **32**(4), 662–667 (2009)
15. Aslam, M., et al.: Hadcc: hybrid advanced distributed and centralized clustering path planning algorithm for WSNs. In: IEEE 28th International Conference on Advanced Information Networking and Applications (AINA). IEEE (2014)
16. Chen, B., et al.: Novel hybrid hierarchical-K-means clustering method (HK-means) for microarray analysis. In: Computational Systems Bioinformatics Conference, Workshops and Poster Abstracts. IEEE (2005)
17. Liu, W.J., Feng, K.T.: Greedy anti-void routing protocol for wireless sensor networks. IEEE Commun. Lett. **11**(7) (2007)
18. Fang, Q., Gao, J., Guibas, L.J.: Locating and bypassing holes in sensor networks. Mob. Netw. Appl. **11**(2), 187–200 (2006)
19. Yaakob, N., et al.: By-passing infected areas in wireless sensor networks using BPR. IEEE Trans. Comput. **64**(6), 1594–1606 (2015)

Multimodal Biometric Authentication System Using Local Hand Features

Gaurav Jaswal, Amit Kaul and Ravinder Nath

Abstract In this work, the hand-based multimodal biometric system is presented using score-level fusion of hand geometry and local palmprint features. Initially, a palm ROI of fixed size has been cropped on the basis of finger base points. However, these images are not well aligned and reduce the matching accuracy. To better align them, L-K tracking-based palm image alignment method has been presented. Following this, the poor contrast ROI image is enhanced using novel fractional G-L filter. Then, local keypoints of aligned ROI images are extracted using Block–SIFT descriptor. Secondly, a set of novel geometrical features has been computed from Palmer region of hand image. Further, the highly uncorrelated features are selected from palm and hand geometry using Dia-FLD. In order to handle robust classification, a high-performance method Linear SVM has been used. Finally, score-level fusion rule has been employed which has shown the increased performance of combined approach in terms of Correct Recognition Rate (99.34%), Equal Error Rate (2.16%), and Computation Time (2084 ms). The proposed system has been tested on largest publicly available contact based and contactless databases: Bosphorus hand database, CASIA, and IITD palmprint databases to validate the results.

Keywords Fusion · Local features · SIFT · Lucas–Kanade tracking

---

G. Jaswal (✉) · A. Kaul · R. Nath
Signal Processing and Instrumentation Laboratory, Department of Electrical Engineering, National Institute of Technology, Hamirpur, India
e-mail: gauravjaswal@nith.ac.in

A. Kaul
e-mail: amitkaul@nith.ac.in

R. Nath
e-mail: r.nath@nith.ac.in

D. Reddy Edla et al. (eds.), *Advances in Machine Learning and Data Science*, Advances in Intelligent Systems and Computing 705,
https://doi.org/10.1007/978-981-10-8569-7_33

1 Introduction

Developing a personal recognition biometric system based on hand characteristics has recently grown interest of researchers worldwide. A Palmer region basically enriches with most discriminative textural features (lines, creases) and geometrical (length, width, angles) which can be easily clubbed and captured together. All features can be captured in a contactless manner using cheap low resolution cameras with less user cooperation [1]. However, fingerprint provided the basis for initial identification systems. But, fingerprint technology requires high-resolution images (>400 dpi) which limit its role in certain commercial applications. It has been observed that fingerprint can be easily disturbed with injury and wound on the finger surface. On the contrary, the quality of palm surface is literally found good. A palmprint consists of bigger ROI region and more dense features than fingerprint. Likewise, geometrical features of hand/finger do not change much with time and provide a supplementary source of information.

Palmprint A palmprint refers to a unique pattern (principal lines, geometry, creases, and singular points) of human hand lie in between wrist and fingers. In [2], authors used different size kernels and orientations of Sobel operator to detect edge response. In [3], authors suggested to incorporate an Iterative Closest Point (ICP) alignment to improve line detection schemes. In [4], authors proposed SIFT algorithm with RANSAC. The subspace methods like LDA [5] and their variants were well used in the literature. In addition to this, Compcode [6] and other coding methods are well-defined in the literature.

Hand Geometry The finger/palm length, width, thickness, area, curvature, angle, etc., usually involve in determining the geometry of hand. These physical dimensions are measured over several points. In [7], authors computed 31 geometric features and achieved the best accuracy with Gaussian mixture model. In [8], authors extracted several 2-D and 3-D geometric measurements of hand and fused them. In [9], authors combined shape- and geometry-based features at score level using contact/contactless hand images.

Contribution The performance of any single biometric trait is often affected by varying environment situations, sensor accuracy, poor-quality images, spoof attacks, etc. Grouping of palmprint and hand geometry traits takes the advantage to improve the performance of single modality-based hand recognition systems. Palmprint and hand geometry can be acquired simultaneously using a low-cost camera, which makes the proposed system highly user-friendly, economic, and speedy. The key contribution of this paper is summarized as: The ROI of raw hand image is cropped by considering two points at bottom of fingers as a reference and further images are aligned by a novel L-K tracking image alignment method. Next, ROI images are enhanced using novel G-L fractional filter with optimum non-integer order. In addition to this, each set of features is further optimized using Dia-FLD. The linear SVM method is used for classification, and scores of palmprint and hand geometry are fused to reveal the human identity. The proposed system is tested over a benchmark IITD [10], CASIA [11] (contactless palmprint databases), and Bosphorus (contact based) hand database [12], and its performance is evaluated using EER, CRR, and Computation Time.

The remainder of the article is organized into following six main sections: Sect. 2 highlights the ROI extraction, image enhancement procedure, and role of image alignment. The feature extraction and dimension reduction methods are also presented in the same section. The classification and fusion criterion is discussed in next section. Then, experiment results are presented in Sect. 4. Finally, the conclusion is drawn in the last section.

2 Proposed Multimodal Biometric System

A hand-based multimodal biometric system for personal authentication is proposed using palmprint and hand geometry features.

2.1 ROI Extraction

To more efficiently segment the palm region, an improved version of ROI extraction method for IITD contactless palmprint images has been presented. First of all, re-size the input image to 600×800 and contrast adjust the input image. Smooth the input image and obtain gradient image for the Gaussian filtered image. Multiply the gradient image by 2 and add the result to the Gaussian filtered image in order to brighten the image at the boundary of the foreground. Find Otsu thresholded image, and dilate this image to get fingers separated from each other in case of noise in the image. Fill holes in the image, and remove all white connected regions other than the largest one (palm). This removes any other small objects in the background and removes background portions that appear white in binary image. Obtain the cropped palm region by finding two points at the bottom of the ring and middle fingers in the image. Next, find Euclidean distance between centroid and boundary of palm region. Multiply Euclidean distance by -1 to make points in boundary nearest to centroid as peaks. Smooth the distance, and find the peaks and their locations. The peaks correspond to points at the bottom of finger and points obtained due to noise and irregularities in the boundary. Now, measure the angle between the horizontal line and the line joining the two points. Rotate the image to make the line joining the two points horizontal. Find out the largest square inside the palm region such that size of square is larger than the distance between the two bottom points of finger, and all four corner points of finger are in the foreground of the rotated black and white image. Following this, remove 35 pixels from all 4 borders of the ROI image obtained in previous step to remove dark regions. Finally, re-size the ROI to 150×150

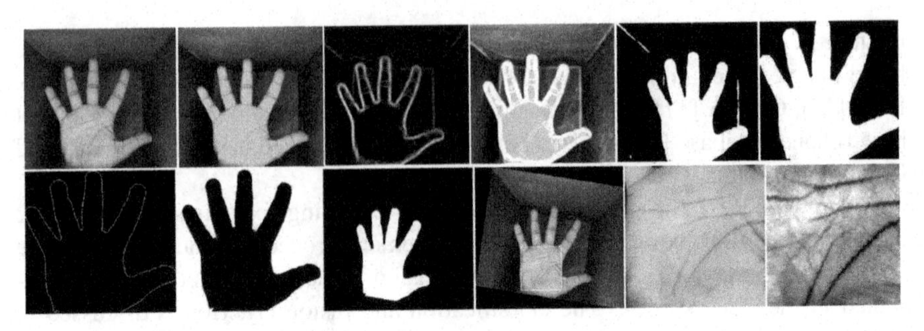

Fig. 1 Palmprint ROI extraction (IITD contactless images)

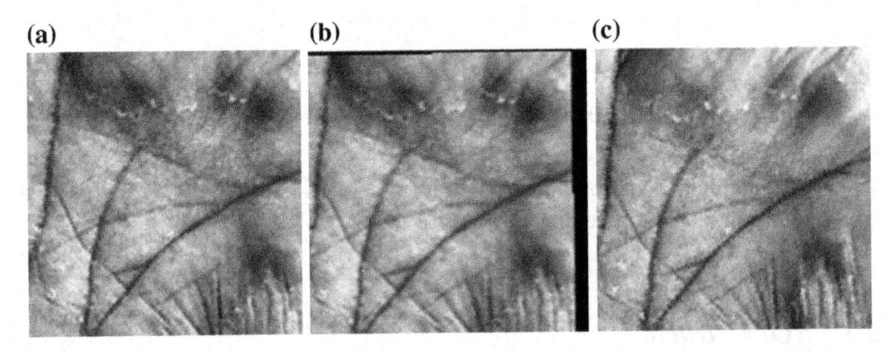

Fig. 2 Alignment results: **a** Reference ROI **b** rotated ROI **c** target ROI

as shown in Fig. 1. The proposed ROI segmentation algorithm is also tested with CASIA palmprint data set with small changes in parameter selection. However, these images are not well aligned and thus affects the matching performance.

Lucas–Kanade (L-K) Image Alignment: L-K is a widely used method for optical flow estimation and feature tracking developed by Bruce D. Lucas and Takeo Kanade in 1981 [13]. In this work, the main objective of the L-K algorithm is to align the target with the reference image as depicted in Fig. 2 and minimize the sum of squared error function between them. Let us assume that $I_{(x,y)}$ and $J_{(x,y)}$ be the reference and target images. $W_{(x,y;p)}$ defines the parametric wrap (geometric mapping) which takes the pixel in $I_{(x,y)}$ and transforms it to the sub-pixel location in $J_{(x,y)}$. Now, estimate the transformation parameter p that minimizes the sum of squared error between $I_{(x)}$ and $I_{(T)}$:

$$min \sum_{x} \sum_{y} [I_{(x,y)} - J_{(x,y)w_{(x,y,p)}}]^2 \tag{1}$$

The minimization of the error function is performed w.r.t. p. Now update the parameter p until *pp* such that:

$$p \leftarrow p + \delta p \tag{2}$$

Now the error function is minimized w.r.t. p as:

$$min \sum_{x} \sum_{y} [I_{(x,y)} - J_{(x,y)w_{(x,y;p+\delta p)}}]^2 \tag{3}$$

The above equation can be modified by first-order Taylor expansion, and further the steepest descent parameter is measured by taking dot product between error image and steepest descent image.

2.2 Image Enhancement

To enhance the overall visual effect of palmprint ROI, a 2-D discrete Grunwald–Letnikov (G-L) fractional derivative is used. For $\alpha > 0$, the αth-order G-L-based mask with respect to x for the duration [a, x] is defined as:

$$D^x_\alpha f(x,y) = \lim_{h \to 0} h^{-v} \sum_{m=0}^{n-1} \frac{1}{\tau(-v)} \frac{\phi(m-v)}{\tau(m+1)} f(x-mh) \tag{4}$$

It is a fractional derivative operator when $\alpha = 0$. The frequency characteristics of this FD mask are shown in Fig. 3b that clearly described that at negative values it behaves as a low-pass filter. Suppose, a ROI image $f(x,y)$ is convolved with a $(z \times t)$ moving window function which applies the mask to every pixel in the image, then the response to signal $I_{\alpha(x,y)}$ is given by:

$$I_{\alpha(x,y)} = \sum_{z=-a}^{a} \sum_{t=-b}^{b} w(z,t) f(x+z, y+t) \tag{5}$$

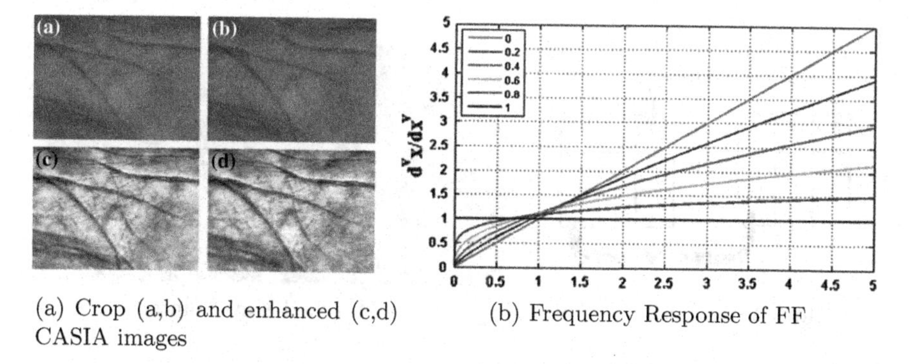

(a) Crop (a,b) and enhanced (c,d) CASIA images

(b) Frequency Response of FF

Fig. 3 Image enhancement using fractional mask

where $f(x, y)$ represents spatial coordinates of a pixel and $w(z, t)$ is the value of the fractional differential mask. Thus, the image gets overall improvement through edges, contours and retains texture information. The enhanced ROI images are shown in Fig. 3a. The optimal response to several palmprint images is obtained at a different fraction (α) orders as shown in Fig. 3b.

2.3 *Feature Extraction*

In this work, two type of features are computed from palmer region of hand, i.e., local texture and geometrical features. However, it is suspected that some of original extracted features may be noisy or not relevant, and addition of these features may lead to larger computations and deterioration in the classification accuracy. Therefore, a highly robust dimension reduction method known as Dia-FLD has been applied to each feature vector separately, and then matching scores of each modality are fused to take the final decision that the recognized result is impostor or genuine.

Block-SIFT-Based PalmPrint Recognition: In the proposed method, a 2-D palmprint image is first divided into several sub-blocks using non-overlap window. Each of these sub-blocks is convolved with Gaussian kernels of different scales and produce multi-scale SIFT features as shown in Fig. 4. Since, blocking the image before feature extraction is essential such as to get the full image information and to match corresponding blocks between two images so that more effective results can be realized. In this analysis, the size of sub-block window is selected (8×8) empirically to obtain the best representation of local features. Thereafter, the keypoints with optimize orientations are extracted and 128-dimensional descriptor for each block is created [14]. Any two pairs of SIFT-based image descriptors (Xi, Yi) are said to be matched if the distance between keypoints inside the descriptors is less than the selected threshold value.

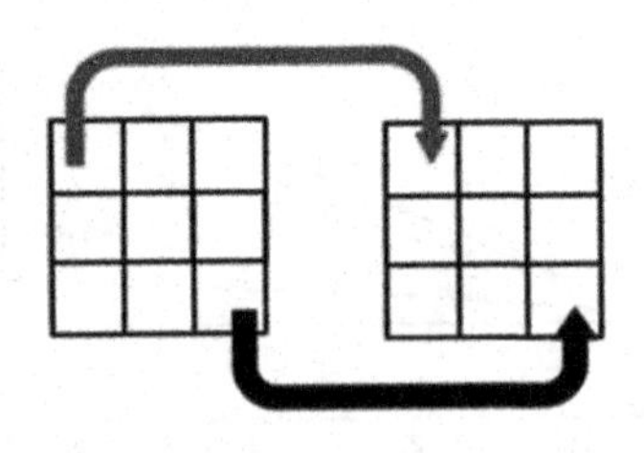

(a) Block Wise SIFT Matching

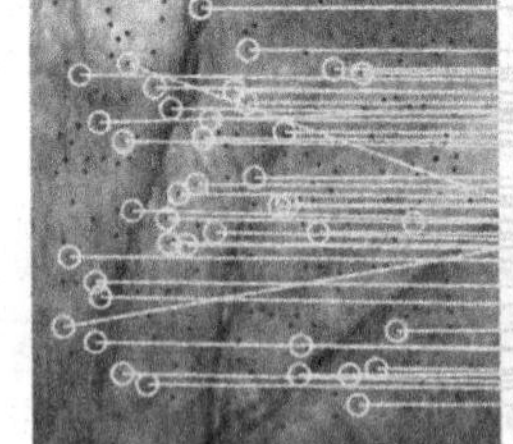

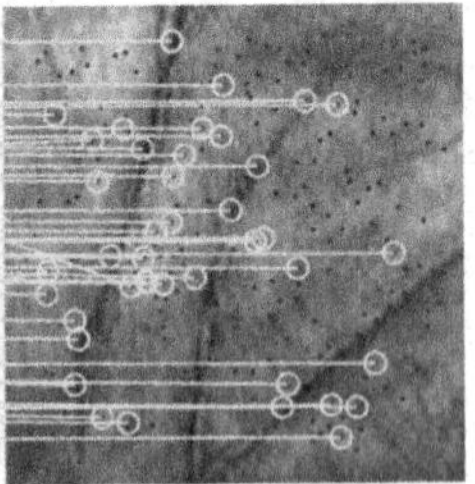

(b) Matched Key Points

Fig. 4 Correct and false matching with B-SIFT

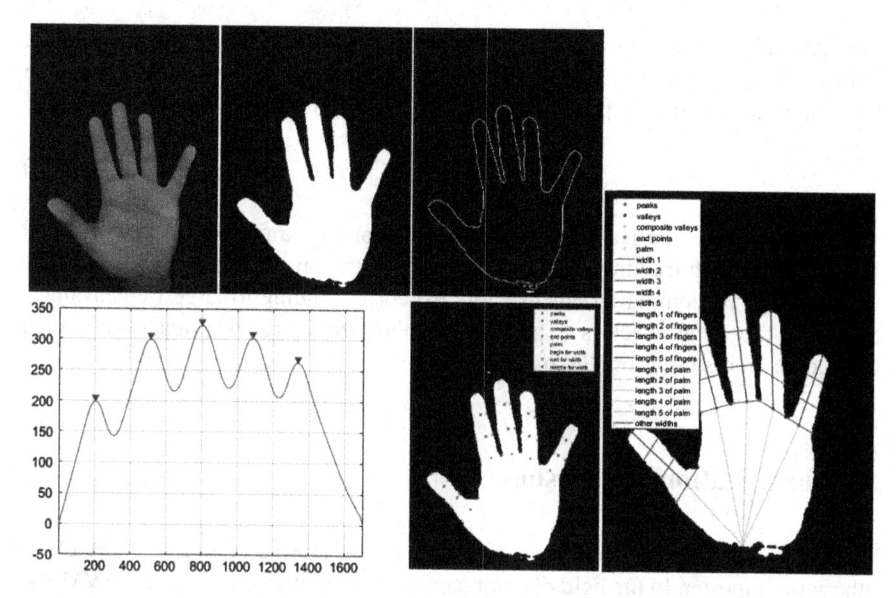

Fig. 5 Finger peak, reference point, valleys, contour detection, and geometrical measurements over Bosphorus hand database

Hand Geometry Recognition: There might be deviations in the spatial location of the fingers during hand capture. Hence, a 2-D coordinate system is developed to find the peaks, valleys, reference point, and the boundary of fingers/palms as shown in Fig. 5. This method finds 29 features from each of five fingers including thumb that characterizes the geometry of hand very well. These geometrical features include: (a) finger/palm heights: A total of 10 measurements are considered, out of which 5 are taken between fingertips and midpoint of left–right valleys, and other 5 distances are measured between mid-valley points to reference point (near wrist). (b) Finger widths: A total of 15 width measurements are taken (3 widths from each finger). (c) Angles: Four angles are measured in between consecutive fingers. All these measurements are combined to form a joint feature vector.

2.4 *Diagonal Fisher Linear Discriminant Analysis*

In Dia-FLD method, the optimal projection vectors are obtained from diagonal image coefficients rather than row or column direction of images. By doing this, Dia-FLD maintains the correlations between variation of both rows and columns of images without image to vector transformations. Suppose, we have N training samples per subject, $I_j; (j = 1, 2, \ldots, N)$ and M total number of available subjects ($M * I_j$ samples). Let us suppose that width (w) is greater than height (h) of image, and then the diagonal image block (D_j) is selected using the method as given in [15]. Following this, the within class (S_w) and between-class (S_b) scatter matrices are defined with the knowledge of mean diagonal (D) image. Let $Y = [y_1, y_2, \dots, y_d]$ represents the optimal projection matrix projecting the training hand images I_j onto Y such that:

$$C_{j(training)} = I_j Y \tag{6}$$

where $C_{j(training)}$ is the feature matrix of corresponding training samples obtained by projecting each training image onto X. Since, the optimal projection vectors X can be found by computing eigenvectors (d) corresponding to largest eigenvalues. Similarly, the weight vectors $C_{j(test)}$ can be computed for the test images, and then SVM classifier is used for classification.

3 Classification and Fusion

The score is considered genuine match if both the images belong to same class, otherwise imposter. In the field of computer vision and object recognition, SVM has emerged as a powerful tool for binary classification. It separates the two classes by constructing an optimal separating hyperplane and ensures that the margin between two classes is maximized. Support vectors are the bounds between data sets and the optimal separating hyperplane. Hence, the objective in support vector is to maximize the distance or the margin between the support vectors. The choice of suitable fusion rule to combine multiple traits is very important for better performance. In this work, a weighted sum rule is employed to integrate the normalized scores. Weights are provided as per the individual performance of the biometric trait.

4 Experimental Analysis

In distinguishing experiments, the individual and integrated performance of two modalities (palmprint and hand geometry) have been evaluated in identification and verification mode using IITD [10], CASIA [11], and Bosphorus hand [12] databases. All the experiments are performed using intersession-based testing protocol. The results are determined by selecting same number of training/test samples, and varying projection vectors to check the behavior of CRR because accuracy is greatly affected by number of principal components. For better justification of the proposed method, the following experiments are conducted:

Experiment-1: In the first test, each category of CASIA Palmprint database: Right Hand (RH-276), and Left Hand (LH-290), is assumed separately, and the proposed palmprint and hand geometry algorithms are tested. The respective EER and CRR curves are shown in Fig. 6a, b. The objective of this test is to check the baseline

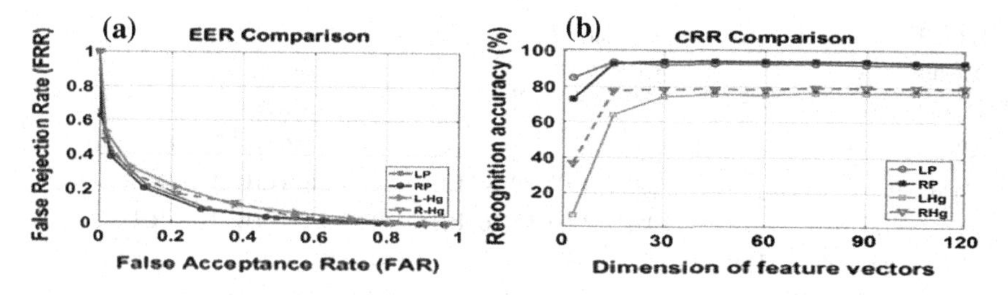

Fig. 6 Unimodal recognition performance: **a** ROC analysis, **b** CRR analysis

performance of each individual method. The average of genuine (4640) and impostor (1,340,960) matchings is obtained for left-hand data set (LH-290). Likewise, in case of right hand (RH-276) the average number of genuine (4416) and imposter (1,214,400) matchings are obtained. The conclusions from the first test illustrate that: (a) The palm-print schemes have shown significant improvement in overall results than hand geometry. (b) Among individual performance, palmprint (RH)-based SVM signifies the superior results (CRR-99.15%; EER-1.276%; speed-1575 ms) than all other methods. (c) In case of performance of hand geometry methods, the distance measurements on hand get affected due to unstable gap between the fingers, specifically over contactless images. The hand geometry accuracy results (CRR-88.58%; EER-11.62%; speed-2094 ms) with SVM classification over RH data set are somewhat better than LH data set. (d) The use of Dia-FLD plays an important role to improve the recognition accuracy as well as computation speed of these methods. Table 1 summarizes the overall results of proposed multimodal biometric system using three databases.

Experiment-2: In the second test, the whole 566 (LH + RH) number of subjects and their corresponding poses (5668 = 4528) are included for performance evaluation of each one of proposed palmprint and hand geometry. A similar methodology is adopted for selection of training and test images per class. Thus, 5,116,640 number of impostor and 9056 genuine matching scores are computed for CASIA palmprint and CASIA hand geometry alone. The results over complete data sets are described in Table 1. Further, in the next case the fusion of full CASIA-1 (palmprint) and full CASIA-2 (hand geometry) is performed. Thus, a 566 number of subjects of each modality are considered for a multimodal virtual data set which results into 5,116,640 number of impostor and 9056 genuine matching scores. Also, the EER and CRR plots are shown in Fig. 7a, b. It is clear that the optimum results (CRR-99.34%; EER-2.16%; speed-2084 ms) are achieved with proposed multimodal scheme over full (palmprint + hand geometry) data sets which is significantly higher than the individual modality-based schemes. It largely depends upon the easiness in allocating score weights to individual methods. Hence, it can be concluded that our experimental results on CASIA database using proposed multimodal (palmprint + hand geometry) fusion with SVM are highly robust for hand-based access control

Table 1 Comparative performance analysis

Method	CRR (%)	EER (%)
CompCode (CASIA) [6]	–	GAR-98.4
SIFT-RANSAC (CASIA) [4]	–	0.3969 (LH); 0.489 (RH)
Hand geom. (self) [16]	89.81 (LH); 89.74 (RH)	3.61 (LH); 3.61 (RH)
Palm (self) [16]	95.77 (LH); 95.46 (RH)	1.97 (LH); 2.02 (RH)
Shape + geometry (IITD) [9]	–	0.52
Left palm (CASIA)	98.86	1.387
Right palm (CASIA)	99.65	1.276
All palm (CASIA)	99.44	0.688
Left-hand geom. (CASIA)	86.78	13.05
Right-hand geom. (CASIA)	88.58	11.62
All hand geom. (CASIA)	89.12	8.76
All fusion (CASIA)	99.34	2.16
All palm (IITD)	99.56	1.54
All hand geom. (IITD)	95.18	5.02
All fusion (IITD)	99.18	2.78
Hand geom. (Bosphorus)	96.34	4.22

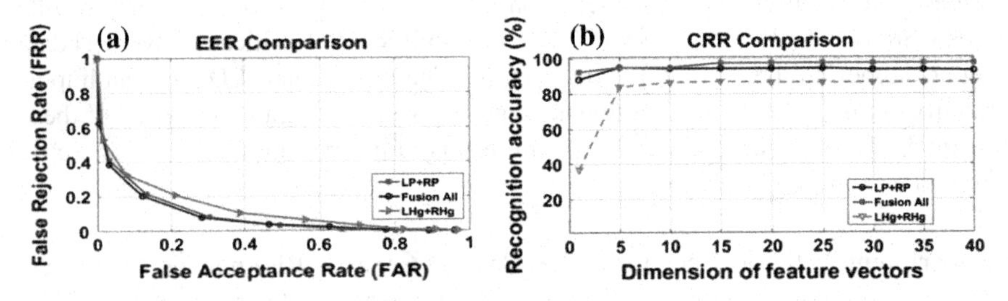

Fig. 7 Unimodal recognition performance: **a** ROC analysis, **b** CRR analysis

biometric systems. Likewise, the proposed multimodal scheme has been tested over IITD data set and achieves significant results (CRR-99.18%; EER-2.78%; speed-2266 ms).

Experiment-3: Apart from this, the performance of proposed methods has been evaluated with IITD and Bosphorus databases and also compared with reported state-of-the-art systems [4, 6, 9, 16]. We have constructed two separate virtual image multimodal data sets using CASIA and IITD images and chosen a same number of test and training images per person for recognition. Table 1 presents a comparison of four state-of-the-art methods with our proposed work on the basis of CRR, and EER. Experiment results obtained by Block–SIFT-based texture analysis show significant

CRR/EER values than other Unimodal state-of-the-art palm recognition methods [4, 6, 16] and also superior to proposed geometric analysis approach. There is need to include robust shape features for performance improvement of geometric analysis.

5 Conclusion and Future Scope

This work has been carried out with the motive of developing a hand recognition system which can be used for various access control applications. In this work, the proposed multimodal recognition system based on the fusion of scores of features such as palm texture and hand geometry is tested on benchmark CASIA, IITD palmprint, and Bosphorus hand data sets. Based on proposed ROI extraction, alignment and image enhancement procedures, the SIFT palm and geometrical features over front region of hand are classified using SVM. Finally, the scores of two modalities are fused using weighted sum rule which significantly improves the overall recognition performance. This work can be further extended to focus on more robust and deep hand features.

References

1. Jaswal, G., Kaul, A., Nath, R.: Knuckle print biometrics and fusion schemes-overview, challenges, and solutions. ACM Comput. Surv. **49**(2), 34 (2016)
2. Wong, K.Y., Chekima, A., Dargham, J.A., Sainarayanan, G.: Palmprint identification using Sobel operator. In: 10th IEEE International Conference on Control, Automation, Robotics and Vision, pp. 1338–1341 (2008)
3. Zhang, L.W., Zhang, B., Yan, J.: Principal line-based alignment refinement for palmprint recognition. IEEE Trans. Syst. Man Cybern. **42**(6): 1491–1499 (2012)
4. Wu, X., Zhao, Q., Bu, W.: A SIFT-based contactless palmprint verification approach using iterative RANSAC and local palmprint descriptors. Pattern Recogn. **47**(10): 3314–3326 (2014)
5. Wu, X., Zhang, D., Wang, K.: Fisherpalms based palmprint recognition. Pattern Recogn. Lett. **24**(15): 2829–2838 (2003)
6. Kong, A.K., Zhang, D.: Competitive coding scheme for palmprint verification. In: 17th IEEE International Conference on Pattern Recognition, pp. 520–523 (2004)
7. Sanchez-Reillo, R.: Hand geometry pattern recognition through Gaussian mixture modelling. In: 15th IEEE International Conference on Pattern Recognition, Vol. 2, pp. 937–940 (2000)
8. Kanhangad, V., Kumar, A., Zhang, D.: Combining 2D and 3D hand geometry features for biometric verification. In: IEEE Computer Society Conference on Computer Vision and Pattern Recognition Workshops, pp. 39–44 (2009)
9. Sharma, S., Dubey, S.R., Singh, S.K., Saxena, R., Singh, R.K.: Identity verification using shape and geometry of human hands. Expert Syst. Appl. **42**(2), 821–832 (2015)
10. IIT Delhi Touchless palmprint Database. http://www4.comp.polyu.edu.hk/~csajaykr/IITD/Database_Palm.htm
11. CASIA palmprint Database. http://biometrics.idealtest.org/
12. Bosphorus Database. http://bosphorus.ee.boun.edu.tr/hand/Home.aspx
13. Baker, S., Matthews, I.: Lucas-Kanade 20 years on: a unifying framework. Int. J. Comput. Vision **56**(3), 221–255 (2004)

14. Lowe, D.G.: Distinctive image features from scale-invariant keypoints. Int. J. Comput. Vision **60**(2), 91–110 (2004)
15. Noushath, S., Kumar, G., Kumara, P.: Diagonal fisher linear discriminant analysis for efficient face recognition. Neurocomputing **69**(13), 1711–1716 (2006)
16. Michael, G.K.O., Connie, T., Teoh, A.B.J.: A contact-less biometric system using multiple hand features. J. Vis. Commun. Image Represent. **23**(7), 1068–1084 (2012)

Automatic Semantic Segmentation for Change Detection in Remote Sensing Images

Tejashree Kulkarni and N Venugopal

Abstract Change detection (CD) mainly focuses on the extraction of change information from multispectral remote sensing images of the same geographical location for environmental monitoring, natural disaster evaluation, urban studies, and deforestation monitoring. While capturing the Landsat imagery, there may occur data missing issues such as occlusion of cloud, camera sensor, and aperture artifacts. The existing machine learning approaches do not provide significant results. This paper proposes a DeepLab Dilated convolutional neural network (DL-DCNN) for semantic segmentation with the goal to occur the change map for earth observation applications. Experimental results reveal that the accuracy of the proposed change detection results provides improved results as compared with the existing algorithms and maps the semantic objects within the predefined class as change or no change.

Keywords Change detection · Remote sensing · Multispectral
Deep learning

1 Introduction

Recently, CD [1] on urban remote sensing plays a significant role for researchers for the evaluation of images from multi-temporal image data scene. Monitoring and detection of land cover changes are the crucial steps for continuous monitoring of application such as forest change detection, land management, natural hazard

T. Kulkarni (✉)
Electrical and Electronics Engineering [Embedded System], PES University, Bengaluru, India
e-mail: tejashree2668@gmail.com

N. Venugopal
Department of EEE, PES University, Bengaluru, India
e-mail: venugopaln@pes.edu

D. Reddy Edla et al. (eds.), *Advances in Machine Learning and Data Science*, Advances in Intelligent Systems and Computing 705,
https://doi.org/10.1007/978-981-10-8569-7_34

analysis, disaster management, area planning, policy-making, and grazing land management.

CD is normally classified into supervised and unsupervised based on the availability of the training samples [2]. The supervised CD [3] is moderately quick and requests generally with less computational power than some other training strategies that are used as a part of machine learning. This approach has critical downsides of genuine applications.

Machine learning programs rely upon supervised learning. A man is expected to mark the information, and naming procedure is tedious and costly. Deep learning strategy stays away from this downside since they exceed expectations at unsupervised learning. The key contrast among supervised and unsupervised learning is that the information is not named in unsupervised learning. Deep learning can be connected effectively to huge information for information application, information-based forecast [4].

Segmentation is the process of dividing the image into several parts but not specifies the detailed description of the images. The semantic segmentation provides the detailed information of the meaningful parts and classifies the each and every parts other than low-level features according to the already defined classes [5]. In the change detection approach, there is a need for the detailed segmentation and accurate predictions in order to improve the accuracy [6]. The existing authors proposed numerous techniques for change detection as mentioned in [7–17]. Although some of the researchers introduced deep learning approach for change detection, it lacks in the overall accuracy measurement due to the fixed size input and output, and training did not occur in the end-to-end manner, time-consuming, and restricted or small field of view. Also, there may occur a missing data problem which is a serious issue in the change detection.

A novel DeepLab Dilated CNN learning approach significantly identifies changes in the multispectral remote sensing images which overcome the drawbacks of the existing approaches. The proposed work identifies the changed trajectories of land cover types and produces the accuracy evaluation for the change detection results.

2 Related Work

The recent related work corresponds to the change detection of remote sensing images and is given as follows:

Romero et al. [18] developed a deep convolutional network with single-layer for data analysis of the remote sensing images. The sparse features are extracted with the unsupervised pretraining process. The greedy layer-wise training process was coupled with the unsupervised learning for CD.

Leichtle et al. [19] introduced an unsupervised approach for CD as object based on VHR remote sensing imagery. Their technique mainly focused to detect the changes in buildings. Su et al. [20] proposed a concept or unbalanced images using deep learning and mapping (DLM) framework.

Zhang et al. [21] presented a framework in light of multi-spatial and multi-determination change identification system, which consolidates profound design-based unsupervised element learning and mapping based on component change examination. Furthermore, the denoising autoencoder was stacked to learn nearby and abnormal state portrayal/highlight from the neighborhood of the given pixel in an unsupervised design. Thirdly, inspired by the way that multi-determination picture match shares a similar reality in the unaltered districts, they endeavor to investigate the internal connections between them by building a mapping neural system. The picked-up mapping capacity can connect the distinctive portrayals and feature changes. At long last, we can assemble a vigorous and contractive change outline that highlights closeness investigation, and the change location result is obtained through the division of the last change delineate. Regardless of the possibility, multi-spatial and multi-determination change location issue was contemplated and the exactness was restricted because of its element vectors.

Lu et al. [22] developed a concept based on improving sparse coding method for change detection and that unchanged pixel in different images could be well reconstructed by the joint dictionary for multispectral remote sensing. Initially, the query image pair was inserted into the joint dictionary to know the knowledge of changed pixels. The reconstructed error between changed pixels and unchanged pixels of different images is shown. Although numerous researchers introduced various concepts, there may be some drawbacks in their works.

3 Proposed DeepLab Dilated Convolutional Neural Network for Semantic Segmentation

The proposed work is divided into preprocessing, semantic segmentation, and accuracy evaluation stages. In the preprocessing step, first the Landsat images are corrected for geometric, radiometric and atmospheric disturbances. Then, the ResNet-101 model will extract the features automatically with its pretrained classes. The segmented change map with the detailed semantic representation is reproduced at the output of this model. Finally, the dilated convolutional neural network (DCNN) predicts the results as change or no change.

Preprocessing: First, the temporal images get corrected with correction radiometric correction, atmospheric-correction, and mosaicking to remove radiometric, geometric, and atmospheric distortions.

DeepLab Dilated Convolutional neural network (DL-DCNN) for Semantic Segmentation: The proposed approach automatically identify the meaningful map from two different images of the same area and identifies the multiple changes at the same location as change or unchanged regions.

(a) Initialization—Xavier initialization is performed to attain convergence.
(b) Activation function—To speed up training and achieve improved results than leaky rectifier linear unit.

    Leaky rectifier linear unit is used for nonlinearly transforming the data to achieve better results than the more classical sigmoid or hyperbolic tangent functions and speed up the training process.
(c) DeepLabv2 (ResNet-101) Architecture—The architecture of the DeepLabv2 involves a dilated spatial pyramid pooling and the repurposed ResNet-101 for semantic segmentation by dilated convolution.
(d) Weight optimization—A derivative-free optimization algorithm called coordinate descent is used to find the local minima of a function. Also, it takes the advantage of the normally used gradient descent optimization.

Accuracy Assessment: Finally, the segmentation results show the accuracy assessment of the proposed approach while segmenting the change area.

4 Results

The proposed feature extraction-based change detection is implemented in the MATLAB 2017a software. The resulting analysis of the proposed work is tested on the USGS scene. This dataset was taken in the years 2008 and 2010 to perform change detection techniques.

4.1 Result Analysis

The performance analysis of the proposed technique is described in the following graphical representation to show the high accuracy detection results.

Figure 1 represents the original image USGS dataset for the years 2008 and 2010 with its semantic segmented results. The land area, wetlands, and lakes are labeled in the segmented image. The change detection result in Fig. 2 shows that the blue color indicates the no change, green color indicates the lake, and white color indicates the wetlands.

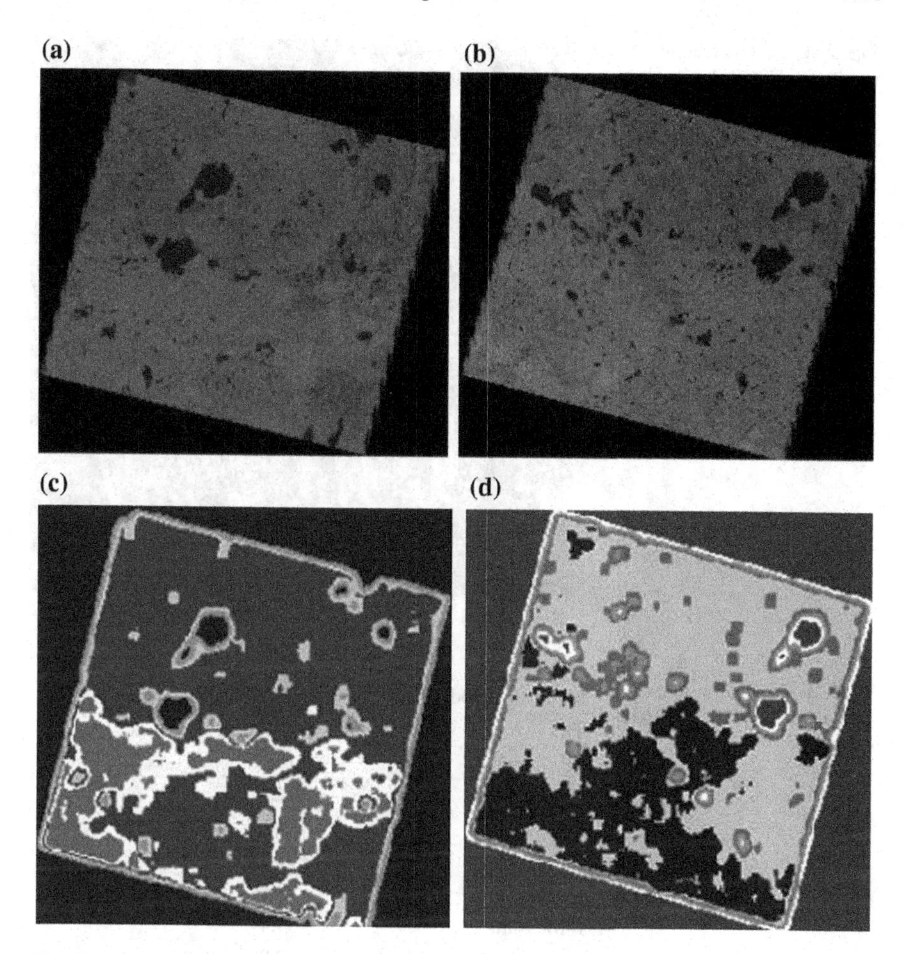

Fig. 1 Semantic segmentation results for USGS dataset **a** original 2009, **b** 2010, **c** segmentation results 2009 and **d** 2010

4.2 Evaluation Metrics

The performance is evaluated in terms of TP, FP, TN, FN, precision, recall, F-Measure, TNR, FPR, FNR, accuracy, kappa coefficient, overall accuracy, and error rate.

The analysis of the various performances is evaluated in Table 1. The table evaluates the TP, TN, FP, FN, precision, recall, F-Measure, TNR, FPR, FNR, PCC, kappa, overall accuracy, and overall error of the proposed method.

Table 2 represents the proposed method to achieve higher kappa value and AUC related to SSND, SSFA, and CDT. If the performance of the accuracy is higher, then we clearly show that the change detection segmentation results are also get improved to measure the details of the images.

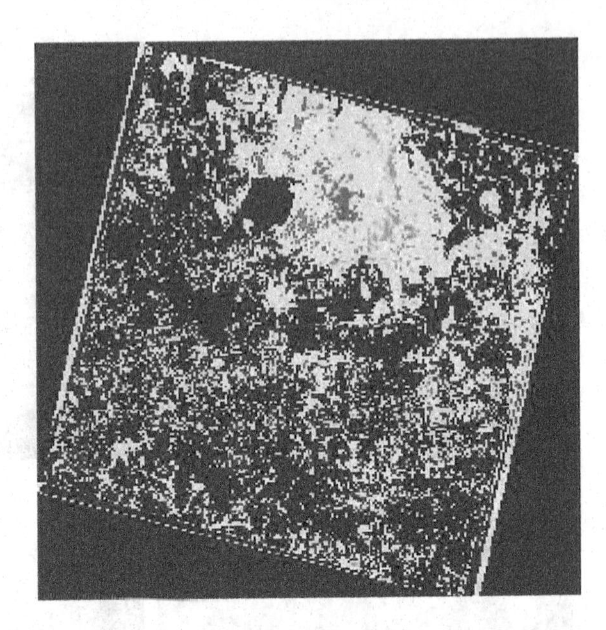

Fig. 2 Change detection results for USGS dataset

Table 1 Evaluation metrics over state-of-the-art methods on dataset

Metrics/method	Proposed method
TP	159,375
FP	1687
TN	24,375
FN	187
Precision	98.95
Recall	99.88
F-Measure	99.41
TNR	99.88
FPR	93.52
FNR	64.74
PCC	95.76
Kappa	80.23
Overall accuracy	98.98
Overall error	1.87

Table 2 Overall accuracy (AUC) and kappa coefficients over state-of-the-art methods on experiment dataset

	SSND [23]	SSFA [24]	CDT [25]	Proposed
Kappa	75.36	93.12	0.9665	80.23
AUC	60.225	83.138	89.681	98.4375

5 Conclusion

Thus, the paper presents the novel DeepLab Dilated learning approach to identify area change and trajectories of changes in the land cover types and evaluates the accuracy of the change detection maps. Accurate predictions and detailed segmentation are the key roles of this paper. In order to evaluate such results, this paper introduces a novel deep learning approach for CD in the remote sensing images. Compared with the existing approaches, our system yields improved performance.

References

1. El-Kawy, O.A., Rød, J.K., Ismail, H.A., Suliman, A.S.: Land use and land cover change detection in the western Nile delta of Egypt using remote sensing data. Appl. Geogr. **31**(2), 483–494 (2011)
2. Fichera, C.R., Modica, G., Pollino, M.: Land Cover classification and change-detection analysis using multi-temporal remote sensed imagery and landscape metrics. Eur. J. Remote Sens. **45**(1), 1–18 (2012)
3. Jin, S., Yang, L., Danielson, P., Homer, C., Fry, J., Xian, G.: A comprehensive change detection method for updating the national land cover database to circa 2011. Remote Sens. Environ. **132**, 159–175 (2013)
4. Zhu, Z., Woodcock, C.E.: Continuous change detection and classification of land cover using all available Landsat data. Remote Sens. Environ. **144**, 152–171 (2014)
5. Hao, M., Shi, W., Zhang, H., Li, C.: Unsupervised change detection with expectation-maximization-based level set. IEEE Geosci. Remote Sens. Lett. **11**(1), 210–214 (2014)
6. Wu, C., Zhang, L., Zhang, L.: A scene change detection framework for multi-temporal very high resolution remote sensing images. Sig. Process. **124**, 184–197 (2016)
7. Neagoe, V.E., Stoica, R.M., Ciurea, A.I., Bruzzone, L., Bovolo, F.: Concurrent self-organizing maps for supervised/unsupervised change detection in remote sensing images. IEEE J. Sel. Top. Appl. Earth Obs. Remote Sens. **7**(8), 3525–3533 (2014)
8. Liu, Z.G., Dezert, J., Mercier, G., Pan, Q.: Dynamic evidential reasoning for change detection in remote sensing images. IEEE Trans. Geosci. Remote Sens. **50**(5), 1955–1967 (2012)
9. Hussain, M., Chen, D., Cheng, A., Wei, H., Stanley, D.: Change detection from remotely sensed images: from pixel-based to object-based approaches. ISPRS J. Photogram. Remote Sens. **80**, 91–106 (2013)
10. Du, P., Liu, S., Xia, J., Zhao, Y.: Information fusion techniques for change detection from multi-temporal remote sensing images. Inf. Fusion **14**(1), 19–27 (2013)
11. Chen, G., Hay, G.J., Carvalho, L.M., Wulder, M.A.: Object-based change detection. Int. J. Remote Sens. **33**(14), 4434–4457 (2012)
12. Gong, M., Zhou, Z., Ma, J.: Change detection in synthetic aperture radar images based on image fusion and fuzzy clustering. IEEE Trans. Image Process. **21**(4), 2141–2151 (2012)
13. Bovolo, F., Marchesi, S., Bruzzone, L.: A framework for automatic and unsupervised detection of multiple changes in multitemporal images. IEEE Trans. Geosci. Remote Sens. **50**(6), 2196–2212 (2012)
14. Mishra, N.S., Ghosh, S., Ghosh, A.: Fuzzy clustering algorithms incorporating local information for change detection in remotely sensed images. Appl. Soft Comput. **12**(8), 2683–2692 (2012)
15. Gu, W., Lv, Z., Hao, M.: Change detection method for remote sensing images based on an improved Markov random field. Multimedia Tools Appl. 1–6 (2015)

16. Du, P., Liu, S., Gamba, P., Tan, K., Xia, J.: Fusion of difference images for change detection over urban areas. IEEE J. Sel. Top. Appl. Earth Obs. Remote Sens. **5**(4), 1076–1086 (2012)
17. Ghosh, A., Subudhi, B.N., Bruzzone, L.: Integration of Gibbs Markov random field and hopfield-type neural networks for unsupervised change detection in remotely sensed multitemporal images. IEEE Trans. Image Process. **22**(8), 3087–3096 (2013)
18. Romero, A., Gatta, C., Camps-Valls, G.: Unsupervised deep feature extraction for remote sensing image classification. IEEE Trans. Geosci. Remote Sens. **54**(3), 1349–1362 (2016)
19. Leichtle, T., Geiß, C., Wurm, M., Lakes, T., Taubenböck, H.: Unsupervised change detection in VHR remote sensing imagery–an object-based clustering approach in a dynamic urban environment. Int. J. Appl. Earth Obs. Geoinf. **54**, 15–27 (2017)
20. Su, L., Gong, M., Zhang, P., Zhang, M., Liu, J., Yang, H.: Deep learning and mapping based ternary change detection for information unbalanced images. Pattern Recogn. **66**, 213–228 (2017)
21. Zhang, P., Gong, M., Su, L., Liu, J., Li, Z.: Change detection based on deep feature representation and mapping transformation for multi-spatial-resolution remote sensing images. ISPRS J. Photogram. Remote Sens. **116**, 24–41 (2016)
22. Lu, X., Yuan, Y., Zheng, X.: Joint dictionary learning for multispectral change detection. IEEE Trans. Cybern. **47**(4), 884–897 (2017)
23. De Morsier, F., Tuia, D., Borgeaud, M., Gass, V., Thiran, J.P.: Semi-supervised novelty detection using SVM entire solution path. IEEE Trans. Geosci. Remote Sens. **51**(4), 1939–1950 (2013)
24. Wu, C., Du, B., Zhang, L.: Slow feature analysis for change detection in multispectral imagery. IEEE Trans. Geosci. Remote Sens. **52**(5), 2858–2874 (2014)
25. Lu, X., Yuan, Y., Zheng, X.: Joint dictionary learning for multispectral change detection. IEEE Trans. Cybern. **47**(4), 884–897 (2017)

A Model for Determining Personality by Analyzing Off-line Handwriting

Vasundhara Bhade and Trupti Baraskar

Abstract Handwriting analysis is the scientific method or way of determining or understanding or predicting the personality or behavior of a writer. Graphology or graph analysis is the scientific name of handwriting analysis. Handwriting often called as brain writing or mind writing, since it is a system of studying the frozen graphic structures which have been generated in the brain and placed on paper in a printed or cursive style. Many things can be revealed from handwriting such as anger, morality, fears, past experience, hidden talents, mental problems. Handwriting is different from person to person. People are facing various psychological problems. Teenagers also face so many mental problems. Criminals can be detected by using handwriting analysis. Counselor and mentor can also use this tool for giving advice to clients. Proposed work contains three main steps: image preprocessing, identification of handwriting features, and mapping of identified features with personality traits. Image pre-processing is the technique in which the handwriting sample is translated into a format which can be easily and efficiently processed in further steps. These steps involve noise removal, grayscale, thresholding, and image morphology. In feature identification, there is an extraction of handwriting features. Three features of handwriting are extracted that are left margin, right margin, and word spacing. Lastly, extracted features are mapped with personality using the rule-based technique. The personality of a writer with respect to three handwriting features is displayed. The proposed system can predict 90% accurate personality of the person.

Keywords Feature identification · Handwriting analysis
Personality prediction · Rule-based technique

V. Bhade (✉) · T. Baraskar
Department of Information Technology, Maharashtra Institute of Technology, Pune, India
e-mail: vasubhade@gmail.com

T. Baraskar
e-mail: trupti.baraskar@mitpune.edu.in

D. Reddy Edla et al. (eds.), *Advances in Machine Learning and Data Science*, Advances in Intelligent Systems and Computing 705,
https://doi.org/10.1007/978-981-10-8569-7_35

1 Introduction

Handwriting analysis is often called as graphology. It is the science of prediction or recognition or identification of behavior or personality by examining various characteristics, strokes, features, and patterns of an individual's handwriting. Many things can be disclosed through handwriting such as anger, morality, fears, past experience, hidden talents, mental problems. Strokes, characteristics, and pattern of handwriting are different so handwriting is peculiar to a specific person [1].

Handwriting analysis would be used in the recruitment process for assessment of the student's personality. Handwriting analysis also used in forensic for detecting criminals. It is a very good tool for child development and for personal development also. It is an essential skill for both children and adults. It is important for counselors and mentor also for giving psychological advice to clients. It is an important tool for historical profiling, security checking, and career guidance [2].

In this work, we propose a model to determine personality from writer's handwriting. Personality or behavior of a person is analyzed from features of handwriting that are left margin, right margin, and spacing between words. Process contains three main steps: image preprocessing, identification of handwriting features, and mapping of identified features with personality traits. Image pre-processing is the technique in which the handwriting sample is translated into a format which can be easily and efficiently processed in further steps. Three features of handwriting are extracted that are left margin, right margin, and word spacing. Lastly, extracted features are mapped with personality using the rule-based technique.

The rest of the paper is organized as follows: First, in Sect. 2, we describe the existing various techniques for personality prediction. In Sect. 3, there is description of system architecture. Section 4 provides implementation and result. Finally, Sect. 5 devoted to conclusion and future work.

2 Related Work

In 2016, Abhishek Bal and Rajiv Saha proposed a method for an off-line handwritten document analysis from segmentation, skew recognition, and pressure detection for cursive handwriting. This method is based on orthogonal projection. The proposed method was tested on IAM database [3].

In 2015, Laurence Likforman-Sulem et al. proposed a method for extracting style and emotion from online and off-line handwriting. This method proposed that anxiety can be recognized through handwriting using random forest technique [4].

In 2013, Esmeralda Contessa Djaml, Risna Darmawati, and Sheldy Nur Ramdlan proposed a technique that personality can be predicted from handwriting and signature. This method uses structure algorithms and multiple artificial neural network (ANN). Some features are processed using multistructure algorithm, and remaining features are processed using ANN algorithm [5].

In 2010, Champa H. N., K. R. Anandakumar proposed a technique that the personality can be predicted from handwriting features such as baseline, pen pressure, the letter "t," the lower loop of letter "y," and the slant. These features are inputs to a rule based which give personality of writer [6].

In 2008, Sofianita Mutalib et al. proposed a method for emotion recognition through handwriting using fuzzy inference. This technique uses a baseline feature of handwriting in determining the level of emotion through Mamdani inference [7].

3 System Architecture

Steps of system architecture are: (1) input of the system, (2) image preprocessing, (3) feature selection, (4) identification of features, (5) mapping of identified features with personality, and (6) output of system (Fig. 1).

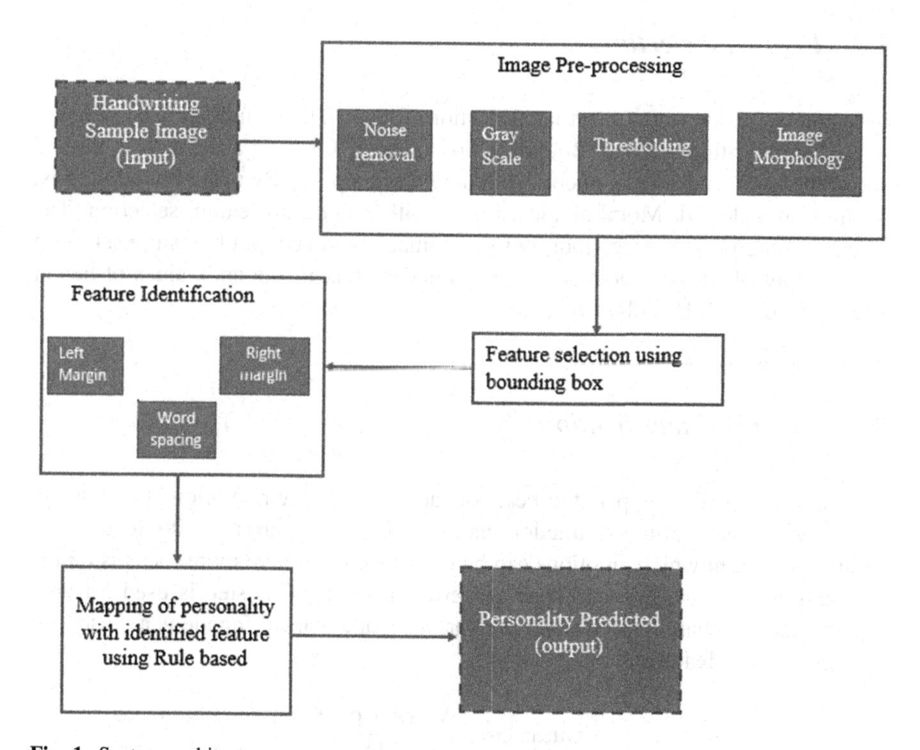

Fig. 1 System architecture

3.1 Input of the System

In this system, manual dataset is used. Eleven different handwriting paragraphs have been taken. Handwriting image is input to the system. Users have to upload his/her handwriting image. The size of the image should be from 100 to 700 KB. For accurate personality, handwritten paragraph must be of minimum 50 words. The user has to click on select image for uploading image into the system.

3.2 Image Preprocessing

Handwriting image will be input to the image preprocessing step. In case of preprocessing, the first method is noise removal, second is grayscale, third is thresholding, and fourth is image morphology. Output of image preprocessing will be morphological image [8].

3.3 Feature Selection

In feature selection step, object identification is done. The bounding box formation around handwriting is done. Bounding box is formed around a word. It is formed according to X- and Y-axis of corresponding word's pixel. By using bounding box, features are selected. Morphological image will be input to feature selection. The output of preprocessing, i.e., morphological image, will be input to feature selection step. Output of feature selection will be bounding box image and values of height and width of each bounding box [5].

3.4 Feature Identification

Feature identification step is the heart of the overall system. Various patterns are recognized using feature extraction method. There are around 300 features of handwriting from which emotions can be recognized, but here more focus is on the left margin, right margin, and spacing between words. This step is used for analyzing data from handwritten samples. Various mathematical functions are used for identification of features:

$$\text{Value in cm} = \frac{\text{Value in pixel}}{37.8} \tag{1}$$

Min(w) = Minimum starting width of bounding box
Max(w) = Maximum ending width of bounding box
Min(h) = Minimum starting height of bounding box
Max(h) = Maximum ending height of bounding box

A. Margin

$$\text{Left Margin} = \text{Min(w)} \tag{2}$$

$$\text{Right Margin} = \text{Image width} - \text{Max(w)} \tag{3}$$

$$\text{Top Margin} = \text{Min(h)} \tag{4}$$

$$\text{Bottom Margin} = \text{Image height} - \text{Max(h)} \tag{5}$$

As shown in Fig. 2, "Still" is leftmost word whose dimensions are {36.0, 228.0} {84.0, 244.0} means starting width of "still" is 36 which is minimum width and ending width is 84 while starting height is 228 and ending height is 244.

According to Eq. (2), left margin is Min(W). So the left margin of this handwriting paragraph is 36 pixels. Here, we want result in centimeter. So, according to Eq. (1), left margin = 36/37.8 = 0.95238 cm.

"Paper" is rightmost image whose dimensions are {569.0, 114.0} {632.0, 135.0} means ending width of "Paper" is 632 which is maximum width. And image width is 640. So, according to Eq. (3), 640 – 632 = 8 Pixels and 8/37.8 = 0.211 cm.

"Which" and "Compression" are top and bottom words, respectively. According to Eqs. (4) and (5) and (1), top margin is 0.582 cm and bottom margin is 2.830 cm.

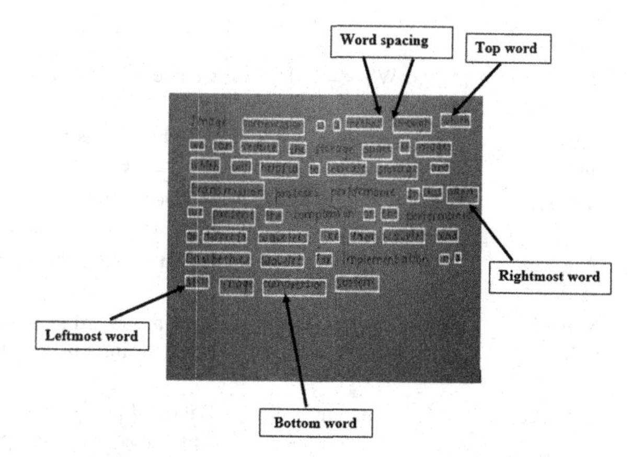

Fig. 2 Word spacing and margin

B. Word Spacing

$$\text{Avg diff} = \sum \left(\frac{\text{starting width of next word} - \text{ending width of previous word}}{No.\, of\ Bounding\ box - 1} \right). \quad (6)$$

Word spacing means how many spaces are between two words. As shown in Fig. 2, suppose we have to find out the word spacing of "method" and "through," then according to Eq. (6), starting width of "through" – ending width of "method" will be word spacing of "method" and "through."

Here total words means number of bounding box words are 45. According to Eq. (6), summation of 45 words (starting width of next word – ending width of first word)/44. Here by (6) and (1), word spacing is 0.52096 cm.

3.5 *Mapping of Identified Features with Personality Traits and Emotions*

After the image is processed and cleansed, it undergoes feature extraction where the peculiar features of the handwriting are identified. These features act as input to the classifier. Using these features, the classifier recognizes the emotions and the state

Table 1 Threshold value

Features	Threshold value	Personality
Narrow word spacing	0.0 <= WS < 0.3	Need of constant contact with people
Normal word spacing	0.3 <= WS < 0.7	Capable of building healthy relationship
Wide word spacing	0.7 < WS <= 2.0	Unable to establish steady relationship
Narrow left margin	0.0 <= margin < 0.5	Attached to his past, decision may depend on past experience
Balanced	0.5 <= margin < 1.5	Balanced person when it comes to risk taking and being attached to the past
Wide left margin	1.5 <= margin < 3.0	Not attached to his past
Narrow right margin	0.0 <= margin < 0.5	Take a forward step and that he may not be experiencing uncertainties at that period of his life
Balanced	0.5 <= margin < 1.5	Balanced person when it comes to risk taking and being attached to the past
Wide right margin	1.5 <= margin < 3.0	To take a future step as he may still be making his mind about it

of the inner mind of the writer. The classifier categorizes these features and accordingly presents the output of the nearest matching emotions. Table 1 shows thresholding value of mapping of personality to feature extraction. According to thresholding value, features are mapped with personality.

3.6 *Output of the System*

The output of the system is personality of the writer. Personality is predicted from extracted features. Using various mathematical equation, features are extracted. Extracted features are mapped with personality using rule-based technique. Extracted features are mapped with personality traits using thresholding.

4 Implementation and Result

Table 2 describes values of three features of 11 handwriting paragraph, and graphs related to features values are shown below.

Table 2 Result analysis of 11 handwriting paragraphs

Images	Word spacing	Related features	Left margin	Related features	Right margin	Related features
1	0.90828	Wide word spacing	2.116	Wide left margin	0.4497	Narrow right margin
2	0.52096	Normal word spacing	0.9523	Balanced	0.211	Narrow right margin
3	0.36816	Normal word spacing	0.634	Balanced	1.3492	Balanced
4	0.49823	Normal word spacing	0.3968	Narrow left margin	1.6402	Wide right margin
5	0.56613	Normal word spacing	1.5079	Wide left margin	2.1428	Wide right margin
6	0.3848	Normal word spacing	1.428	Balanced	2.037	Wide right margin
7	0.43503	Normal word spacing	1.0582	Balanced	1.746	Wide right margin
8	0.56277	Normal word spacing	1.19	Balanced	2.6455	Wide right margin
9	0.18518	Narrow word spacing	0.8465	Balanced	0.26455	Narrow right margin
10	0.39241	Normal word spacing	1.5873	Wide left margin	2.1957	Wide right margin
11	0.53791	Normal word spacing	0.26455	Narrow left margin	0.39682	Narrow right margin

Figure 3 is word spacing analysis. The graph shows values of word spacing of 11 different writers. In 11 writer, 9 persons writing style is normal word spacing the related personality is capable of building healthy relationship, 1 is narrow word spacing (image 9) related personality is need of constant contact with people and 1 is wide word spacing (image 1) related personality is unable to establish steady relationship.

Figure 4 is left margin analysis. The graph shows values of left margin of 11 different writers. In 11 writer, 6 persons writing style is normal left margin the related personality is balanced person when it comes to risk taking and being attached to the past, 2 are narrow left margin (image 4 and image 11) related personality is attached to his past, decision may depend on past experience and 3 are wide left margin (image 5, 10, 1) related personality is not attached to his past.

Figure 5 is right margin analysis. The graph shows values of right margin of 11 different writers. In 11 writer, only 1 persons writing style is normal right margin (image 3) the related personality is balanced person when it comes to risk taking and being attached to the past, 4 are narrow right margin (image 1, 2, 9, 11) related personality is take a forward step and that he may not be experiencing uncertainties at that period of his life, and 6 are wide right margin (image 4, 5, 6, 7, 8, 10) related personality is to take a future step as he may still be making his mind about it.

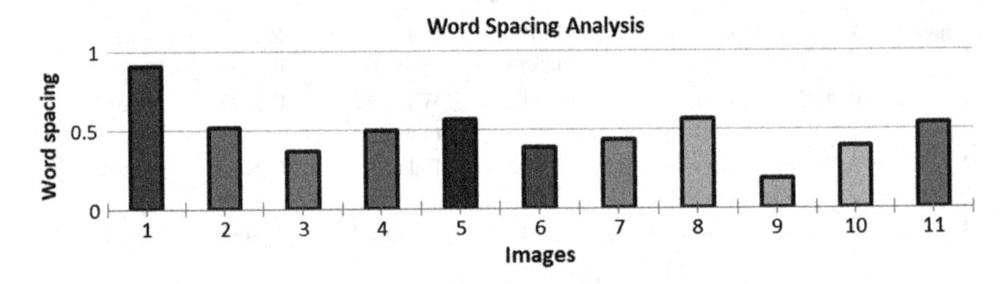

Fig. 3 Word spacing analysis

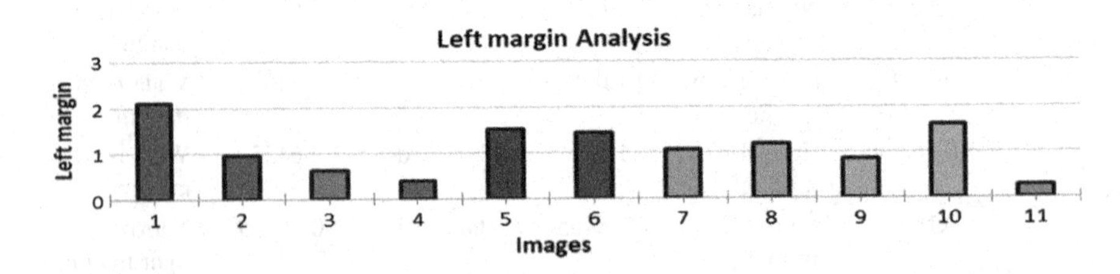

Fig. 4 Left margin analysis

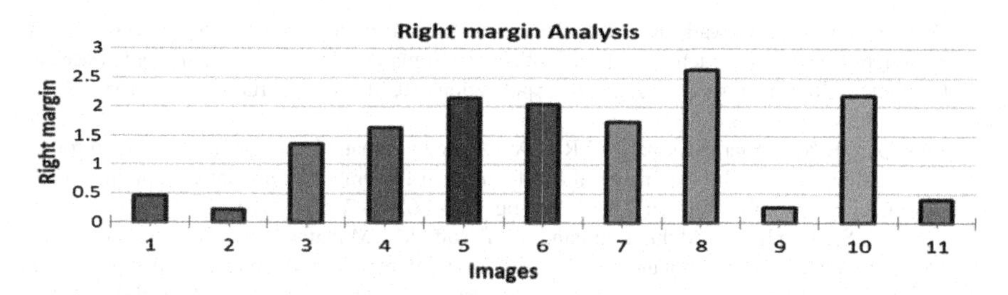

Fig. 5 Right margin analysis

5 Conclusion and Future Work

This system can identify the type of personalities from a variety of features using rule-based technique. Proposed work extracts three features, and it predicts personality related to five features efficiently. Proposed system identified three features that are left margin, right margin, and word spacing. Out of three features, left margin is the most efficient implemented feature as compared to others because left margin is extracted 95% accurately through bounding box. Secondly, right margin extracted 90%. Cordially word spacing feature extracted 85%. The analysis of 11 different handwritings has been done. Out of 11 persons, 50–70% write by normal feature style.

The current system has limitations in terms of number of features and accuracy. Future work includes acquiring more accuracy in predicting personality. The current system can be extended to smart system in which most of the handwriting features and also signature features can be extracted, so that personality can be recognized from the corresponding extracted features.

References

1. Kedar, S.V., Bormane, D.S., Dhadwal, A., Alone, S., Agarwal, R.: Automatic emotion recognition through handwriting analysis: a review. In: International Conference on Computing Communication Control and Automation (ICCUBEA), Pune, pp. 811–816, 26–27 Feb 2015
2. Coll, R., Fornes, A., Llados, J.: Graphological analysis of handwritten text documents for human resources recruitment. In: International Conference on Document Analysis and Recognition, pp. 1081–1085, 26–29 July 2009
3. Bal, A., Saha, R.: An improved method for handwritten document analysis using segmentation, baseline recognition and writing pressure detection. In: International Conference on Advances in Computing and Communications, Cochin, India, pp. 403–415, 6–8 Sept 2016
4. Likforman-Sulem, L., Esposito, A., Faundez-Zanuy, M., Clemencon. S.: Extracting style and emotion from handwriting. In: Advances in Neural Networks: Computational and Theoretical Issues. Smart Innovation, Systems and Technologies, vol. 37, pp. 347–355. Springer International Publishing, Switzerland (2015)

5. Djamal, E.C., Darmawati, R., Ramdlan, S.N.: Application image processing to predict personality based on structure of handwriting and signature. In: International Conference on Computer, Control, Informatics and Its Applications (IC3INA), Jakarta, pp. 163–168, 19–21 Nov 2013
6. Champa, H.N., AnandaKumar, K.R.: Automated human behavior prediction through handwriting analysis. In: International Conference on Integrated Intelligent Computing, IEEE Computer Society Washington, DC, USA, pp. 160–165, 5–7 Aug 2010
7. Mutalib, S., Ramli, R., Abdul Rahman, S., Yusoff, M., Mohamed, A.: Towards emotional control recognition through handwriting using fuzzy inference. In: International Symposium on Information Technology, Kuala Lumpur, Malaysia, vol. 2, pp. 1–5, 26–28 Aug 2008
8. Fallah, B., Khotanlou, H.: Identify human personality parameters based on handwriting using neural network. In: Artificial Intelligence and Robotics (IRANOPEN) (IEEE), Iran, pp. 120–126, 9–9 April 2016

Wavelength-Convertible Optical Switch Based on Cross-Gain Modulation Effect of SOA

Sukhbir Singh and Surinder Singh

Abstract All-optical switching based on wavelength conversion using cross-gain modulation (XGM) effect of semiconductor optical amplifier (SOA) has been proposed and demonstrated for 10 Gbps NRZ modulated data signals. Error-free operation is successfully achieved for converted signal with Q-factor of >28.96 at optimum input probe power of −8 dBm. The proposed simple and cost-effective structure of optical switch can be utilized for future ultra-fast optical switching circuit and to expand the optical network.

Keywords Cross-gain modulation (XGM) · Semiconductor optical amplifier Wavelength conversion · Optical switching

1 Introduction

With exponential growth in demand of more bandwidth due to internet traffic, ultra-fast optical transmission system needs ultra-fast photonics devices to provide efficient service to the users. For such applications, photonic devices such as all-optical logic gates, optical switches, optical amplifier, wavelength converter, optical time-division multiplexer and demultiplexer are widely used [1–5].

SOA-based all-optical interferometric switches and wavelength converter are widely used and attractive candidate for future high-speed all-optical signal processing because SOA has much advantageous as compared to other nonlinear devices. These devices are mainly based on the four-wave mixing (FWM), XGM, and cross-phase modulation effect (XPM) of SOA [6–11].

In this contribution, the proposed wavelength-convertible optical switch is based on XGM effect on SOA which can be utilized for optical switching and wavelength conversion to expand the optical network. The performance of optical switch for

---

S. Singh (✉) · S. Singh
Department of Electronics and Communication Engineering, Sant Longowal Institute of Engineering and Technology, Longowal 148106, Punjab, India
e-mail: Sukhbir.mrar@gmail.com

D. Reddy Edla et al. (eds.), *Advances in Machine Learning and Data Science*, Advances in Intelligent Systems and Computing 705,
https://doi.org/10.1007/978-981-10-8569-7_36

10 Gbps NRZ modulated signals has been evaluated at variable input signal power and different parameter of SOA in terms of Q-factor of converted wavelength signal. Brief introduction about SOA-based optical devices has been given in Sect. 1 and architectural setup in Sect. 2. Section 3 describes the performance evaluation of designed wavelength-convertible optical switch and in Sect. 4, conclusion has been made.

2 Architecture of Wavelength-Convertible Optical Switch

Figure 1 shows the architecture of wavelength-convertible optical switch utilizing the XGM effect of SOA. In XGM, a control pulse of probe signal at center wavelength of 1550 nm and pump signal of wavelength 1540 nm are injected to

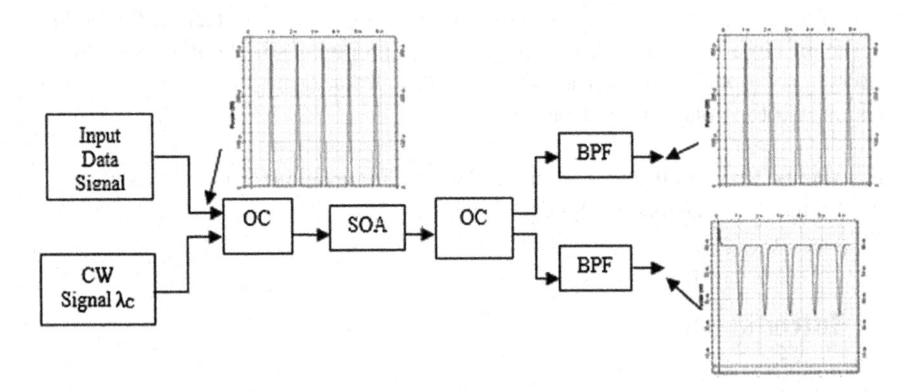

Fig. 1 Architecture of wavelength-convertible optical switch based on XGM effect of SOA

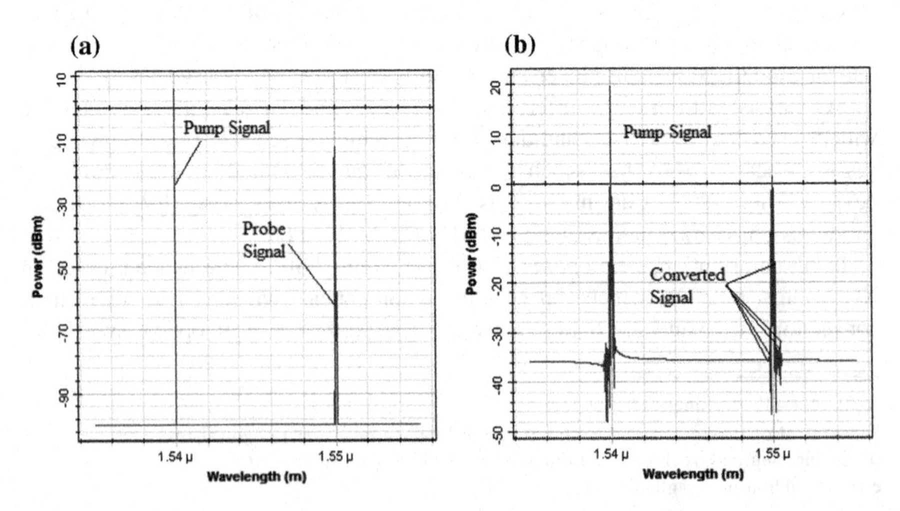

Fig. 2 Received optical spectrum at **a** input of SOA and **b** output of SOA after wavelength conversion

Table 1 SOA parameter considered to design wavelength-convertible switch

Parameter	SOA
Bias current (mA)	150
Active layer length (m)	500×10^{-6}
Active layer width (m)	3×10^{-6}
Active layer thickness (m)	0.2×10^{-6}
Confinement factor	0.3
Spontaneous carrier life time (s)	0.19×10^{-9}
Transparency carrier density (cm^{-3})	1×10^{18}
Material gain constant (m^2)	2.3×10^{-16}
Line width enhancement factor	3
Material loss (m^{-1})	10.5

active region of SOA and phase of probe signal is modulated due to refractive index change via carrier depletion. Hence at the output of SOA, converted wavelength signal is achieved as shown in Fig. 2.

Parameters of SOA are considered to design optical switch given in Table 1. Optical band-pass filters are utilized to recover original signal and converted signal. Switching action of converted wavelength signal is clear in timing diagram of Fig. 1. To achieve fast speed of operation highest extinction ratio, large gain, low saturation output power, and gain recovery time of SOA are desirable. The typical value of gain recovery time is less than 20 ps of the SOA.

3 Performance Analysis of Wavelength-Convertible Optical Switch

Performance of converted wavelength has been analyzed for different parameter of SOA at variable probe input signal power. Figure 3 shows the quality of converted wavelength signal of 1540 nm for various value of injection current. From the curves, it can be observed that at low injection current (100 mA) quality of converted signal is much degraded due insufficient power and gain saturation of SOA. At higher input probe signal, the SOA is saturated which significantly degrade the quality of converted signal. But after gain recovery, the signal quality improves with increase in SOA injection current. The optimum value of Q-factor is 33.56 at 200 mA for –8 dBm input power.

The optimized SOA length of designed wavelength-convertible optical switch for better Q-factor performance is illustrated in Fig. 4. For small active region of SOA, incremental portion does not reach the saturation and quality of converted signal is degraded for 400 μm length. As increase in active length of SOA quality has been improved as compared with 400 μm length but increasing input power, gain coefficient starts decreasing hence degradation in quality. But after gain

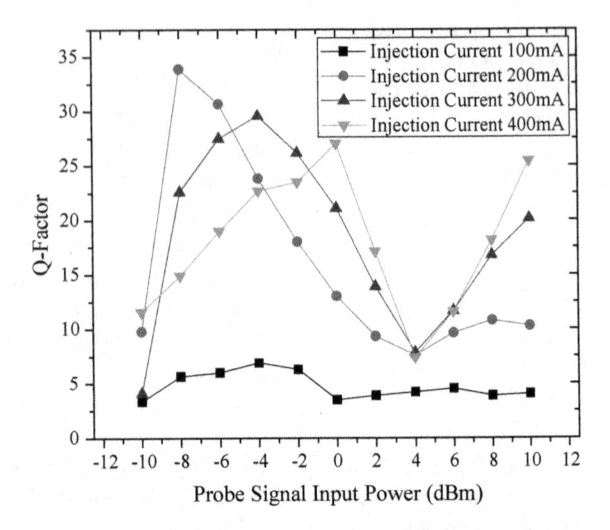

Fig. 3 Q-factor measurement of converted wavelength of 1540 nm at variable probe signal input power for different value of injection current to SOA

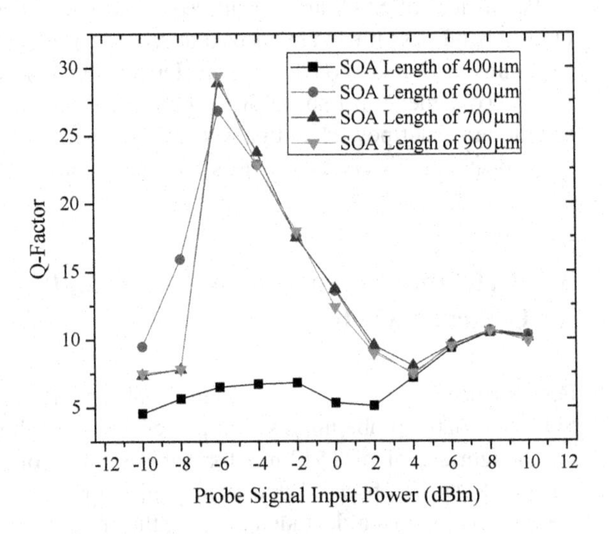

Fig. 4 Q-factor measurement of converted wavelength of 1540 nm at variable probe signal input power for different length of SOA

recovery, quality also improves and optimum value of Q-Factor is 28.96 for 900 μm length of SOA.

The converted signal Q-factor performance for variable input probe signal at different value of confinement of SOA is shown in the Fig. 5. It is observed form the curves that increased value of confinement degrades the quality of converted signal due to more probability of gain saturation. Threshold current density is inversely proportional to the confinement factor. At higher value of confinement factor decrease the threshold current density and decrease the gain of SOA. So, for this investigation confinement factor value is 0.4 considered.

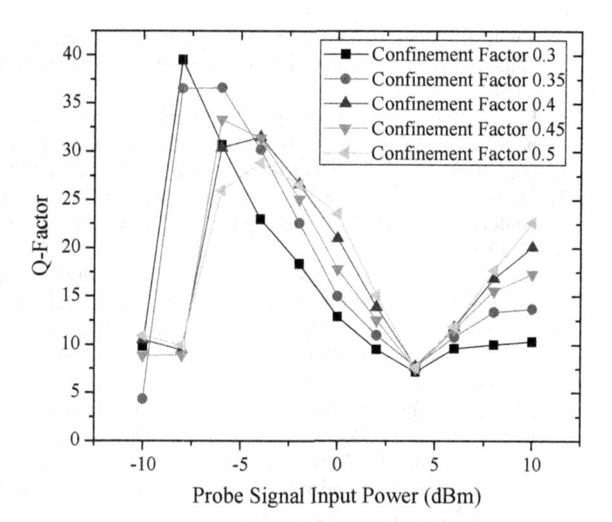

Fig. 5 Q-factor performance of converted wavelength of 1540 nm at variable probe signal input for different value of confinement factor of SOA

4 Conclusion

In the presented paper, wavelength-convertible optical switch based on cross-polarization effect of semiconductor optical amplifier has been proposed and analyzed the Q-factor performance of converted 10 Gbps NRZ modulated signals for different value of injection current, SOA active layer length and confinement factor of SOA. It can be concluded that gain recovery and input probe signal power play a vital role in operation of wavelength-convertible optical switch. The optimum value of Q-factor for injection current of 200 mA, 900 μm active layer of SOA, and confinement factor of 0.4 is 28.96. It is also observed that designed optical switch will be more reliable in switching and wavelength conversion for future optical network.

References

1. Castoldi, P., Raponi, P.G., Andriolli, N., Cerutti, I., Liboiron-Ladouceur, O.: Energy-efficient switching in optical interconnection networks. In: Proceedings of the ICTON 13th International Conference on Transparent Optical Networks (ICTON) (2011)
2. Durhuus, T., Mikkelsen, B., Joergensen, C., Danielsen, S.L., Stubkjaer, K.E.: All-optical wavelength conversion by semiconductor optical amplifiers. J. Lightwave Technol. **14**(6), 942–954 (1996)
3. Schaafsma, D.T., Bradley, E.M.: Cross-gain modulation and frequency conversion crosstalk effects in 1550-nm gain-clamped semiconductor optical amplifiers. IEEE Photon. Technol. Lett. **11**, 727–729 (1999)
4. Singh, S., Singh, S.: Performance analysis of hybrid WDM-OTDM optical multicast overlay system employing 120 Gbps polarization and subcarrier multiplexed unicast signal with 40 Gbps multicast signal. Opt. Commun. **385**, 36–42 (2016)

5. Singh, S., Singh, S.: Investigation on four wave mixing effect in various optical fibers for different spectral efficient orthogonal modulation formats. Opt. Laser Technol. **76**, 64–68 (2016)
6. Fu, S., Wang, M., Zhong, W.-D., Shum, P., Wen, Y., Wu, J.J., Lin, J.: SOA nonlinear polarization rotation with linear polarization maintenance: characterization and applications. IEEE J. Sel. Top. Quantum Electron. **14**(3), 816–825 (2008)
7. Singh, S., Ye, X., Kaler, R.S.: All optical wavelength conversion based on cross polarization modulation in semiconductor optical amplifier. J. Lightwave Technol. **31**(11), 1783–1792 (2013)
8. Contestabile, G., Presi, M., Ciaramella, E.: Multiple wavelength conversion for WDM multicasting by FWM in an SOA. IEEE Photonic Technol. Lett. **16**, 1775–1777 (2004)
9. Liu, H., Li, Y., Peng, H., Huang, J., Kong, D.: Multicast contention resolution based on time-frequency joint scheduling in elastic optical switching networks. Opt. Commun. **383**, 441–445 (2017)
10. Zhang, D., Guo, H., Yang, T., Wu, J.: Optical switching based small-world data center network. Comput. Commun. (Article in Press) 1–12 (2017)
11. Zhang, D., Guo, H., Chen, G., Zhu, Y., Yu, H., Wang, J., Wu, J.: Analysis and experimental demonstration of an optical switching enabled scalable data center network architecture. Opt. Switch. Netw. **23**, 205–214 (2017)

Contrast Enhancement Algorithm for IR Thermograms Using Optimal Temperature Thresholding and Contrast Stretching

Jaspreet Singh and Ajat Shatru Arora

Abstract IR thermography is a noninvasive and non-contact type radiometric technique which creates the 2D thermal images based on infrared radiations. Usually, these are gray-level images which provide poor color contrast. However, various pseudo-coloring algorithms are available to transform these images into RGB space, but contrast enhancement is still required for better visualization of thermograms. In this study, the non-training contrast enhancement algorithm is proposed for IR thermograms. The contrast enhancement in this proposed methodology is achieved by: (i) eliminating the background interference using optimal temperature thresholding and (ii) color enhancement using decorrelation contrast stretching. The performance of proposed methodology has been evaluated based on variations in entropy values. The increasing trend in entropy values indicates the contrast enhancement achieved by using this method.

Keywords Color enhancement · Optimal temperature thresholding
Decorrelation contrast stretching

1 Introduction

1.1 Infrared Radiations

In nineteenth century, a British astronomer and musician, William Herschel discovered the invisible electromagnetic radiations [1], by means of its heating effect on thermometer. These radiations cover the wavelengths (0.75 μm to 1 mm) between microwave band and visible band of electromagnetic spectrum known as infrared

J. Singh (✉) · A. S. Arora
Electrical & Instrumentation Engineering Department, Sant Longowal Institute of Engineering & Technology, Longowal, Punjab, India
e-mail: jassi.mehrok330@gmail.com

A. S. Arora
e-mail: ajatsliet@gmail.com

D. Reddy Edla et al. (eds.), *Advances in Machine Learning and Data Science*,
Advances in Intelligent Systems and Computing 705,
https://doi.org/10.1007/978-981-10-8569-7_37

(IR) radiations. All animate and inanimate objects above −273.15 °C temperature radiate IR energy from its surface due to molecular rotational–vibrational motion. Basically, IR radiations emanate from the surface of an object are a function of its surface temperature and spectral emissivity. The radiant energy emitted by an object and its wavelength distribution is given by Planck's radiation law [2]. The total radiant energy per unit area is given by the Stefan–Boltzmann law as follows:

$$E = \sigma \varepsilon T^4 \quad \mathrm{J\,m^{-2}\,S^{-1}} \tag{1}$$

where

E	Radiant energy
σ (Stefan–Boltzmann's constant)	1.38054×10^{-23} $\mathrm{WsK^{-1}}$
ε	Emissivity factor
T	Temperature in Kelvin

1.2 IR Thermography

IR thermography is a non-contact type radiometric technique which measures the temperature values in the captured field of view based on IR radiations and represents it in the form of false color image. The false color image generated by thermography is known as thermogram. Usually, thermograms are gray-level images which provide poor color contrast among adjacent gray levels. However, gray-level images require less storage space. With the advent of digital image processing, different pseudo-coloring algorithms are available such as iron bow, arctic, hot, lava, rainbow, and many more [3, 4]. This color mapping gives better visualization of patterns in thermograms which lead to better human interpretation. It has been remarked in many studies that the IR thermography has potential to diagnose the abnormal conditions related to superficial organs including head and neck tumor, skin cancer, colorectal cancer, musculoskeletal disorders, diabetes mellitus, paranasal sinusitis [5, 6], and cutaneous perforator mapping in different phases of plastic surgery [7]. With the advancement in couple of related fields, the utility of thermal imaging steadily increases in every field, where clinical environment is not exceptional. The noninvasive and non-ionizing nature of thermography is highly preferable in clinical applications where the patient health is of highest priority.

2 Related Work

Zahedi et al. and Aghdam et al. proposed different types of pseudo-coloring algorithms for thermograms [3, 4]. Furthermore, they compared the outcomes of different proposed algorithms based on objective and subjective evaluation criteria with each other. In 1992, a study from University of Washington, Seattle, proposed an enhancement algorithm for multispectral IR images using decorrelation contrast stretching [8]. In this algorithm, three processing stages were included transformation of data into basic components (decorrelation), independent contrast stretching of data from decorrelated image bands, and retransformation of stretched data based on inverse of the principle component rotation. Using empirical mode decomposition (EMD), a novel approach for image fusion and enhancement has been proposed by Hariharan et al. [9]. In this study, EMD was used to decompose the thermal and visual images into their intrinsic mode functions which were further used to perform image fusion for enhancement. Choi et al. proposed a convolutional neural network-based contrast enhancement method especially for low-resolution thermal images [10].

Apart from image processing for thermal image enhancement, Villasenor-Mora et al. applied different substances on the region of interest to improve visualization of subcutaneous veins [11]. The authors tested various innocuous substances, such as Vaseline, baby oil, ultrasound imaging gel, sunscreen, ethylic alcohol, and penicillin cream, where sunscreen provided better visualization of veins.

3 Methodology

The proposed methodology is divided into five sections: (i) dataset, consists of 20 thermograms, (ii) pseudo-coloring algorithm, (iii) optimal temperature thresholding, (iv) color contrast enhancement, and (v) performance evaluation, by comparing enhanced thermograms with acquired thermograms.

3.1 Dataset

The FLIR E60 thermal imaging camera has been used to acquire the thermograms of different body parts, such as hand, face, foot, back, shoulder, and abdominal regions. These anatomical sites were captured with anterior, posterior, and lateral views at room temperature of 22 ± 3 °C with 50 ± 10% humidity. During thermal imaging, the camera settings kept constant, such as reflected temperature at 20 °C and emissivity at 0.98 [12]. In this way, the dataset of 20 thermograms has been created.

3.2 Pseudo-coloring Algorithm

In this study, the non-iterative smooth color map has been proposed to transform the gray-level thermograms into RGB color space for better visualization of thermal data. Firstly, the thermal data was normalized by using Eq. 2:

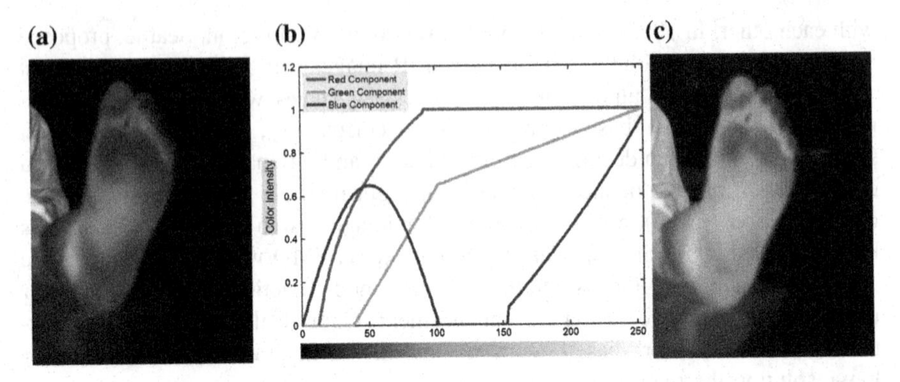

Fig. 1 Depicts the implementation of proposed pseudo-coloring algorithm on plantar view of foot thermogram. **a** Normalized gray-level image, **b** RGB component profile of proposed colormap, and **c** false color image of gray-level image

$$T_{ij} = \frac{I_{ij} - I_{min}}{I_{max} - I_{min}} \tag{2}$$

where I_{ij} is the acquired thermal data, i represents rows, j represents columns, I_{max} and I_{min} are maximum and minimum temperature of thermogram, respectively. Basically, T_{ij} is a gray-level image with intensity values between 0 and 1. Then, the pseudo-coloring process (Eqs. 3–5) was implemented to map the false colors on this normalized thermal data.

$$R_{ij} = \begin{cases} 0, & 0 \le T_{ij} < \frac{1}{20} \\ \frac{3 + \log T_{ij}}{2}, & \frac{1}{20} \le T_{ij} < \frac{9}{25} \\ 1, & \frac{9}{25} \le T_{ij} \le 1 \end{cases} \tag{3}$$

$$G_{ij} = \begin{cases} 0, & 0 \le T_{ij} < \frac{3}{20} \\ \frac{13}{5}\left(T_{ij} - \frac{3}{20}\right), & \frac{3}{20} \le T_{ij} < \frac{2}{5} \\ \frac{7}{12}\left(T_{ij} - \frac{2}{5}\right) + \frac{13}{20}, & \frac{2}{5} \le T_{ij} \le 1 \end{cases} \tag{4}$$

$$B_{ij} = \begin{cases} \frac{13}{20}\left(\sin\left(\frac{5}{2}\pi T_{ij}\right)\right), & 0 \le T_{ij} < \frac{2}{5} \\ 0, & \frac{2}{5} \le T_{ij} < \frac{3}{5} \\ \exp\left(T_{ij}\right) - \frac{7}{4}, & \frac{3}{5} \le T_{ij} \le 1 \end{cases} \tag{5}$$

Here, R_{ij}, $G_{ij,}$ and B_{ij} are red, green, and blue components of false color image, respectively. Figure 1 illustrates the acquired gray-level image, color composition of proposed colormap, and false color image after colormap implementation.

3.3 Optimal Temperature Thresholding

Image preprocessing is required to get the better visualization of geometry of region of interest (ROI) by eliminating the background interference from thermograms. In this study, the iterative method for optimal threshold is proposed to convert the gray-level thermogram into binary image [13]. This process is performed in the following steps:

Algorithm: Optimal temperature thresholding

Require: T_{max} and T_{min} are maximum and minimum temperature of thermogram respectively
Require: Maximum iteration number $N = 100$, Tolerance $T_{ol} = 0.001$
Require: Initial value for temperature threshold, $T_{\cap}^{1} = \frac{(T_{max} + T_{min})}{2}$ and $T_{\cap}^{2} = T_{\cap}^{1} + 1$
Require: Variance of thermogram σ^2, probability of foreground $P_{\gamma} = 0.6$, probability of background $P_{\beta} = 0.4$
Step 1: for i = 1 **to** N **do**
Step 2: if $(T_{\cap}^{i+1} - T_{\cap}^{i}) < T_{ol}$; **Terminate**
Step3: Background image, $I_{\beta}^{i} = \begin{cases} 1, & I < T_{\cap}^{i} \\ 0, & else \end{cases}$
Foreground image, $I_{\gamma}^{i} = \begin{cases} 1, & I \geq T_{\cap}^{i} \\ 0, & else \end{cases}$
Step 4: Average temperature μ_{β}^{i} and μ_{γ}^{i} of regions I_{β}^{i} and I_{γ}^{i} respectively
Step 5: $T_{\cap}^{i+1} = \frac{(\mu_{\beta}^{i} + \mu_{\gamma}^{i})}{2} + \frac{\sigma^2}{(\mu_{\beta}^{i} - \mu_{\gamma}^{i})} + ln\left(\frac{P_{\gamma}}{P_{\beta}}\right)$
Step 6: end for

By using temperature threshold $T_{\cap}^{i+1}$, the gray-level thermograms were converted into binary images, where the ROIs were depicted with white color as shown in Fig. 2. It shows that the thresholding provides better visualization by eliminating the background interference.

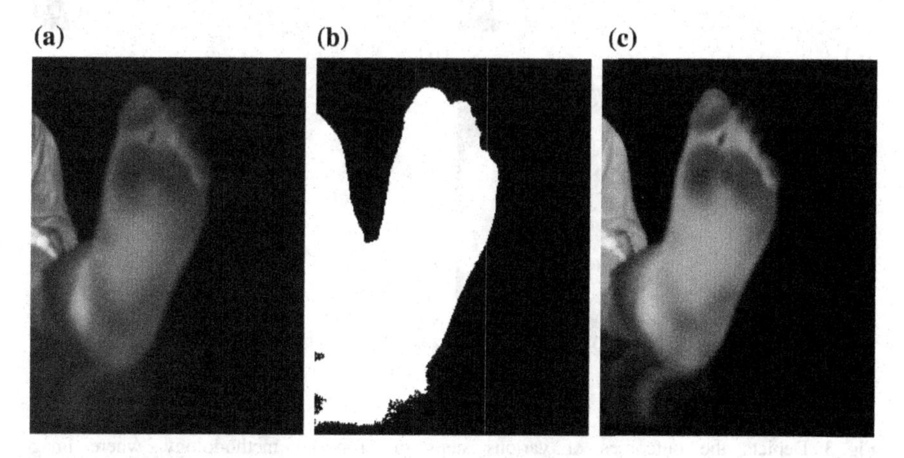

Fig. 2 Illustrates the outcome of optimal temperature thresholding method. **a** Gray-level thermogram of plantar view of foot, **b** binary mask of thermogram obtained after optimal thresholding, and **c** false color image corresponding to binary mask region

3.4 Color Contrast Enhancement

In this section, decorrelation contrast stretching (DCS) has been used to enhance the color distribution only over the segmented region of a thermogram. MATLAB has been used to perform the DCS on thermograms. The outcome of proposed methodology after color contrast enhancement is shown in Fig. 3.

3.5 Performance Evaluation

The performance of this proposed methodology for color enhancement has been evaluated based on entropy value. The entropy values were found at three different stages, (i) P: pseudo-color the thermogram, (ii) TP: thresholding then pseudo-coloring, and (iii) TPEN: thresholding pseudo-coloring then contrast

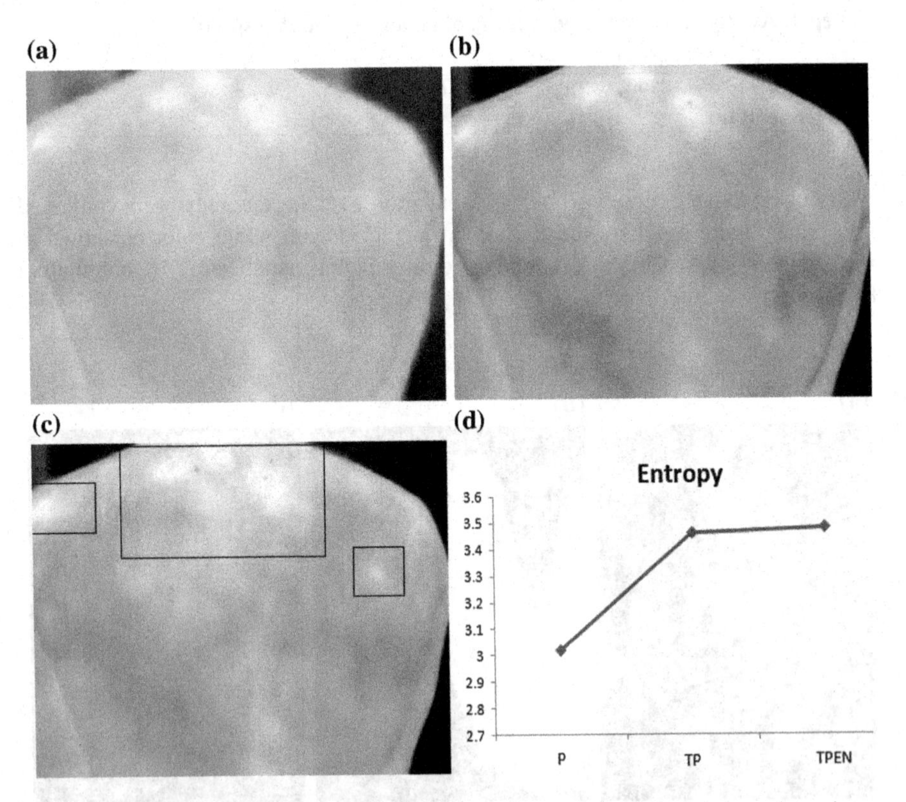

Fig. 3 Depicts the outcomes of various steps of proposed methodology, where image enhancement is shown in (**c**). **a** Thermogram after pseudo-coloring, **b** pseudo-coloring after thermogram thresholding, **c** image enhancement on previous thermogram, and **d** variation in entropy values of these three thermograms

enhancement. The largest entropy value at TPEN stage depicts that the proposed methodology enhances the color contrast which leads to better visualization of thermal patterns (see Fig. 3).

4 Results and Discussion

The proposed methodology has been implemented on acquired thermograms. The entropy values for thermograms at different stages of proposed methodology have been calculated and provided in Table 1. The entropy value of a thermogram gives measure of randomness in color contrast. So, large entropy value means that the thermogram delivers more information than the thermograms having less entropy values.

The increasing trend in entropy values from P to TP and TP to TPEN clearly shows the color enhancement performance of proposed methodology. More the entropy difference between the two stages, better the color enhancement is achieved. In contrast, two cases out of twenty show degradation in color contrast

Table 1 Entropy values for thermograms at different stages of proposed methodology

Thermogram	Entropy		
	P	TP	TPEN
1	4.3	4.52	4.69
2	3.98	3.99	4.42
3	3.98	4.08	4.08
4	4.16	4.20	4.67
5	2.49	2.75	2.76
6	4.76	4.78	5.48
7	3.62	4.31	4.68
8	3.98	4.09	4.24
9	5.27	4.78	6.62
10	4.54	4.58	5.69
11	4.81	4.95	5.56
12	4.82	4.99	5.20
13	3.85	4.00	3.97
14	3.75	3.89	3.93
15	2.66	2.68	2.85
16	3.91	3.97	4.00
17	3.02	3.46	3.47
18	3.68	3.85	3.83
19	3.97	3.77	4.34
20	3.72	3.73	3.76

(highlighted in Table 1). Apart from this statistical measure, the performance of proposed methodology can be analyzed visually.

5 Conclusion

The results of proposed methodology show that it improves the visual perception of thermograms by enhancing the color contrast. In addition, the proposed methodology is comprised of simple and non-training algorithms which require less storage space and computation time. However, the performance of this methodology has been evaluated on various superficial organs of human subjects, but it is also flexible to use in other applications of IR thermography.

References

1. Liew, S.C.: Electromagnetic Waves. Centre for Remote Imaging, Sensing and Processing, pp. 10–27 (2006)
2. Planck, M.: On the theory of the energy distribution law of the normal spectrum. Ann. Phys. **4** (553), 237–245 (1901)
3. Aghdam, N.S., Amin, M.M., Tavakol, M.E., Ng, E.Y.K.: Designing and comparing different color map algorithms for pseudo-coloring breast thermograms. J. Med. Imag. Health Inform. **3** (4), 487–493 (2013)
4. Zahedi, Z., Sadri, S., Moosavi, A.: Breast thermography and pseudo-coloring presentation for improving gray infrared images. In: Photonics Global Conference (PGC), pp. 1–5. IEEE (2012)
5. Gauthier, E., Marin, T., Bodnar, J., Stubbe, L.: Assessment of the pertinence of infrared thermography as a diagnostic tool in sinusitis-Cases study. In: The 12th International Conference on Quantitative Infrared Thermography, vol. 41 (2014)
6. Arora, A.S., Singh, J.: Paranasal sinusitis detection using thermal imaging. In: Science and Information Conference (SAI), pp. 184–188. IEEE (2015)
7. Anbar, M., Milescu, L., Naumov, A., Brown, C., Button, T., Carty, C., AlDulaimi, K.: Detection of cancerous breasts by dynamic area telethermometry. IEEE Eng. Med. Biol. Mag. **20**, 80–91 (2001)
8. Gillespie, A.R.: Enhancement of multispectral thermal infrared images: decorrelation contrast stretching. Remote Sens. Environ. **42**(2), 147–155 (1992)
9. Hariharan, H., Gribok, A., Abidi, M.A., Koschan, A.: Image fusion and enhancement via empirical mode decomposition. J. Pattern Recognit. Res. **1**(1), 16–32 (2006)
10. Choi, Y., Kim, N., Hwang, S., Kweon, I.S.: Thermal image enhancement using convolutional neural network. In: IEEE/RSJ International Conference on Intelligent Robots and Systems (IROS) 2016, pp. 223–230. IEEE (2016)
11. Villasenor-Mora, C., Sanchez-Marin, F.J., Garay-Sevilla, M.E.: Contrast enhancement of mid and far infrared images of subcutaneous veins. Infrared Phys. Technol. **51**(3), 221–228 (2008)
12. Steketee, J.: Spectral emissivity of skin and pericardium. Phys. Med. Biol. **18**(5), 686–694 (1973)
13. Sahoo, P.K., Soltani, S., Wong, A.K.C.: A survey of thresholding techniques. Comput. Vision Graph. Image Process. **41**(2), 233–260 (1988)

Data Deduplication and Fine-Grained Auditing on Big Data in Cloud Storage

RN. Karthika, C. Valliyammai and D. Abisha

Abstract The computing expedient and indulgence are made available in cloud servers by redistributing innumerable resources over the cyberspace. The utmost hefty on-demand services in cloud are data storage. In the world of technocrats, there is a colossal use of national information infrastructure from where an immense amount of data is produced in day-to-day life. To handle those prodigious data on demand is a challenging chore for current data storage systems. The prominence of data deduplication (dedupe) is pointed out by data explosion and colossal slew in redundant data. In the proposed scheme, source-based deduplication is used to eliminate duplicate data, where the client check for the unique data in local (or) remote index through the backup of lower network bandwidth with fast and lower computation overhead. Firstly, in source-based deduplication the data are stored in the physical memory and the fragments of the data are cuckoo hashed before storing the data their physical memory. Secondly, the cloud correctness of data and security is a prime concern, and it is achieved by signing the data block before sending it to the server. And the proposed scheme guarantees the data integrity by fine-grained auditing using Boneh–Lynn–Shacham (BLS) algorithm for signing process, which is one of the secured algorithms. The homomorphic authentication with random masking technique is used to attain privacy-preserving and public auditing.

Keywords Deduplication · Cloud storage · Cyberspace · Infobahn
Prodigious data · Fine-grained auditing · Cuckoo hashing

RN. Karthika (✉) · C. Valliyammai · D. Abisha
Department of Computer Technology, MIT Campus, Anna University, Chennai, India
e-mail: rannar@gmail.com

C. Valliyammai
e-mail: cva@annauniv.edu

D. Abisha
e-mail: abisha.abiya@gmail.com

D. Reddy Edla et al. (eds.), *Advances in Machine Learning and Data Science*,
Advances in Intelligent Systems and Computing 705,
https://doi.org/10.1007/978-981-10-8569-7_38

1 Introduction

In this technological era, bigdata herald an epoch of computing in the cloud. Cloud is a resource pool which provides applications and services on demand. Cloud computing is an Internet technology that exploits both central remote servers and the Internet to handle the data and applications. Computing is being transformed into a model consisting of service that is commoditized and delivered on demand. Due to the massive growth, the generated big data is a challenging and time-demanding task that requires a large computational framework to ensure efficacious data processing or sharing and analysis. Data auditing is especially recommended that the users who have high-level security demands over their data. Data auditing schemes on the existing system have various properties of hidden risks and inefficiency of unauthorized audit requests and in processing small updates still exist. In this paper, proposed work focuses on the benefits of scalability and efficiency in the cloud server. Source-based deduplication is implemented with three components: fragmentation, cuckoo hashing, and chunking. Thus, a key advantage of source-based deduplication is that no additional hardware is required for backup. The integrity of the data is checked in source-based deduplication, and fine-grained auditing (FGA) is carried out with the help of linear homomorphic authentication to dynamically check the small updates, which benefits the scalability and efficiency.

2 Related Work

The interactive paths to improve the efficacy of fully randomized secure dedupe have two collaborating schemes; they are constructed based on static and active dedupe decision tree. The static dedupe decision tree is formulated by the random nodes from source, which does not let the tree bring up-to-date. The dynamic deduplication decision tree is constructed based on the designed self-generation tree, which allows the server to conduct tree update and other optimization techniques [1]. Rekeying-Aware Encrypted Deduplication (REED) storage system with deduplication capability enables secure and lightweight rekeying. REED enables dynamic access control by controlling the group of users access to a file. The stored files are re-encrypted with low overhead using both active and lazy revocations [2]. POD is primarily designed for storage and reduces small write traffic, improving cache efficiency, and read performance. Active RAID reconstruction is an important and integral part of parity-based RAID systems of the hard disk drive (HDD). Extra power is consumed while generating the hash value. The experimental analysis of the online RAID reconstruction demonstrates that the data dedupe can condense the reconstructed time, by improving the system reliability. At this level, a file is checked for deduplication and hash values are used for the file checking function where tag consistency is guaranteed and allows full advantage to be taken over encrypted data of deduplication [3]. This attains secure and fine-grained possession, management in cloud storage for secure, and efficient data deduplication. The efficient deduplication in a distributed storage (DEDIS) performs cluster-wide offline deduplication, and DEDIS does not rely on remote data and novel optimization for reducing the deduplication overhead and its increasingly reliable and the fine-grained performance tuning that limits the request size variability and provides better IO control timelines for storage bandwidth [4, 5].

The fragmentation drops down the efficiency of restoring in dedupe based on backup systems. The fragmentation is categorized into two; they are sparse and out-of-order containers. Sparse containers ascertain the utmost rebuild performance, while inoperative containers ascertain the rebuild performance under limited restore cache. History-Aware Rewriting (HAR) precisely identifies and rewrites sparse containers via exploiting historical information. Optimal restores caching scheme and a hybrid rewriting algorithm as foils of HAR to diminish the pessimistic bearings of inoperative containers are also implemented [6, 7]. The existing problem in data dedupe is information outflow which provides an origination of the problem based on natural graphical modeling. The proposed scheme of the above problem is tackled in an effectual, in an inelastic way by using fragmentation to protect secrecy of sensitive connotations. The fragments of the attributes are independent where the values depend on the other attribute values in which they are related [8]. The cuckoo hashing was carried out under random-walk insertion method with polylogarithmic bound which holds a large d value with high probability for insertion time for graph structure which results in the cuckoo graph in regardless of vertex. The directions for improvement include reducing the value of d for which this type of result holds and reducing the time-bound exponent [9]. For precise, the average case analysis of cuckoo hashing and the probability of cuckoo hashing have no conflicts and are produced in the upper bound of the construction time, where the cuckoo graph is a random bipartite graph used in the application of double saddle point method to obtain asymptotic expansion. The asymptotic results converge in the number of cycles and trees of given size including limiting distributions [10].

The auditing and file-level deduplication is done in SecCloud system which includes file uploading protocol, integrity auditing protocol, and proof of ownership. This protocol is designed based on the Merkle hash tree (MHT). The block-level auditing and sector-level auditing are done for verifying the data, and the performance evaluation is done by using 64-bit t2 Micro Linux servers in Amazon EC2 platform as the auditing server and storage server [11, 12]. The privacy-preserving public auditing protocol allows a TPA to ensure the veracity of subcontracted data on behalf of the users lacking of encroach on the privacy of the data by using the MHT authentication which is developed to assure the correctness of the partial signatures in the enhanced protocol. Privacy-preserving public auditing protocols in the existing system assume that the users' end devices are powerful enough to calculate the computation efficiently when a file is to be outsourced. When the trade off time, t = 300, the auditing protocol achieved the trade-off between efficiency and error detection probability. Because of the low-performance, it is more practically used in end devices [13]. The auditing scheme is multi-replica dynamic public auditing (MuR-DPA). A new scheme authenticated data structure (ADS) based on the MHT called as MR-MHT (multi-replica-MHT) [11]. To support full dynamic updates and authentication of block indices, they include rank and level values in the computation of MHT nodes. The theoretical and experimental results of this scheme not only provide verification, integrity, communication overhead but also provide security against the deceitful cloud service provider (CSP) [7].

3 Proposed Architecture

The cloud user checks for the duplicate files using source-based deduplication as illustrated in Fig. 1 where the uploaded files are stored in cloud server. The duplicate files are verified by using their physical memory address which checks using byte level. Then the deduped data are fragmented, and each fragment is cuckoo hashed, and the deduplicated data are stored in the cloud server. The solution repossesses disk space by obliterating the duplicate data. Before space repossession is carried out, an integrity check can be achieved to warrant that the deduplicated data equal to the original data objects. The integrity check is done by the TPA, and the data owner requests for the auditing. The third-party auditor transmits the audit report to the data owner. Thus, the internal verification and auditing are done.

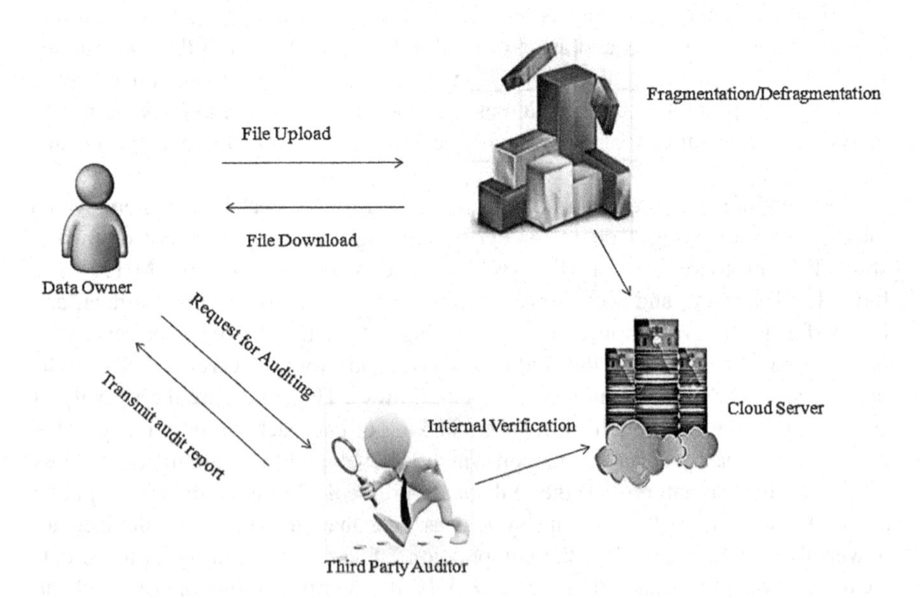

Fig. 1 Proposed architecture of deduplication and auditing

4 Source-Based Data Deduplication

Source-based deduplication takes place at the client side or where the data originates, and it consists of backup application at the client side and some form of disk storage to serve as the repository for backup data. The data are then sent over the LAN/WAN to a disk storage for protection. Single instance storage compares the entire files based on a hash value of a file. If the hash already exists, the data is deemed to be a duplicate and eradicated. If the hash does not exist, then the data is stored in the new hash index. The algorithm for source-based deduplication is as follows

```
Algorithm 1: Deduplicate (string HAshKey, Sting data)

Input:  Files containing raw data
Output: Deduplicated data are stored in the database

dedupeFile ( string HAshKey, string data ) {
bool isUnique = false;
if( exist in CuckooHash(fileHAshKey))
{
if ( ! is Dupe In Cache (fileHAshKey))
{
    isUnique = true;
    }
}
Else
    {
    IsUnique = true;
}
If ( isUnique )
{
saveInCache ( FileHAshKey );
bs. set Buffered Data ( fileHAshKey, data );
}
}
End deduplicate
```

4.1 Deduplication Using Cuckoo Hashing

The technique is based on the number of bits of the hash function which is directly proportional to its computational complexity. It creates a unique mathematical hash representation of all uploaded files and if there is a match in the hash already generated, then the files are removed.

```
Algorithm 2: insert (word key, word value)

Input:  Fragmented Files containing raw data
Output: Hash value for each fragmented file

//hash1(key)=h1; hash2(key)= h2
HashEnt *neNode (key,value)
While(word key)
    int result ←find(key, <ht1,hb1>, <ht2,hb2>)
     if (result = EFIRST  ∨ result = ESECOND)
          Update
     if (ht1=NULL)

    if(¬CAS(&table[0][h1],<ht1, hb1>,<neNode,hb1>))
           While (word key)
        insert
    if (ht2=NULL)

    if(¬CAS(&table[1][h2],<ht2,hb2>,<neNode,hb2>))
        insert
          end if
end While
end insert
```

5 Evaluation

5.1 Experimental Setup

The evaluation is based on a number of duplicate copies bypassed on saving data to the storage server. The evaluation is performed on a system equipped with Intel Core i5 CPU 2.00 GHz with 8 GB RAM and installed Windows 7 Professional, 64-bit Operating System. The machine is connected to Cloud server hosted with domain name DriveHQ via the Internet (1.0 Gbps). The dataset has been taken from clue Web (http://www.lemuproject.org 2012).

5.2 *Data Deduplication*

The proposed system is evaluated for different file set for checking the duplication in the file, and after deduplication, the storage space has been reduced and it is illustrated in Fig. 2.

5.3 *Auditing*

In Fig. 3, the auditing time is computed with and without dedupe. The auditing time without dedupe data is stored and tagged independently in the cloud.

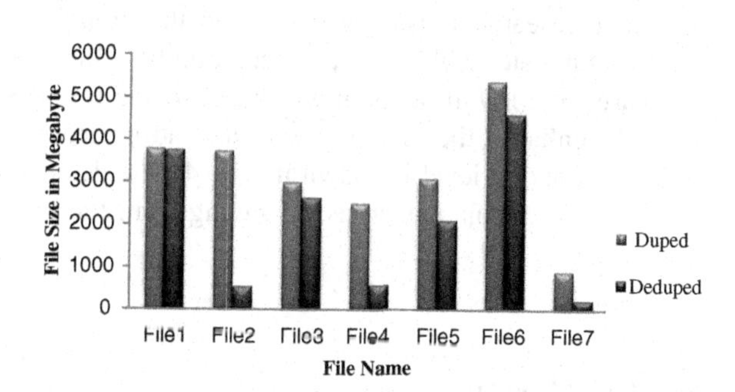

Fig. 2 Data size before/after dedupe

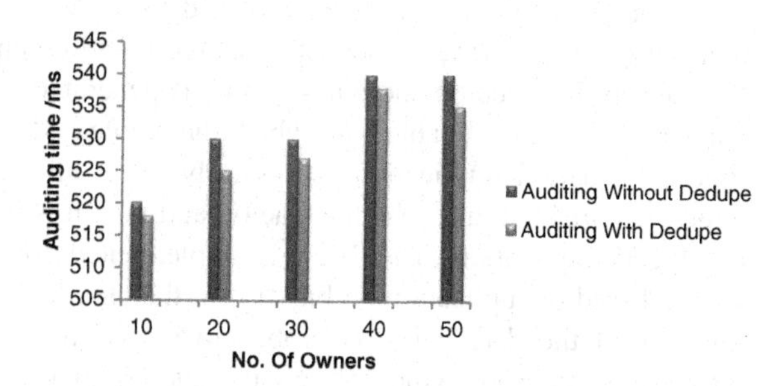

Fig. 3 Auditing time comparison with/without dedupe

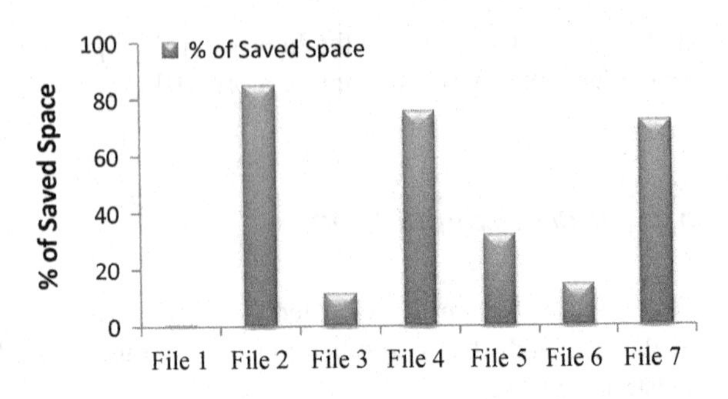

Fig. 4 Deduplication ratio

5.4 *Deduplication Rate*

The deduplication rate is an important indicator of the effect of deduplication, which is defined as follows:

$$\text{deduplication rate} = \frac{\text{Original bytes of data} - \text{stored bytes of data}}{\text{original bytes of data}}$$

Dedupe rate depends primarily on dedupe algorithm and deduplication datasets. As some research illustrates that when the granularity of dedupe increases, the duplication rate will decrease accordingly. But the granularity deduplication decreases, the I/O throughput will increase and it will also cut down the efficiency of CPU, enhance the memory demand, and decrease overall system performance, and the variation level is shown in Fig. 4. The deduplication ratio in percentage is obtained by calculating projected saving from the observed value of the deduplication rate.

6 Conclusion

In the proposed work, the deduplicated data save the space on the disk. Among them, the most important concept is source-based deduplication, which refers that the files are fragmented and a hash value is generated with each fragmented chunk and stored in the hash table with which the whole deduplication system works. The process of data deduplication successfully stores all the data into the storage disk which requires the disk a high capacity, and then it is possible for the user to carry out the deduplication. The HLA is implemented for fine-grained auditing with source-based deduplication on big data in the cloud which is used for checking the integrity of the data. HLA is probed to reduce the overheads of reckoning and information transfer, which allows the auditor to verify the integrity of the data without recuperating the whole data from the cloud storage.

References

1. Jiang, T., Chen, X.: Secure and efficient cloud data deduplication with randomized tag. IEEE Trans. Inf. Foren. Secur. **12**(3) (2017)
2. Qin, C., Li, J., Lee, P.P.C.: The design and implementation of a rekeying-aware encrypted deduplication storage system. ACM Trans. Storage **13**(1), Article 9, 30 p. (2017)
3. Mao, B, Jiang, H., Wu, S., Tian, L.: POD: performance oriented I/O deduplication for primary storage systems in the cloud. In: IEEE International Parallel and Distributed Processing Symposium, vol. 10, pp. 767–776 (2014)
4. Zhou, B, Wen, J.-T: Metadata feedback and utilization for data deduplication across WAN. J. Comput. Sci. Technol. **31**(3), 604–623 (2016)
5. Paulo, J., Pereira, J.: Efficient deduplication in a distributed primary storage infrastructure. ACM Trans. Storage **12**(4), Article 20, 35 p. (2016)
6. Fu, M., Lee, P.P.C., Feng, D., Chen, Z., Xiao, Y.: A simulation analysis of reliability in primary storage deduplication. In: IEEE International Symposium on Workload Characterization (IISWC), pp. 1–10 (2016)
7. Zhou, B., Wen, J.T.: Improving metadata caching efficiency for data deduplication via in-RAMmetadata utilization. J. Comput. Sci. Technol. **31**(4), 805–819 (2016)
8. De Capitani, S., di Vimercati, S., Foresti, S.J.: Fragmentation in presence of data dependencies. IEEE Trans. Dependable Secur. Comput. **1**(6), 510–522 (2014)
9. Frieze, A., Mitzenmacher, M.: An Analysis of Random-Walk Cuckoo Hashing. Published in LNCS, pp. 490–503 (2009)
10. Drmota, M., Kutzelnigg, R.: A Precise analysis of cuckoo hashing. ACM Trans. Algorithm **8** (2), 1–40 (2012)
11. Liu, C., Ranjan, R., Yang, C., Zhang, X., Wang, L., Chen, J.: MuR-DPA: top-down levelled multi-replica merkle hash tree based secure public auditing for dynamic big data storage on cloud. IEEE Trans. Comput. **64**(9) (2015)
12. Li, J., Li, J., Xie, D., Cai, Z.: Secure auditing and deduplicating data in cloud. IEEE Trans. Comput. **65**(8) (2016)
13. Zhanga, J., Dong, Q.: Efficient ID-based public auditing for the outsourced data in cloud storage. Elsevier, Inf. Sci. (2016)

Retraction Note to: In-silico Analysis of LncRNA-mRNA Target Prediction

Deepanjali Sharma and Gaurav Meena

Retraction Note to: Chapter "In-silico Analysis of LncRNA-mRNA Target Prediction" in: D. Reddy Edla et al. (eds.), *Advances in Machine Learning and Data Science*, Advances in Intelligent Systems and Computing 705, https://doi.org/10.1007/978-981-10-8569-7_28

The authors have retracted this chapter [1] because upon re-review of the methodology, errors have been identified in the presented algorithm. Therefore, the results are incorrect. All authors agree to this retraction.

[1] Sharma D., Meena G. (2018) In-silico Analysis of LncRNA-mRNA Target Prediction. In: Reddy Edla D., Lingras P., Venkatanareshbabu K. (eds) Advances in Machine Learning and Data Science. Advances in Intelligent Systems and Computing, vol 705. Springer, Singapore; https://link.springer.com/chapter/10.1007/978-981-10-8569-7_28.

The retracted online version of this chapter can be found at https://doi.org/10.1007/978-981-10-8569-7_28

D. Reddy Edla et al. (eds.), *Advances in Machine Learning and Data Science*, Advances in Intelligent Systems and Computing 705,
https://doi.org/10.1007/978-981-10-8569-7_39

Author Index

D. Reddy Edla et al. (eds.), *Advances in Machine Learning and Data Science*,
Advances in Intelligent Systems and Computing 705,
https://doi.org/10.1007/978-981-10-8569-7

Printed in the United States
By Bookmasters